Process Instrumentation Applications Manual

Other McGraw-Hill Chemical Engineering Books

CHOPEY • *Fluid Movers*

CHOPEY • *Handbook of Chemical Engineering Calculations*

CONSIDINE • *Process/Industrial Instruments and Controls Handbook*

COOK AND DUMONT • *Process Drying Practice*

CROOM • *Filter Dust Collectors*

DATTA-BARUA • *Natural Gas Measurement and Control*

DEAN • *Lange's Handbook of Chemistry*

DESHOTELS AND ZIMMERMAN • *Cost-Effective Risk Assessment for Process Design*

DILLON • *Materials Selection for the Process Industries*

FITZGERALD • *Control Valves for the Chemical Process Industries*

HARPER • *Handbook of Plastics, Elastomers, and Composites*

KISTER • *Distillation Design*

KISTER • *Distillation Process Applications and Operations*

MANSFIELD • *Engineering Design for Process Facilities*

MCGEE • *Molecular Engineering*

MILLER • *Flow Measurement Engineering Handbook*

PALLUZI • *Pilot Plant and Laboratory Safety*

PALLUZI • *Pilot Plant Design, Construction, and Operation*

POWER • *Steam Jet Ejectors for the Process Industries*

REIST • *Aerosol Science and Technology*

RHINE AND TUCKER • *Modelling of Gas-Fired Furnaces, Boilers, and Other Industrial Processes*

SATTERFIELD • *Heterogeneous Catalysis in Industrial Practice*

SCHWEITZER • *Handbook of Separation Techniques for Chemical Engineers*

SHINSKEY • *Feedback Controllers for the Process Industries*

SHINSKEY • *Process Control Systems*

SHUGAR AND BALLINGER • *Chemical Technician's Ready Reference Handbook*

SHUGAR AND DEAN • *The Chemist's Ready Reference Handbook*

SIKICH • *Emergency Management and Planning Handbook*

SMALLWOOD • *Solvent Recovery Handbook*

SMITH • *Chemical Process Design*

TATTERSON • *Mixing and Gas Dispersion in Agitated Tanks*

TATTERSON • *Scale-Up and Design of Industrial Mixing Processes*

YOKELL • *A Working Guide to Shell and Tube Heat Exchangers*

Process Instrumentation Applications Manual

Bob Connell

McGraw-Hill
New York San Francisco Washington, D.C. Auckland Bogotá
Caracas Lisbon London Madrid Mexico City Milan
Montreal New Delhi San Juan Singapore
Sydney Tokyo Toronto

Library of Congress Cataloging-in-Publication Data

Connell, Bob.
Process instrumentation applications manual / Bob Connell.
p. cm.
ISBN 0-07-012353-5
1. Chemical process control. 2. Engineering instruments.
I. Title.
TP155.75.C663 1996
660'.283—dc20 95-22624
CIP

McGraw-Hill

A Division of The McGraw-Hill Companies

1 2 3 4 5 6 7 8 9 0 DOC/DOC 9 0 0 9 8 7 6 5

ISBN 0-07-012353-5

The sponsoring editor for this book was Harold Crawford, the editing supervisor was Joseph Bertuna, and the production supervisor was Pamela A. Pelton. It was set in Century Schoolbook by Estelita F. Green of McGraw-Hill's Professional Book Group composition unit.

Printed and bound by R. R. Donnelley & Sons Company.

Contents

Preface

There is no shortage of textbooks about process measurement and control these days. There are so many that for a long time I had the distinct feeling that there was nothing new or useful that I could possible add to the large fund of theory and advice that was already available in printed form.

Many years ago I actually had a contract to write a textbook on process instrumentation. At that time, I thought that I had enough material in note form to produce a useful reference book without a laborious effort on my part. Such was not the case. Writing a book took a great deal more time than I had anticipated. Furthermore, people with prior experience warned me that I would never get rich from the proceeds from any textbook. All that I could expect would be a certain amount of recognition if I happened to turn out a popular product. Thus work on the book was put off and put off, until my contract became null and void.

However, today circumstances are somewhat different. Being in a condition of retirement, I have more time to devote to writing. Furthermore, I also have my own word processor with which to crank out the material. These facts, plus the added encouragement from my former supervisor at Northern Alberta Institute of Technology, led me to reactivate the project.

The program head for the Instrumentation Engineering Technology Program at the Northern Alberta Institute of Technology (NAIT) has been prodding me for some time to put together a book that discusses process instrumentation from a more facts-of-life point of view. After working for 30 years on measurement and control systems for Imperial Oil Limited and Syncrude Canada Limited, I should have learned some things that could be helpful to others who want to avoid problems of the type I have encountered. Many of these problems were the result of my own naivety, or my failure to reason things out in a objective manner.

Please note that the words I just used—"After working for 30 years on instrumentation systems"—were chosen advisedly. I might have said, "Having had 30 years of experience." Yet statements of years of experience in process control now tend to leave me skeptical. The process control field has developed at an incredible rate. What is commonplace today in control systems would have been considered fantasy 10 years ago. Thus I have to conclude that while it is all right for someone to claim that he has had 30 years of experience in the instrumentation field, the fact remains that if he has not learned anything in the last 5 years, the knowledge of which he may be proud could be virtually useless today.

In any event, as I have told my students on numerous occasions, 80 percent of what engineers like to refer to as "practical experience" is really knowing what *won't* work. These days, solutions to process control problems seem to be developing faster than we can find problems to which we can usefully apply them. This inevitably leads to solutions being applied where there really are no problems. In other words, change for the sake of change.

I imagine there is no other technology field in which theorists and manufacturers have so consistently come up with new methods, hardware, and systems that can solve one's problems. In most cases, the new ways work. However, any salesperson's manual will state, "Don't ever be dishonest, but on the other hand, don't give a prospective customer any reason for not buying your product." As a result, you often find out that the new system you bought because you believed it would solve all the problems inherent in your old system does in fact do this, but it brings along some new problems of its own. We saw this when we changed from pneumatic to electronic, from steam tracing to electric tracing, and from analog to digital.

There are probably no perfect solutions. The decision with which we are continually faced, therefore, is should we opt for the new system, which eliminates all our old problems but which has an assortment of new problems we have to get accustomed to? Or would we be better off to stick with the old system and its problems, which we have learned to deal with?

One can get a lot of advice about what is new and promising and what is old and therefore unsuitable in process control. All this is presumably based on a knowledge of the latest process control theory and on a sometimes tilted analysis of the quality of the hardware that has been developed. Yet, in my experience, the biggest single factor in which course of action is finally selected is *people*. People with all their attitudes, or biases, or preferences, or dislikes, or convictions of the moment.

I have tried to impress on the students I have taught that rules of thumb, and you see many of them in process instrumentation, are adopted from ignorance. The only instrumentation rule of thumb that I know of that you can bet your reputation on is that simplicity and reliability go hand in hand. Therefore, one does not select a computer to do every control job, even though a computer could probably do every control job. It is our responsibility as process instrumentation specialists to be aware of all the useful methods that are currently available to us and to use the simplest one that will produce the required results.

Furthermore, one should be on guard against the advice of those who tell us that any of our current ways are obsolete. These are the people whose attitude is that hardware or systems are obsolete if they are not the latest model. Applying this principle, pneumatic instrumentation should now be considered obsolete. What nonsense! Nothing is obsolete as long as it is doing the job that you want it to do. This applies to your car, your TV set, your personal computer, and anything else for which you have paid a lot of money. It also applies to process instrumentation systems. It is only when your machine or system will not produce the results you want, because it does not have the necessary features, or because maintenance has become intolerable, or because repair parts are no longer available, that it is reasonable to replace it with a newer model.

My career in this remarkable field of process instrumentation technology is all but over. For myself, I could not have chosen a better one. Not the least of the factors which have made it so worthwhile for me are the associations I have had with people who have helped me along the way, especially at those points in my career when I was the most naive.

My first boss, when I joined Imperial Oil, was Bill Graeb. Bill was the finest supervisor I ever had, and he remained one of my closest friends. Greg Ziegler, of the famous Ziegler and Nichols team, on many occasions took the trouble to show me, a rookie engineer working for Taylor Instruments, what automatic process control was really about. And finally, D. Mykitiuk, my supervisor at NAIT, whom I admire greatly for his more than 20 years of dedication to developing a top-notch process instrumentation program for his students.

In the writing department, I have to acknowledge the talent of Gregory McMillan and Stanley Weiner in producing their book, *How to Become an Instrument Engineer*. This work has all the appearances of being a spoof, but it contains a great deal of real field experience that should not be ignored by any conscientious instrument engineer or technologist.

Those who are students of control theory will probably not be turned on by my book. Much of it is philosophical rather than theoretical. Occasionally, I have broken the unwritten rule that forbids use of the first person pronoun. For this I apologize to the reader in advance.

Where measurement or control loops have been shown, the ISA conventions have not been used. Transmitters, current-to-air converters, valve positioners, and other peripheral instrument components whose use is implied have been left out to simplify the sketches. In addition, since none of the measurement or control systems shown are specifically pneumatic or electronic, the ISA symbols for electronic or pneumatic signal lines have not been used either. Instead, the general signal line symbol directly below has been adopted:

Not ----------------- **or** —//————//— **but** ——⋀⋁——⋀⋁——

As much as I am dedicated to the process control field, I have had to learn over a period of time that process control is not an end unto itself. It is, to be sure, a very important component in the machinery of production. The safety and profitable operation of a plant depend on process control, but not on process control alone. As a technology, process control has now become important enough that process control engineering should eventually become an engineering discipline in its own right. Even so, it is really the quantity and quality of what the plant produces that are the key priority, and we must not lose sight of the fact that our expertise in this field, while important, is but one spoke in the wheel.

Finally, advanced control theory notwithstanding, the control systems that will be designed for the new plants of the future are going to be feedback control systems in 90 percent or more of the applications. Feedback control systems have the characteristic closed loop, and their overall performance depends on the individual performance of each of the three components of the loop: the controller, the control valve, and the process. Advanced control algorithms make their contribution through the automatic controller, which is fine, except that the controller is rarely the dominant component in the loop.

In almost every case it is the process and its characteristics, both static and dynamic, that will determine more than anything else how well the control system as a whole performs. Unfortunately, the chances that any process equipment manufacturer would alter the design of his or her product so as to improve its process control characteristics are not good at all. The reason: Any changes that would contribute to better control also would increase the cost of the equipment and make it less competitive.

Each time I began another of my Instrumentation Application courses with a new class, I tried to impress on them an important fact which I left with them in the form of a sign to be hung on the wall of each one's first office after graduation. I also add the comment, "If you don't learn anything else from me, learn this." The sign reads: THE ESSENTIAL INGREDIENT OF ALL SUCCESSFUL INSTRUMENTATION SYSTEMS IS ATTENTION TO DETAIL.

One of my favorite comics is Calvin and Hobbs. In one episode, Calvin's father has taken the family out for a drive. They come to a bridge that has a sign that says, "Load Limit 10 Tons." This prompts Calvin to ask, "How do they know the load limit, Dad?"

After due consideration, Calvin's father comes up with the answer. "Well son, they drive heavier and heavier trucks over it until the bridge breaks; then they weigh the last truck and put the bridge back together." Calvin's response to this is, "I guess I should have thought of that." But his mother comments, rather dryly, "If you don't know the answer, why don't you just admit it?"

I trust that in this book I have not answered too many questions about measurement and control in the way that Calvin's father answered his question.

Bob Connell

Process Instrumentation Applications Manual

Chapter

1

Physical Properties

Physical properties are those properties of a substance which do not change irrespective of the size of the sample that is examined. This precludes mass and volume, both of which are affected by the sample size. It does include properties such as color, odor, texture, and in fact, many other things. Physical properties are often specific to a particular substance and, as such, can sometimes be used to identify the substance.

From the instrumentation point of view, the physical properties of materials are often involved in the design of instrument components. They are also an important factor in the calibration of some meters and in the sizing of devices such as flowmeters, control valves, and safety valves. This being the case, it is worthwhile to devote some space to a discussion of certain of the more prominent physical properties and their meaning.

1.1 Density

Density is probably the most fundamental of all the physical properties. It applies to substances in any of the three states: solid, liquid, or gas. *Density* is defined as the mass of a unit volume of a substance. It is usually identified by the Greek lowercase letter rho (ρ). Most substances have their own peculiar value of density; hence it is often possible to identify a substance if its density can be measured.

To obtain the density of a sample, both its mass and its volume must be determined. Then the density can be calculated by dividing the mass of the substance by its volume to obtain the mass per unit volume. Symbolically,

$$\rho = \frac{m}{V}$$

The units of density are kilograms per cubic meter (kg/m^3) in the SI system. In the foot-pound-second system, the basic units for density are pounds per cubic foot (lb/ft^3). The conversion relation is

$$1 \text{ pcf} = 16.018 \text{ kg/m}^3$$

Note that density is determined by dividing the mass of an object (not its weight) by its volume. Another consideration is that the volume of the sample probably will be measured at its prevailing temperature. Since the volume of most substances changes with temperature, then the density of a substance also must change with temperature. Increasing the temperature of a sample of a substance generally increases its volume and therefore decreases its density. To be valid, data tables that give the densities of substances must state the temperature at which the measurement was made.

1.1.1 Specific gravity

In industry, the density of liquids is not always specified in units of kilograms per cubic meter or pounds per cubic foot. Specific gravity, which is the density of a liquid relative to the density of water, is used widely. The specific gravity of a liquid is determined by dividing the density of the liquid at its prevailing temperature by the density of water *at a specified standard temperature.* The densities of both the liquid and the water must have the same units so that the specific gravity value will be a dimensionless ratio.

It is worth noting that since specific gravity has no units, it does not belong in any system of units. Therefore, it is not for the proponents of the SI system, or any other system of units, to decree that specific gravity should not be used. It will continue to be used, if for no other reason than that it is a useful concept. In the SI system, the term *relative density* has been introduced. It means the same as specific gravity.

Unfortunately, opinion as to what constitutes the "specified standard temperature" of water is not consistent. Scientists and academics tend to use the density of water at either 0°C, which is the lowest temperature water can have as a liquid, or 4°C, which is the temperature at which water has its highest density value. The density values at 0 and 4°C are 999.84 and 999.97 kg/m^3, respectively.

In the petroleum industry in North America, the density value for water that is used is the 15°C value, namely, 999.10 kg/m^3.

1.1.2 American Petroleum Institute gravity

The petroleum industry uses, in addition, a peculiar density scale, which is *API gravity.* This scale is a tradition among petroleum people, and it makes sense to them because their more valuable products, which usually have lower densities, have higher values on the API gravity scale. The relationships are

$$\text{API gravity} = \frac{141.5}{\text{specific gravity}} - 131.5 \text{ degrees}$$

or

$$\text{Specific gravity} = \frac{141.5}{\text{API gravity} + 131.5}$$

On the API gravity scale, water has a value of 10 degrees, bitumen is about 6 degrees, and motor gasoline is about 60 degrees.

1.1.3 Baumé scale

The *Baumé scale* is another scale used to specify the density of certain chemicals, caustic solutions being a common example. The relationship, for chemicals having a specific gravity greater than 1, is

$$\text{Degrees Baumé} = 141.5 - \frac{141.5}{\text{specific gravity}}$$

1.1.4 Specific weight

Another basis for defining the density of a substance is the concept of specific weight, which is the weight of a sample of the substance (not its mass) divided by the volume of the sample. As such,

$$\text{Specific weight} = \text{density} \times \text{gravitational constant at the location}$$

Specific weight is not a very practical concept. Fortunately, it is seldom used.

1.1.5 Other density scales

Other density scales that are specific to particular industries are the *Brix scale,* which is used for syrup density in sugar refining, and the *proof scale,* which is used for the alcohol content in alcoholic beverages.

1.1.6 Density considerations

The calculations involved in sizing flowmeters and control valves require the density of the fluid at the operating temperature. This raises the question of whether the density at the operating temperature can be calculated if the density at some other temperature is known. For water, light hydrocarbon liquids, and certain other liquids, the increase in volume for a 1°C rise in temperature is not the same at all temperatures. If the density of the liquid were plotted over a range of temperatures, the graph would not necessarily be a straight line.

This means that calculating the density of a liquid at some specified temperature by using the known density at some other temperature and a coefficient of density change with temperature is not a reliable procedure. The only way to be certain of the density of a liquid at a particular temperature is to actually measure its density at that temperature in the laboratory. Published coefficients of density change with temperature will be averaged values at best and should only be used for rough estimates when no better method of determining the density is available.

1.2 Viscosity

Viscosity is a property of both liquids and gases. It is usually used as an indication of how freely a fluid will flow. Fluids that have a low viscosity are relatively "fluid" and will flow freely through conduits. Fluids that have a high viscosity flow sluggishly.

The coefficient of viscosity μ of a fluid is the quantitative measure of the viscousness of the fluid. In other words, it is a method of putting a number on the viscosity. Consider Fig. 1.1, in which a solid cylindrical rod is at the center of a tube filled with a liquid. The objective is to cause the rod to move along the axis of the tube with a velocity v. This will require a force F to be applied to the rod, since the viscous nature of the liquid imposes a drag on the rod.

The factors that determine the force F (in newtons) that must be applied to the rod are

- The desired velocity v (m/s). The greater the velocity, the greater is the required force.
- The surface area A (m^2) of the rod. The greater the area of contact between the liquid and the rod, the greater is the force.

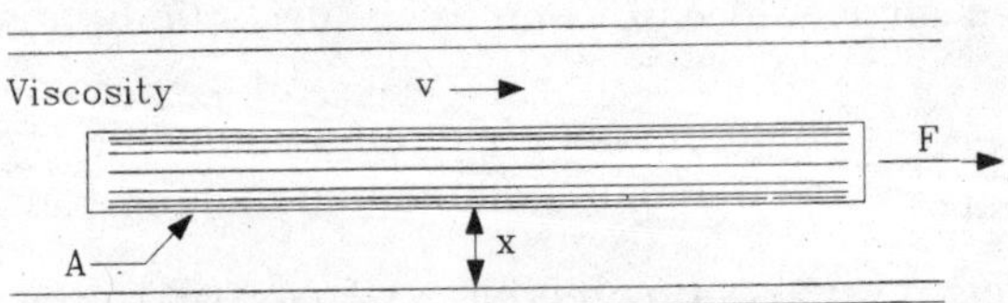

Figure 1.1 The force required to move a cylindrical rod down the center of a circular conduit is determined by four separate factors.

- The viscosity μ of the liquid. The greater the coefficient of viscosity, the greater is the force.
- The distance x (m) between the surface of the rod and the inside wall of the tube. The greater this distance, the less the liquid will tend to stretch or shear on itself and therefore the less the force.

Putting these factors down as an equation, we get

$$F \propto \frac{vA\mu}{x}$$

Let the units of μ be chosen so that $F = \dfrac{(vA\mu)}{x}$. Thus,

$$\mu = \frac{Fx}{Av}$$

The units of F/A are pressure units, or pascals. The units of x/v are seconds. The units of μ are, consequently, pascal-seconds (Pa•s) in the SI system.

However, 1 Pa•s is a very large increment of viscosity. Accordingly, it is more practical to measure viscosities in millipascal-seconds (mPa•s). It turns out, conveniently, that

$$1 \text{ mPa•s} = 1 \text{ centipoise (cP)}$$

The coefficient of viscosity μ (mPa•s) is called the *absolute viscosity,* or the *dynamic viscosity,* of a fluid. There is also an allied property called the *kinematic viscosity,* which is defined as

$$\text{Kinematic viscosity} = \frac{\text{absolute viscosity (Pa•s)}}{\text{density (kg/m}^3\text{)}}$$

The units of kinematic viscosity are square meters per second (m^2/s). A proper question is, Why do we need kinematic viscosity as well as absolute viscosity? It would be nice if we did not need it, but unfortunately, certain instruments used to measure the viscosity of liquids respond to the kinematic viscosity rather than the absolute viscosity.

The coefficient of viscosity of liquids varies greatly with temperature. Raising the temperature of a liquid decreases its viscosity. The viscosity of gases is also affected by temperature, but in the opposite way. Increasing the temperature of a gas increases its viscosity. The result is that any statement of the viscosity of a liquid or a gas is meaningless unless the temperature is also stated.

1.3 Vapor Pressure

Whether a substance is in its liquid or vapor state, the molecules of the substance will be in constant motion. It is the bombardment of the molecules of the substance against the inner walls of a containing vessel that creates the pressure against these surfaces. If a closed vessel is completely evacuated and then partly filled with a sample liquid, the velocity of the molecules in the liquid will cause some molecules to leave the liquid and migrate up into the space above the liquid. This will continue until the vapor space above the liquid saturates with molecules, at which time the reverse effect begins. Then some of the molecules in the vapor space return to the liquid.

Eventually, an equilibrium condition is reached in which the number of molecules leaving the liquid phase per unit time exactly matches the number of molecules returning from the vapor to the liquid. Since molecular bombardment creates pressure, there will be a measurable pressure in the vapor space. This pressure is the *vapor pressure* of the substance. The following points are significant:

- Vapor pressure is always reported in absolute pressure units; otherwise, a different table of vapor pressure values for liquids would be required for every different location.
- The vapor pressure of a liquid increases rapidly as the temperature of the liquid increases. The relationship is not linear. Thus any vapor pressure reading for a liquid is meaningless unless the temperature of the liquid is also given.
- If the temperature of the liquid is raised to the point where the vapor pressure of the liquid becomes equal to the surrounding pressure, that is when the liquid boils. Consequently, a liquid can be caused to boil either by raising its temperature or by lowering the ambient pressure. This explains why water will boil at a lower temperature at higher elevations where the atmospheric pressure is lower.
- Liquids having relatively high vapor pressures will evaporate faster when the container is left open. This is due to the higher molecular activity in liquids that have higher vapor pressures.

For pure liquids, the relationships between temperature and vapor pressure are well known and are predictable. Thus it is possible to build a thermometer that is activated by the vapor pressure of a liquid and which can indicate temperature through the use of a proper correlation.

1.4 The Kinetic Theory of Gases

Figure 1.2 shows a cube of a gas. Each side of the cube measures s meters. In addition, three assumptions are made at the outset:

1. The size of the molecules of the gas inside the cube is negligible compared with the distances between the molecules.
2. The motion of the molecules is due to the internal energy of the gas (i.e., its temperature), not to any mutual forces of attraction.
3. Collisions between the molecules themselves and between the molecules and the walls of the cube are purely elastic; that is, there is no loss in energy as a result of a collision.

Consider first a single molecule. Its mass is m kilograms. Its velocity is u meters per second. If the molecule struck any wall of the cube dead on and dropped to the bottom, the change (loss) in its momentum would be mu. However, in reality, it rebounds in the opposite direction with the same velocity. Therefore, the net (vector) change in the momentum of the molecule will be 2 mu.

Although it can be assumed that a molecule travels in a straight path, it will not necessarily move perpendicular to any one wall of the cube. Its velocity can be resolved into three components parallel to each of the three axes that define the walls of the cube. Call these the x, y, and z velocity components. Thus

$$u^2 = x^2 + y^2 + z^2$$

Now let us examine any one of the component velocities (say, the z component). The walls at each end of the z axis are s meters apart. Therefore, the travel time of the molecule from one wall to the other and back to the starting point will be

$$\frac{\text{Distance}}{\text{Velocity}} = \frac{2s}{z}$$

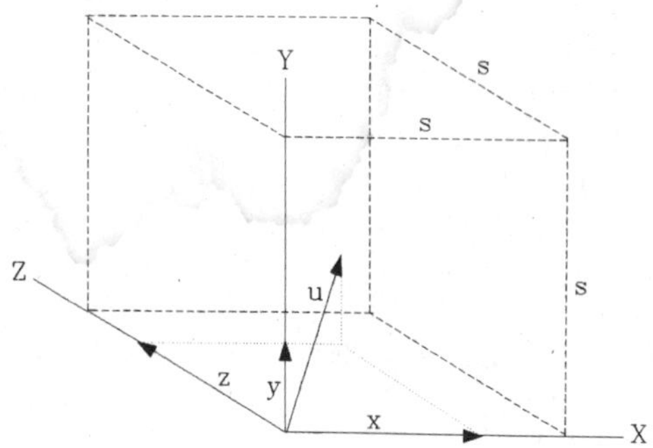

Figure 1.2 Motion of a molecule of gas inside a cube enclosure.

In $2s/z$ time units, the molecule makes two impacts. Therefore, the number of impacts per unit time will be

$$\frac{2}{2s/z} = \frac{z}{s}$$

The change in momentum for this component of the motion of the molecule per unit time will be

$$\frac{z}{s} \times 2mz = \frac{2mz^2}{s}$$

For the x and y components, the change in momentum per unit time will be $2mx^2/s$ and $2my^2/s$. Therefore, the total change in momentum for a single molecule, per unit time, will be

$$\frac{2mx^2}{s} + \frac{2my^2}{s} + \frac{2mz^2}{s} = \frac{2mu^2}{s}$$

If all the molecules in the cube are numbered 1, 2, 3, 4 . . . , then the change in momentum per unit time for all molecules will be

$$\frac{2mu_1^2}{s} + \frac{2mu_2^2}{s} + \frac{2mu_3^2}{s} + \frac{2mu_4^2}{s} + \cdots$$
$$= \frac{2m}{s}(u_1^2 + u_2^2 + u_3^2 + u_4^2 + \cdots)$$

For the squares of the velocities of the molecules, that is, u_1^2, u_2^2, u_3^2, u_4^2, . . . , there will exist some average value U^2. Let the number of molecules inside the cube be N. Then the total change in momentum per unit time for all the molecules in the cube will be

$$\frac{2m}{s} \times NU^2$$

Since force is equal to the rate of change of momentum, this expression will be equal to the net force at any instant on all the walls of the cube.

The cube has six inside walls, each with an area s^2. The pressure P inside the cube will be equal to

$$\frac{\text{Total force}}{\text{Total area}} = \frac{2mNU^2}{s} \times \frac{1}{6s^2} = \frac{mNU^2}{3s^3} = P$$

Since s^3 is the volume V of the cube, then

$$P = \frac{1}{3}\frac{mNU^2}{V}$$

This is the *kinetic theory of gases.*

1.4.1 The kinetic theory of gases and the gas laws

While the development of the kinetic theory of gases relationship may seem a bit tedious, it is nevertheless useful because all the well-known gas laws can be developed from the one fundamental relation. In general, the kinetic energy possessed by any object having a mass m in motion with a velocity v will be $\frac{1}{2}mv^2$. For the molecules of gas inside the cube, their *average* kinetic energy will be $\frac{1}{2}mU^2$, and the sum total of their kinetic energy will be $\frac{1}{2}mNU^2$. The kinetic theory of gases holds that the kinetic energy of the molecules is directly dependent on the temperature of the gas.

Suppose that some of the gas is withdrawn from the cube, and it is then replaced by a different gas until the pressure in the cube is back up to the original value. The temperature is held constant while this is happening. Since the mixture has only one pressure and one temperature, all the molecules in the container will have the same average kinetic energy, irrespective of the gas they represent. Furthermore, it will be the same average kinetic energy that was present when the first gas was by itself in the cube at the same temperature and pressure.

If the temperature is increased, the average kinetic energy of all the molecules of the mixture in the cube will be increased by the same amount. Certainly, if the temperature were lowered to absolute zero, then the average kinetic energy of every gas molecule should decrease to zero. In practice, this cannot be proven because all gases change to their liquid state before their temperature becomes this low.

This simply supports the conclusion that molecules of different gases at the same temperature will have the same average kinetic energy. If there are two gases numbered 1 and 2, the average kinetic energy of a molecule of gas 1 will be $\frac{1}{2}m_1U_1^2$, and for gas 2 it will be $\frac{1}{2}m_2U_2^2$. While m_1 may not be equal to m_2 and U_1^2 may not be equal to U_2^2, nevertheless, if gas 1 and gas 2 are at the same temperature,

$$\tfrac{1}{2}m_1U_1^2 = \tfrac{1}{2}m_2U_2^2$$

If T is the absolute temperature of the gas, then $\frac{1}{2}mNU^2$, and consequently, mNU^2, must be proportional to T. If the $\frac{1}{3}$ factor is combined with the proportional constant to create a constant C, then

$$P = \frac{1}{3}\frac{mNU^2}{V} \qquad \text{becomes} \qquad P = \frac{CT}{V}$$

The kinetic theory relationship accordingly can be rearranged to

$$PV = CT$$

For a fixed quantity of gas, with the temperature constant,

$$PV = CT \qquad \text{becomes} \qquad PV = \text{constant}$$

which is *Boyle's law.*

> If the temperature of a gas is constant, the product of its absolute pressure and its volume will be a constant.

Similarly, if the quantity of gas is fixed and the volume is held constant, then

$$PV = CT \qquad \text{becomes} \qquad P = \text{constant} \times T$$

which is *Charles' law.*

> If the volume of a gas is held constant, the absolute pressure of the gas will be proportional to its absolute temperature.

Let's consider the situation in which there are two containers, both the same size, filled with two different gases, gas 1 and gas 2. Furthermore, the gases in both containers are at the same pressure and temperature.

The original relationship for the kinetic theory of gases is

$$PV = \tfrac{1}{3}mNU^2$$

Consequently, for gas 1, $P_1V_1 = \tfrac{1}{3}m_1N_1U_1^2$. For gas 2, $P_2V_2 = \tfrac{1}{3}m_2N_2U_2^2$.

Since the containers are the same size, the volumes of both gases will be the same, and it has been stipulated that their pressures are also the same. Therefore, $P_1V_1 = P_2V_2$, and since this is so, then

$$m_1N_1U_1^2 = m_2N_2U_2^2$$

Since the average kinetic energy of a molecule of gas 1 will be the same as the average kinetic energy of a molecule of gas 2, then

$$\tfrac{1}{2}m_1U_1^2 = \tfrac{1}{2}m_2U_2^2$$

Considering $\tfrac{1}{2}m_1U_1^2 = \tfrac{1}{2}m_2U_2^2$ and $m_1N_1U_1^2 = m_2N_2U_2^2$ together leads to the conclusion that N_1 and N_2 also must be equal. This is *Avogadro's hypothesis.*

> Equal volumes of different gases at the same pressure and temperature contain the same number of molecules.

The relation $PV = CT$ is the basis for still another gas law. If there is a mixture of gases in a single container, each of the component gases will have the same volume and the same temperature, and all the molecules will have the same average kinetic energy. Therefore, the pressure contributed by each of the component gases will be in proportion to the number of molecules of each gas that are present. This is *Dalton's law* of partial pressures.

> The pressure of a mixture of gases inside a closed container will be equal to the sum of the partial pressures of the individual gases.

The significance of Dalton's law is that each component gas will behave as if it were by itself in the container at a pressure equal to its partial pressure in the mixture. This fact is most important in processing operations such as fractionation.

Finally, the value of the so-called constant C will depend on a number of factors, but not on the kind of gas. In most thermodynamics texts, this constant is known as the *universal gas constant* and is usually designated R. The universal gas law is, accordingly,

$$PV = RT$$

The value of R depends on the quantity of gas and on the units that are used for P, V, and T, but not on the gas itself. For 1 gram mole of gas, with P in pascals absolute, V in cubic meters, and T in degrees R, the value of R is 8.314. For 1 pound mole of gas, with P in pounds per square inch absolute, V in cubic feet, and T in degrees K, the value of R is 10.73.

1.4.2 The universal gas law problem

Each of the gas laws, Boyle's, Charles', Avogadro's, and Dalton's, was developed originally from experimental observations. For a long time they were all accepted as being valid, as it was considered that any disagreement observed between calculated values and measured values was within the limits of experimental error. As accuracy in measurement improved, however, it eventually became obvious that none of the known gases actually obey the universal gas law $PV = RT$ or the other aforementioned laws exactly, although conformance is approached as the pressure of the gas decreases.

Recognition of this deviation is generally made by writing the universal gas law in the form

$$PV = zRT$$

The factor z is called the *gas law deviation factor.* It is a dimensionless quantity and is equal to PV/RT at the particular P and T at which the measurements are made. The true value of z for different gases at various pressures and temperatures can only be determined by laboratory experiment.

Graphs in which z is plotted on a base of reduced pressure at various reduced temperatures show that for many gases the value of z deviates significantly from 1.00 for temperatures lower than about twice the critical temperature of the gas, unless the pressure of the gas is less than about one-fifth its critical pressure. For air, this would mean a temperature of −8°C or a pressure of 650 kPag (95 psig). Thus air is one gas that conforms quite well to the universal gas law at relatively reasonable temperatures and pressures.

Conditions for conformance with the universal gas law, that is, high temperatures and low pressures, are actually what one would expect if one were to take a critical view of the assumptions that were made before the fundamental relation for the kinetic theory of gases was developed. The assumption that the size of the molecules is negligible compared with the distances between the molecules would be reasonable when the pressure is low. At higher pressures, however, the intermolecular spaces decrease, and the assumption becomes questionable.

The assumption that the motion of the molecules is influenced only be their kinetic energy, not by any mutual forces of attraction, is reasonable when the temperature is relatively high and the molecules have substantial kinetic energy. It is believed, however, that there are gravitational forces acting between the molecules. As the temperature decreases and the molecules lose kinetic energy, a point is reached where their motion is influenced by the intermolecular forces as well.

Some confusion exists with respect to the gas law deviation factor owing to the fact that in some texts it is called the *compressibility factor.* In other texts, it has been called the *supercompressibility factor,* which is actually the inverse of the gas law deviation factor, that is, $1/z$. See Appendix for a table of gas law deviation factors at 15°C and 101.325 kPaa for a number of gases.

1.4.3 Density of gases

The corrected gas law is $PV = zRT$, for 1 mol of gas. The value of R depends on the units used for P, V, and T. If the molecular weight of the gas is M, then at the prevailing conditions of P and T, the density of the gas ρ will be M/V, where V is the volume of 1 mol of the gas. Therefore, V is equal to M/ρ. Substituting for V in the gas law equation gives

$$\frac{PM}{\rho} = zRT \qquad \text{and} \qquad \rho = \frac{PM}{zRT}$$

This relation can be useful for calculating the density of a gas. If the pressure is in kilopascals absolute and the temperature is in degrees Kelvin, the value of R will be 8.314, and the density calculated will be in kilograms per cubic meter.

1.4.4 Specific gravity of gases

Some texts and manuals refer to a property called the "specific gravity" of a gas. It is doubtful that when the term *specific gravity* was coined it was ever intended to apply to anything but liquids. Frankly, where gases are concerned, one is on much safer ground if one sticks to the concept of density, which is far more basic, and forgets about specific gravity altogether. This practice is recommended.

Since over a period of time specific gravity for gases has been used in some places, it is probably best to clear up any misunderstanding as to what it really means. The *specific gravity of a gas* is defined as being the ratio

$$\frac{\text{Density of the gas}}{\text{Density of air}}$$

both density measurements being taken *at the same pressure and temperature.*

By definition, both density measurements are to be made at the same pressure and temperature. This being so, it may lead to the following question. Since no specific pressure, or temperature, is specified, only the *same* pressure and temperature, does this mean that the specific gravity of the gas will have the same value at *any* pressure and temperature?

Suppose that at a pressure P_1 and a temperature T_1 a volume of gas V_1 has a mass W_1. Its density at P_1 and T_1 is therefore W_1/V_1. Further, suppose that at the same pressure and temperature the mass of the same volume V_1 of air is W_{A1} so that the density of the air is W_{A1}/V_1. The specific gravity of the gas is then

$$\frac{W_1/V_1}{W_{A1}/V_1} = \frac{W_1}{W_{A1}}$$

Any quantity of gas will be made up of the molecules of the gas, along with the spaces between the molecules. The spaces will not contribute to the mass of the gas. Consequently, the mass of the quantity of gas will be due entirely to the number of molecules in the quantity.

Avogadro's hypothesis states that equal volumes of different gases at the same pressure and temperature will contain the same number of molecules. This implies that the volume V_1 of the gas will contain the same number of molecules as the volume V_1 of air. It also means that the mass W_1 of the gas and the mass W_{A1} of the air represent the masses of two equal numbers of molecules. Therefore, the specific gravity of the gas (W_1/W_{A1}) must be equal to

$$\frac{\text{Mass of an individual gas molecule}}{\text{Mass of an individual air molecule}} = \frac{\text{molecular weight of the gas}}{\text{molecular weight of air}}$$

If another pressure P_2 and another temperature T_2 were chosen, the same process of reasoning could be followed, leading to the same conclusion—that the specific gravity of the gas at P_2 and T_2 is again equal to the ratio of the molecular weight of the gas to the molecular weight of air. The general conclusion, therefore, is that the specific gravity of a gas is equal to the molecular weight of the gas divided by the molecular weight of air (28.97). Furthermore, the specific gravity must have the same value at all pressures and temperatures. This statement actually appears in some texts.

Suppose that the molecular weight of the gas is M, that the gas law deviation factor of the gas at the prevailing pressure and temperature (P and T) is z, and that the density of the gas at P and T is ρ. Suppose also that the corresponding values for air are M_A, z_A, and ρ_A. Using the definition for the specific gravity of a gas and the relationship for gas density developed on page 13, we get

$$\frac{\rho}{\rho_A} = \frac{PM/zRT}{PM_A/z_ART} = \frac{M}{M_A} \times \frac{z_A}{z}$$

The P and T both cancel in this case, since the definition of specific gravity specifies that both density measurements be made at the same pressure and temperature.

This last relation shows that the specific gravity of a gas is not equal to the ratio of the molecular weight of the gas to the molecular weight of air but is actually equal to this ratio modified by the factor z_A/z. Two points regarding this factor are important. First of all, the individual values of z and z_A vary with pressure and temperature, so the value of z_A/z also must depend on these conditions. Thus the value of the specific gravity of a gas cannot be independent of pressure and temperature.

Second, for many gases, especially hydrocarbons, the ratio z_A/z is usually not close enough to unity that one can assume that the specific gravity is equal to M/M_A without a significant error. For practical purposes, the loss in accuracy will likely be unimportant when doing control valve sizing calculations. The error could be significant, however, if $z_A/z = 1$ is assumed when doing gas flowmeter calculations.

Chapter

2

How Instruments Are Selected

The U.S. astronaut Wally Schirra became famous for a comment he made just prior to being launched into space on one of his missions. As he climbed into the space capsule and fastened his seat belt, he reflected, "Just think, Wally, everything in here was supplied by the lowest bidder."

The comment was probably true, as far as it went. What was not strictly correct about it was that it did not tell the whole story. While everything may have been provided by the lowest bidder, every component would have been required to meet a specification that was written for it by people who understood the function that the component was required to perform. Nevertheless, this shows, in a way, how the personal element often can enter into the selection of hardware for vehicles, machines, plants, or instrumentation systems.

It is worth the time, and the paper, to consider just what determines the selection of process instruments for process plants. The list that is laid out here is only roughly in order of the frequency in which the particular item is an influential factor.

2.1 Size or Range

The fact that most instruments and control system components come only in definite ranges and sizes puts a limitation on their use in some applications. A differential pressure (DP) transmitter equipped with integral seals, a useful device for level measurement, has a minimum range of 0 to 5 kPa, or 0 to 20 inches of water. If the application requires a shorter measurement range, this DP transmitter cannot be used.

In the same way, the maximum size in which cage-type control valves are made is about 8 in. If the application requires more capacity than an 8-in cage valve can provide, then some other type of control valve must be used.

2.2 Operating Conditions

Operating conditions that affect the selection of instruments include temperature, pressure, and stream velocity. Measurement of liquid level in a vessel using a sonic beam has some attractive features, particularly the linearity of the measurement and the fact that nothing contacts the liquid. However, at the present time, the transducers used in sonic level systems have an upper temperature limit of about 100°C.

Some plant stream analyzers require that the pressure of the fluid under analysis be held below a specified upper limit. In addition, vortex shedding, turbine, and some other types of flowmeters will not measure accurately unless the stream velocity is greater than a low limit value.

2.3 Stream Properties

In certain cases, the properties of the stream itself will put a restriction on the type of instrument selected. A high viscosity will lower the Reynolds number, which may prohibit the use of some types of flowmeters. Some sonic meters cannot tolerate solids in the stream, whereas others will not work unless solids are present. A magnetic flowmeter cannot function unless the electrical conductivity of the stream is above a minimum value. And the acidity or alkalinity of the stream may rule out a particular type of measurement sensor.

In food processing plants, measuring instruments may have to conform with a sanitation code, which can limit the choice of measurement sensors. These are only a few of the many examples that could be cited of stream properties placing a limitation on the selection of instruments.

2.4 Cost

All too often cost is considered to mean only what it takes to *purchase* an instrument or system. To many project managers, this is the only cost that is relevant. Unfortunately, there is the more insidious matter of what it costs to get the results that you wanted from the instrumentation, as well as what it costs in worker hours to keep it performing at the level required.

This second aspect of cost is sometimes called the *cost of ownership*. It is very real, yet it is frequently ignored by short-sighted individuals who have control over the money. How many project managers will buy the argument that if you spend 20 percent more for a better system, you will get it back, and then some, in more reliable performance, which in turn will result in more profitable plant operations and reduced maintenance costs? The response one usually gets is, "If it meets the specifications, which, incidentally, you wrote for it, then we are going to accept the lowest bid."

A curious fact about instrumentation costs is that for large-scale projects involving hydrocarbon plants, the cost of instrumentation turns out to be just about 5 percent of the overall project cost. What this means is that if we were to spend 20 percent more than the lowest bid to get the best instrumentation, instead of that which barely meets the specs, it would increase the overall project cost by only 1 percent. This, unfortunately, is another argument for which project managers have little sympathy.

Finally, those who have worked on the instrumentation design for large projects have probably noticed that when management begins to lose sleep over the rising costs of the project, and the inevitable edict to economize wherever possible is handed down, every other area of the job can think up reasons why their part cannot be cut, so the instrumentation gets axed to satisfy management that something is being done.

Since this book will never be read by project managers, a subterfuge that has been known to work will be suggested. Assuming that there is no doubt that the higher priced system is significantly superior and would be a worthwhile investment, the arguments for it can be put before the operations department, which will eventually have the system as part of its equipment. If the personnel in that department can be convinced that the better system will mean a better running plant, then they may be inclined to put pressure on project management to come up with the extra money.

Most operations departments have considerable clout, and if their personnel do not get what they feel is necessary, they have been known to go around project management to senior management. Their traditional argument—"If we don't get this, we can't assume responsibility for the success of the plant"—seems to win out every time.

A company president was once quoted as saying, "Don't worry about the first cost, as long as it is the last cost." Management people with this enlightened attitude seem to be, sad to say, few and far between.

2.5 Compatibility

Measurement and control systems have more than one component. A feedback control system will have a transmitter, a controller, and a control valve. These components must be compatible in the sense that the signals that each one sends out must be understood by the following component that receives the signals.

One of the virtues of pneumatic instruments is that years ago the pneumatic signal range was standardized at 3 to 15 psig. Furthermore, in pneumatic systems, polarity, grounding, impedance matching, and so on have no significance. Thus a user of pneumatic instruments can mix and match brands of components as much as he or she pleases.

The same does not necessarily apply to electronic instruments. Even though the signal range is now standardized at 4 to 20 mA, this does not mean that anyone's transmitter can be used with anyone else's controller or that any controller can be used with any valve I/P (current/pressure) converter. In some cases, different brands will be found to be incompatible, and this will impose a restriction on the selection of instruments.

2.6 Standardization

When a plant is satisfied with one particular brand of instruments, why not standardize? Standardization simplifies purchasing and minimizes the learning process for operations and maintenance personnel. What standardization does not do is guarantee the best prices or deliveries. And it may mean that the plant is not getting the benefits of the latest and best technology in instrumentation.

If the instruments are required for an expansion to an existing plant, then it makes sense to specify the same brand that was supplied for the original plant. In any event, the emphasis that should be placed on standardization is a decision for the instrument user, taking into account the circumstances. It is mentioned here because sometimes it is a factor that influences whose instruments are bought.

2.7 Certification by Government Agencies

This item has to do with instruments that are electrically powered. The question of safety is involved, particularly in plants that are processing flammable materials. If there are flammable vapors present, either normally or due to a leak, then an electrical fault can provide the ignition required to create an explosion. It has happened.

There is a definite difference in the approaches taken to this problem in Canada and in the United States. In the United States, the question of the safety of electrically powered equipment is a matter between the user company and its insurance company. This is why the leading testing laboratories, Factory Mutual and Underwriters, are sponsored by insurance firms.

In Canada, however, provincial governments are involved. It is against the law in Canada to power up any electrically operated equipment unless, and until, it has been certified safe to use by a provincial electrical inspector. This is where the Canadian Standards Association (CSA) often gets involved.

CSA is an agency sponsored by the Canadian federal government. Its directive is to ensure that devices offered for sale to the public at large are in fact safe to be used. Although the bulk of the agency's work seems to concern electrical devices, construction materials, plumbing fixtures, and other products are part of the package. CSA carries out this directive by testing products in its testing laboratories, by sending out qualified inspectors to examine products, and by publishing standards.

Although it would appear otherwise, CSA does not have the authority to write laws. However, it is quite possible for any province to require that products or work conform with a particular CSA standard, which, in effect, would make that standard a law. The Canadian Electrical Code, which was prepared by CSA, is probably the most prominent example.

Most provincial inspectors do not feel qualified to evaluate the safety of process instruments made up of circuit boards filled with solid-state components, microprocessor chips, and the like. Consequently, they usually require that the instrument have CSA certification. If it has a CSA sticker on it, it passes. If it does not, it is turned down.

Obtaining CSA certification for an instrument is not particularly easy. First of all, CSA will make a substantial charge for the certification service. An instrument of the model to be certified will have to be submitted to the laboratory for testing to verify its freedom from safety hazards that could be caused by electrical faults. The instrument may be destroyed during the testing process. Finally, depending on the amount of work that CSA has in hand, the testing and certification may take a year or more.

The bottom line is that U.S. manufacturers of specialty instruments, such as plant stream analyzers, will sometimes turn down an invitation to bid on a Canadian order because the cost and trouble associated with obtaining the CSA approval will never be repaid by the amount of business that results. This means that the potential Canadian user is left to look elsewhere for an instrument that will do the job.

Certification by CSA can actually come in either of two forms: general acceptance or special acceptance. See Appendix for more information.

2.8 Maintenance Preferences

The likes and dislikes of the people who do the maintenance work on the instruments often have a bearing on what should be bought or what should not. Generally, instrument maintenance groups want instruments that can have their readings easily verified and can be zero checked without having to remove them from the process. They want instruments that hold their calibration for long periods of time and that can be installed so that they are easy to reach. They also want instruments they can fix themselves, if the need arises, rather than having to send them back to the factory for repair.

What most instrument maintenance people do not want is to be pioneers in new instrumentation developments, especially if they believe the job can be done with some instrument or system with which they are already familiar.

While the desire to do as much work in house is a compliment to the zeal of instrument maintenance groups, and it minimizes delays, a factor that should be considered is the quality of the product produced in the instrument shop versus that which would be expected from an instrument manufacturer who may well have better facilities, materials, techniques, or know-how.

A case in point is the popular choice of thermocouples for temperature measurement. In the Instrumentation Engineering Technology Program at Northern Alberta Institute of Technology (NAIT), one of the first-year laboratories has as its assignment the assembly of a thermocouple. Students cut two lengths of thermocouple wire, one copper and one constantan, twist them together with pliers, and then braise them with a torch. The students then confirm that the thermocouple is actually a temperature sensor by connecting it to a meter that can read small electromotive forces (emfs). All in all, an interesting exercise.

An overlooked result of this laboratory, however, is that it tends to delude students into believing that making up a thermocouple is a simple task. This attitude may stay with them if these students graduate to be instrument service personnel and assume that making thermocouples is part of their job function. What is often forgotten is that a thermocouple is really a tiny battery whose emf output varies with temperature. It is the emf that the meter measures, not temperature per se. The measured emf has to be converted to temperature using correlation tables, usually those published by the U.S. National Institute of Standards and Technology (NIST).

The NIST tables were derived from experimental data, and there can be no doubt that when they were drawn up, the thermocouples used were made very carefully. The purity of the two dissimilar wires, on which the validity of the tables depends, was carefully protected. Procedures that would tend to change the metallurgical qualities of the wires would be avoided. These would include using pliers to cut wires and twist them, which cold works the metal, and braising them with a torch, which induces oxidation, thermally stresses the wires, and in all likelihood changes their crystalline structure. The technique that is used in the NAIT first-year laboratory, and in some instrument maintenance shops as well, would not be considered conducive to producing a thermocouple whose temperature/emf characteristic closely matched the correlations in the NIST tables.

Accordingly, one aspect of procurement is whether maintenance personnel prefer to buy or to make a particular instrument or component. In the event that an instrument is to be bought, then before any supplier is selected, it is good policy to find out if there are any strong maintenance preferences, bearing in mind that negative preferences are as important as positive ones. When maintenance people approve of what they are getting, they will usually go out of their way to ensure that it works. If something else is bought, the results may be indifferent.

2.9 Maintenance Capability

Whenever the installation of a new measurement or control system is being contemplated, one question that must be asked is, "Whom do we have who can make it work?" Failure to do this usually leads to disillusionment. It is not simply a matter of bringing the new instrumentation on line successfully either. This is often done by high-priced personnel, possibly from the engineering department or from the company that made the system.

Sooner or later these people will have to walk out of the gate and leave the maintenance and care of the system to the plant instrument group. If this group does not have the know-how to get the job done, or if it simply does not have enough time to keep the system tuned up properly, then over a period of time there will be a deterioration in the performance of the system until it eventually becomes virtually useless. The problem will be compounded if production depends on the availability of the system. Frequent and expensive service calls by a service person from the manufacturer's plant usually will only delay the inevitable.

Maintenance capability is made up of knowledge, time, and acceptance of responsibility. If the maintenance group does not have all three of these components, as far as the new instrumentation is concerned, or is not prepared to acquire them, then something else with less stringent requirements should be chosen for the job.

Plants in remote areas where the instrument service person doubles as an electrician, as well as doing the occasional pipefitting, welding, or insulating, will be better off with pneumatic instruments.

2.10 Confidence, Known Past Performance

When something has worked well for you before, it is only natural to give it preference the next time a similar job comes along. Changing to some other system or method requires considerable justification.

Known past performance is not necessarily limited to in-house experience either. It also includes the experience of others who have had to debug similar instrument applications in their own plants and who have hard-earned field experience with the instrumentation or system that you are considering. The question to be asked is, "What did you have to do to get the results you were expecting to get when you decided to buy it?"

Figure 2.1 shows an acceptance pattern that occurs fairly often in plants, especially in operating departments. The horizontal scale is a time scale, whereas the vertical scale shows different levels of acceptance for new operating equipment. When the operations department first hears the news, usually from central engineering, that it will be getting some newly developed operating equipment that has never been used in the plant before, the reaction to the idea will likely lie somewhere between suspicion and indifference. This attitude prevails until startup time, when, owing to unfamiliarity with the new equipment and the usual debugging problems, just about everything that can go wrong does. A steep drop in the acceptance level results.

This continues, along with the consequent production problems, until a top-level meeting is held, at which time the plant manager, or someone of equivalent authority, declares, "Something has got to be done!" At this point a hurried call is made to the factory where the equipment was made, and a qualified service engineer, who really understands the equipment, is on the next plane. He or she quickly assesses the situation, corrects the problems, provides the answers

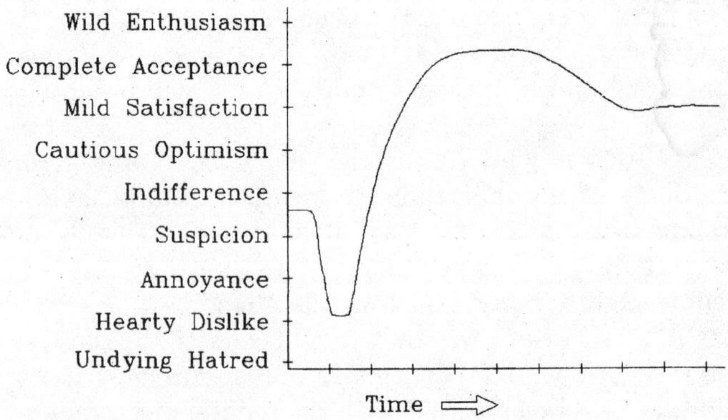

Figure 2.1 Acceptance curve for new operating equipment.

which the instruction manuals did not provide or which the instruction manuals did provide but which were never read, and the new equipment straightens out and starts to do the job it was intended to do all along.

The acceptance level then rises to great heights and remains there for some time. Eventually, however, the natural process of wearing takes place, and some minor problems that require servicing begin to show up. These are corrected easily enough so that the long-term acceptance of the new equipment remains at about the mild satisfaction level, overall, with all concerned.

The point of all this is that if anyone in the plant were asked for his or her assessment of the value of the new equipment, the point that he or she is at on the acceptance curve would have considerable influence on the answer. It is worthwhile to keep this in mind when asking for operating experience on equipment that you have not used yourself and which you are considering buying.

2.11 Design Features

Design features include such things as accuracy, selectivity, serviceability, ambient limitations, and so on. Whereas the preceding consideration, known past performance, is information gleaned from actual experience with the instrument or system by yourself or someone you know who has used it, the design features are the information provided to you by the manufacturer, through sales bulletins or salespeople.

While this information will have its use, it will hardly be complete enough for making a sound decision, since a salesperson will rarely give you a reason not to buy his or her product.

2.12 Delivery

Sometimes an emergency will occur that requires immediate procurement of an instrument so that mechanical work on a plant can be completed and production resumed. The regular instrument supplier can ship in 2 weeks. The competition, which has been trying for years to get one of its instruments into your plant, will ship a comparable instrument that afternoon. Obviously, delivery will have priority over preference in this case.

2.13 Supplier Credibility, Backup

Many instrument manufacturers do not market their products directly but deal through jobbers and supplier representatives. What these people know about the instruments they sell can vary from in-depth knowledge of the strengths and weaknesses at one end of the spectrum to how to fill out customs declaration forms only at the other. A supplier who will tell you that his or her instrument is *not* the one to use in a particular application is a valuable resource indeed.

There is also the question of how much of a spare parts inventory the supplier keeps on hand. It is obviously better, from a business point of view, to have the supplier invest his or her money in spare parts, and provide the space to store them, than to have foot these costs oneself.

2.14 Life Expectancy

How long an instrument should last, relative to the anticipated life of other operating equipment in the plant, is sometimes a factor. It is a well-known fact that in these times, digital data-processing systems can be superseded before the one purchased has been brought on line successfully. On the other hand, some pneumatic instruments that were in vogue in the 1940s can still be obtained today.

When some problem is referred back to the salesperson who sold the instrument, an answer frequently heard is, "That was our old model. You should try our new one."

2.15 Other

Other considerations include reciprocal business, a secrecy agreement, policy that favors business being placed in the home province, state, or country, and the possibility that the sales manager of one of the firms bidding has a brother-in-law in the purchasing department.

2.16 Merits of Pneumatic and Electronic Instruments

Large-scale digital data-processing systems, with their computing power, have received so much publicity that in some circles digital electronics is considered to be the only way to go. Pneumatic instrumentation? It is gone with the 50-cent haircut.

Discerning instrumentation specialists will never take this attitude. They know that even in the 1990s, pneumatic instrumentation has features that electronic instrumentation cannot match. If pneumatic instrumentation is rejected for a project, then it should be for valid reasons, not because misguided individuals consider it to be obsolete.

When comparing pneumatic and electronic instruments, the following points emerge.

2.16.1 Pneumatic

- Pneumatic instruments are *safe* to use in flammable environments. No electrical approvals are required. In gas compressor stations, which normally have a gassy environment, pneumatic instruments may never be replaced with anything else. When pneumatic instruments are used, no safety approvals are required, irrespective of the environment.

- Item for item, pneumatic instruments are less expensive than electronic instruments. At one time, when pneumatic signal lines consisted of ¼-in copper tubes that were individually installed, while electronic signal wires could be pulled in in 36-pair cables, it could by argued that the installation costs for electronic instruments were less than for pneumatic instruments. Today, however, air tubes also can be pulled in in bundles, which lowers their installation cost at least to the level of electronic signal lines.

- Ever since pneumatic signals were standardized at 3 to 15 psig, all pneumatic instrument brands have been compatible. A user can mix and match brands as much as he or she pleases. There are no problems with grounding, impedance matching, polarity, or power supply levels to contend with.

- In pneumatic signal transmission systems, there are never problems with pickup or with superimposing of extraneous signals. The measurement signal line from a pneumatic transmitter can be run alongside power lines or wrapped around transformers or motors, and the signal that is sent along the line will be unaffected.

- How pneumatic instruments function is relatively easy to understand. There are not that many parts involved. The basic detector system in just about all pneumatic instruments is a nozzle and baffle, backed up with a relay valve that amplifies air flow and pressure changes. It is axiomatic that things that are easier to understand are also easier to fix.

  On the other hand, conventional practice in servicing electronic systems with their solid-state components, microprocessors, and circuit boards is to avoid trying to understand what goes on inside them. The usual procedure is to find the location of the problem and install a new component or board.

- Pneumatic instruments are rugged. They can tolerate environments in which electronic instruments will fail. This means, in particular, heat, cold, and high humidity. In fact, just about the only thing that will cause a pneumatic instrument to fail is wet or dirty air.

- Pneumatic instruments have a long service life. Even after an instrument has been withdrawn from the line, parts are generally available for many years. In contrast, the rate at which electronic systems are being superseded, as each manufacturer tries to move one up on the others, is aggravating. The fact that electronic systems are often built up from components obtained from subsuppliers over whom the system supplier has no great control is another potential problem.

If all these factors favor the use of pneumatic instruments, what are their weaknesses? An obvious one is the relatively slow speed with which pneumatic signals are transmitted. The distances, in runs of signal tubing, from the transmitter to the controller in the operations center and back to the control valve can create signal delays that may be great enough to actually cause a deterioration in control. The point at which this is reached is a combined signal line distance, measurement plus control, of about 600 ft. In very large plants, the distances become 1000 ft or more, which pretty well precludes pneumatics.

Although signal speed is a factor, it is not the dominant one, however. The real reason why electronics should be favored over pneumatics is that *electronics can be digital.* This is far more significant than any considerations of signal speed or so-called state-of-the-art technology.

The way that process control has been evolving, the ability of control systems to perform logic, as opposed to conventional feedback control, has become more and more important. Logic cannot be done efficiently with analog components, and this is the only form in which pneumatic instruments come. At one time, fluidics, a form of pneumatic flip-flops with limited logic capability, was thought to have great possibilities. The development of microprocessor chips that were 1/1000 of the size and weight, and had 10,000 times the logic capability pretty well buried fluidics.

The ability to operate in a digital format is also necessary if information is to be displayed on consoles rather than on instrument panels. If it must be digital, pneumatics cannot really do it.

A secondary problem, which is frequently forgotten, is the matter of leaks in pneumatic signal lines. A leak in a measurement signal line means that the receiving instrument sees a measurement signal that is lower than the signal which was transmitted. A leak in a control signal line means that the control valve is not stroked as far as it should be. A leak in a control signal line also can reduce the effectiveness of derivative in the controller.

When pneumatic instruments are originally installed, the signal lines will consist of lengths of tubing spliced together with connectors. These connectors will be tight enough in the beginning, but this does not mean that they are going to be tight forever. In a 10-year-old plant with pneumatic instrumentation, it may be presumed that a perceived deterioration in the performance of the measurement and control systems is due to aging of the instruments themselves. Actually, when a pneumatic instrument wears out, it is generally caused by years of operating its manual controls, or altering its calibration adjustments, or possibly by corrosion due to its environment. It is not caused by the flow of air through its internals.

The real cause of poorer performance is more likely to be deterioration of the signal lines. Yet the plants in which periodic checks on the tightness of signal lines are made seem to be few and far between.

A leak tester for pneumatic signal lines is described in the Appendix.

2.16.2 Electronic

In general, the weaknesses of pneumatic instrumentation are the strong points of electronic, and vice versa.

- Electronic instrument systems can be digital as well as analog. Consequently, their capabilities are much superior when logic is to be performed.
- The digital capability of electronics makes possible the use of console displays.
- The higher signal speed, roughly the speed of light versus the speed of sound, makes electronic instruments superior when measurement and control signals have to transmitted over large distances.
- Digital capability means better compatibility with computers and video display terminals (VDTs).

The key weaknesses are obsolescence, complexity, and requirement for approvals.

Although not necessarily a weakness, there is nevertheless a definite limit to the amount of resistance with which the measurement and control circuits can be loaded up. Intrinsic safety barriers, dropping resistors, and wire all have substantial resistance. Since the power supply for the loop will not have unlimited capability to overcome this, it is possible to run into a situation where there is too much resistance in the loop for proper operation.

2.16.3 The intrinsic safety concept

Intrinsic safety is one aspect of the broader subject of electrical safety. Some general background on electrical safety may prove helpful in putting intrinsic safety in perspective.

In plants that process and handle flammable materials, safety personnel often refer to the *explosion equation:*

$$\text{Flammable gas} + \text{air} + \text{ignition} = \text{explosion!}$$

This equation says that three factors are required to create an explosion. If any one of them is absent, an explosion cannot occur. It would be difficult to eliminate air, since people usually have to work in the potentially hazardous areas. Thus the usual approach is to try to eliminate either the flammable gas or the source of ignition.

The problem of eliminating the flammable gas can be handled by constructing a pressure-ventilated building, within the hazardous area, in which to house any operating equipment that might become a source of ignition. This includes electronic instruments. The building must by virtually airtight. Clean air from a location where there are no flammable vapors is blown in. The ventilation system also keeps the building at a slightly positive pressure so that if there is any leakage, it will be from the inside out, not the other way.

One way to prevent a source of ignition would be to refrain from using any devices that have open flames or electrically powered equipment of any kind, since electrical devices can develop faults that produce sparks and cause ignition. This approach is obviously not very practical, considering modern production methods and the products that are to be made.

A philosophy that eventually emerged, and which is still used, does not adhere to the "remove the source of ignition" approach. It can be applied to instrumentation; thus it deserves some consideration. The concept is to allow the explosion to happen, if conditions are right for it, but to contain it safely. This can be done by mounting electrically powered devices inside explosion-proof enclosures. In order to be effective, an explosion-proof enclosure must have these two features:

- It must be strong enough to resist the buildup of pressure which the explosion will create inside of it without fracturing.
- Whether the cover of the enclosure is screwed on or bolted on, the escape path taken by the hot combustion gases created by the explosion, along the threads or between the mating flange surfaces, must be long enough that by the time the gases reach the outside, they are at too low a temperature to ignite any flammable gases that may be on the outside of the enclosure.

Accordingly, explosion-proof enclosures have to be constructed so that if there is an internal explosion, the combustion products can leak out. There is an implication here that instrumentation people need to recognize. *Explosion-proof enclosures cannot be weatherproof.* To be weatherproof, the cover of the enclosure must be pulled down on a resilient surface, such as a gasket. This will make the enclosure virtually airtight. Therefore, if there were a pressure buildup inside the enclosure, the pressure would not be vented off.

When the first electronic process instruments became available about 1950, the approach taken, if they were to be located in hazardous areas, was to install transmitters, valve converters, and other field-mounted devices in explosion-proof enclosures. The flammable gas, the air, and the potential source of ignition were all present. Consequently, explosion-proof enclosures were the only option available.

All of this notwithstanding, if electrical safety today depended on the use of explosion-proof enclosures for field-mounted electronic instruments, then there would be no electronic instruments in plants that handle flammable materials. The reason: Explosion-proof enclosures simply are not safe when servicing has to be done.

The transmitter that does not need periodic testing, troubleshooting, and adjusting has not been made. In order to do calibration checks or other tests on an electronic transmitter, the cover of the enclosure has to be off, and the transmitter power has to be on. If flammable vapors can be present, this is obviously dangerous.

Another situation that has occurred involves large instruments, such as plant stream analyzers, which require large enclosures with equally large covers. To remove the cover might entail removing 20 or more large bolts. If the instrument has been particularly cranky so that the cover has had to be removed and replaced several times, then it will eventually be found that the cover is being held in place by 2 bolts at the upper corners, while the other 18 bolts are lying on the ground. If an explosion took place under these circumstances, it would not be contained, and any combustion gases would not be cooled down before they reached the outside.

The salvation of electronic instrumentation for plants where hazardous vapors are present has been a development called *intrinsic safety.* The concept originated in Great Britain as a response to the number of explosions that were occurring in coal mines. The hazardous vapors were "fire damp," which is essentially methane, while the source of ignition was electrical equipment, usually lighting.

Explosion-proof lighting was not the answer. One could only imagine what it would be like to wear a miner's helmet equipped with an explosion-proof lamp all day long.

It turns out, however, that even a mixture of flammable gas and air in its most explosive form has to have a minimum amount of energy in an electrical spark if the mixture is to be ignited. Furthermore, the amount of energy in an electrical spark is easily measured in terms of the current that is creating the spark. If the flammable gas is known, an explosive mixture can be created in the laboratory, and the minimum current required to spark and ignite the mixture can be determined with good accuracy.

If the electrically powered equipment is designed so that the currents flowing in the various circuits are always well below the value that could ignite the known flammable gases in the event of a fault, the equipment will be inherently safe. This is the underlying principle of intrinsic safety. Eventually, this principle came to be applied in the design of electronic process instruments.

Considering the advances made to this point, one would deduce that the way had finally been cleared for electronic instruments to be used safely in hazardous areas. Unfortunately, it was not to be, at least not in Canada. Electrical equipment still had to be approved by provincial inspectors, and they were not about to pass intrinsically safe instruments just because the manufacturer claimed that it had put them together in such a way that they could not create a hazard. Some inspectors had not really been tuned in to what intrinsic safety was all about, and none of them felt qualified to look at circuit diagrams and decide whether an instrument was in fact intrinsically safe. Consequently, responsibility for certification was once again passed over to CSA.

In the original stages, the position taken by CSA with respect to electronic instruments and intrinsic safety just about killed any possibility that they could be used. To ensure that there was zero possibility of any hazard, CSA required that process electronic instruments be tested on a loop basis. In other words, the transmitter, controller, and valve converter all had to be connected up and tested together. At that time, the compatibility of different brands of electronic instruments was not particularly good, so this meant that only those manufacturers who made all three components need apply. The manufacturer also had to be one that had the money and the patience to wait out the CSA tests. It was probable that only one of the leading instrument manufacturers had these qualifications. Once again, electronic instruments for hazardous areas appeared doomed.

However, just when everything looked gloomy, the good instrument fairy appeared with another of her priceless gifts. It was called the *intrinsic safety barrier.* Along with it came a general acceptance of the fact that although the transmitter, controller, and valve converter were all connected together electrically, in many cases only the transmitter and the valve converter were out in the hazardous area. The controller usually was located inside an operations center that was pressure ventilated and free of any hazardous vapors. In addition, the power required to operate the transmitter and the valve converter came from the controller.

All that was really necessary for safety, therefore, was to limit the current in the signal lines between the transmitter and the controller and between the controller and the valve converter to an intrinsically safe value. An intrinsic safety barrier, installed in the signal line as it entered or left the operations center, achieved this.

2.16.4 Intrinsic safety barriers

The intrinsic safety barrier changed the rules of the game. It was no longer necessary to have the transmitter, controller, and valve converter tested together. The problem was reduced to having the transmitter or the valve converter tested in combination with the barrier that would be used with it. Although most instrument manufacturers made their own barriers, it was not necessary to buy both the instrument and the barrier from the same company. To obtain certification, however, it was necessary that the combination of the instrument *and* the barrier that would be used with it be approved.

The controller, and in fact, any other components of the loop that were located inside the safe area, did not need approval, at least not for intrinsic safety. This meant that all the peripheral devices such as recorders, alarm switches, function generators, data processors, and so on could be hooked into the loop along with the controller. The only stipulation, from the safety standpoint, was that none of these devices be operating at a voltage level higher than 250 V dc.

The parts that make up an intrinsic safety barrier are shown in Fig. 2.2. Most barriers look something like this one. It consists of a fuse F, along with two zener diodes, D1 and D2, separated by dropping resistors R1 and R2. These are often encapsulated into a block. The current in the signal line is limited in the following way: If the voltage ahead of the dropping resistor R1 exceeds 26 V, the current is shunted to ground by zener diode D1. The second resistor and diode pair backs up the first pair. If the voltage ahead of the dropping resistor R2 exceeds 28 V, zener diode D2 shunts the current to ground. In

Figure 2.2 A basic intrinsic safety barrier.

the event of a large current surge which the diodes cannot handle, the fuse F opens.

Although the principles of intrinsic safety barriers are fairly straightforward, there are some fine points that need further explanation. Electronic instrument signals have a range of 4 to 20 mA dc, which means that there will be a pair of signal wires, one positive and one negative. The barrier illustrated in Fig. 2.2 is for installation in a positive line. If the barrier is to go in a negative line, a different design, which has the zener diodes reversed, is required.

In theory, four barriers are required for each control loop, two on each wire of the measurement signal line and two on each wire of the control signal line. However, a dc circuit usually can be grounded at one place without affecting its operation. If any of the four wires can be grounded, the need for a barrier on that wire is eliminated.

It may actually be possible to ground one wire of the measurement signal pair and one wire of the control signal pair, thus removing the need for two of the barriers. However, this will work only *if the input and the output sides of the electronic controller are electrically isolated.* This is an important detail that must be known about the controller. If the input and output are not isolated, then only one wire can be grounded, and three barriers will be needed for the measurement and control signals.

Since the first barriers were completely encapsulated, a blown fuse meant that the barrier had to be discarded and replaced with a new one. In some locations, screwed connections were not permitted, and replacing a barrier necessitated soldering. The latest barriers have been improved to the point where they are constructed like circuit breakers. If a current surge causes them to open up, they can be reset.

A designer should be aware that an intrinsic safety barrier adds a substantial amount of resistance to the measurement or control loop. The amount of resistance depends on the type of barrier, which in turn depends on the degree of hazard in the hazardous area. A barrier for a Class 1, Group D, Division 2 area (oil refinery type hydrocarbons) will add 175 Ω to the line in which it is installed and 350 Ω to the loop if a barrier is required on each wire. If the hazardous area has hydrogen present and is classified Group B, two intrinsic safety barriers suitable for Group B will add 600 Ω to the loop.

If these resistance loads are combined with a long transmission distance, or if 250-Ω dropping resistors that are used to create 1- to 5-V signals for data processors are added to the circuit, it is quite possible to run out of power to drive the 4- to 20-mA signals around the measurement or control loop.

Chapter

3

Process Measurements

This chapter will be devoted to measurement principles rather than to the way in which specific measurements are made. When the principles are in place, they will become a foundation for the descriptions of the various measurement instruments that will follow.

3.1 General

Process measurements are made in and on operating equipment and on the solid, liquid, or gaseous materials which the process equipment handles. They can be divided into two categories:

1. Measurement of the *condition* of the material or the equipment. Examples are liquid level in vessels, pressure in vessels or piping, flow rates in pipes and ducts, and temperatures of materials or operating equipment.
2. Measurement of the *physical properties,* or *physical makeup,* of the materials handled. These measurements include the measurement of density, viscosity, pH, or the percentage of a component in a mixture.

For a long time the only measurement technology available was in the capability to measure conditions, and plants had to be operated on the basis of this information only. Since 1970, however, great advances have been made through spectroscopy, chromatography, and even mass spectrometry in making measurements of the quality of raw materials, products, and effluents. This capability has proven to be exceedingly valuable.

An important aspect of current process measurement is that the requirement is to make measurements in real time, that is, measurements that are continuous and up-to-date. Thus accurate measurement information must be available at any moment that it may be called for. Taking samples and waiting for a laboratory to analyze them and report back the readings does not qualify as a real-time method.

The type of measurement required will establish the design of the sensor to be used. Obviously, the sensor must produce an output of some type that will correlate predictably with values of the variable being measured. How complicated the sensor has to be in order to do this will have an effect on its cost and probably on how well it will perform in an industrial environment. The output from the sensor will be picked up by some type of readout device that can interpret and display the measurement in terms of the value of the variable being measured. In addition to the usual gauges, meters, and recorders, computer-based readout devices that are available today make it possible to display the value of the variable in many different ways.

The first of the *fundamental laws of process control* reads as follows:

> All process control starts with measurement, and the quality of control obtained can never be better than the quality of the measurement on which it is based.

If the first law means what it says, then measurement technology should get at least as much attention as control theory, yet this does not happen often. It is an unfortunate weakness of present-day instrumentation technology that our ability to process data has far outstripped our ability to obtain good data to process. There is much that needs to be done in the development and use of better sensors.

Webster's Dictionary defines a *meter* as "an instrument for measuring." Accordingly, for the sake of brevity, the term *meter* will be used from here on, rather than measuring instrument. Note, too, that a *meter* is both a measuring instrument and the basic unit of length in the SI metric system.

3.2 Displaying the Measurement

Humans are naturally endowed with five different senses, yet as far as process measurements are concerned, only one of these senses is really useful. This is, of course, the sense of sight. Consequently, no matter what measurement is being made, it must somehow be converted into a visible reading. The more common methods of displaying measurements are

- A pointer moving across a scale
- A pen drawing a graph on a chart

- A meter displaying the measurement as a set of numbers
- A display on the screen of a video terminal, in which the value of the measurement is shown as a digital number or by using any of a number of pictorial methods such as the height of a bar graph.

The first two are analog methods, while the last two are mainly digital. Digital presentations are generally preferred today because the chance that there may be human error in reading the value from the display device is considerably reduced.

In large plant operations, especially those involving continuous processing, it is usual to collect all the important plant measurement information at one central location so that it can be monitored closely by the operating personnel. The alternative would be to have local reading meters. This would require operations personnel to walk through the plant to read the meters. The complexity of processing operations that are run today would not permit this. A dangerous situation could develop without the operators knowing about it in time.

When measurements are to be transmitted over a distance, such as from a measurement point in the process to an operations center, two instruments are required. The first is a transmitter, which will make the measurement required and convert it into a proportional pneumatic or electronic signal that can be transmitted through tubing or along wires. The second is a receiving device, which will take the transmitted signal and convert it back to an actual measurement reading in whatever units are appropriate.

Even though the majority of process meters will measure the variable of concern continuously, nevertheless, current digital processing systems may not read the signal from the meter continuously. Such systems may in fact sample the measurement at a prescribed interval. Whether or not this leads to better information about the variable being measured is a subject of much debate.

To complete the story, there is a class of process instruments designed to transmit or control process variables such as pressure and temperature but without actually displaying their value. The sensor drives the mechanism through which an output signal is obtained but does not display any measurement information. These nonindicating instruments are sometimes referred to as "blind" instruments. If a reading of the variable they are measuring is required, a separate meter must be provided.

3.3 Measurement Range

Any process meter will be able to measure values of a process variable only between fixed lower and upper limits. These limits define the measurement range of the meter. Examples are 0 to 5 meters of liquid level in a vessel, 0 to 1000 kilopascals pressure, 0 to 10 cubic meters per second flow rate, or 0 to 200 degrees Celsius. The ability of the instrument to make any measurement is necessarily limited to values of the measured variable that lie within the measurement range. In other words, the instrument cannot measure any value that is lower than the low limit of its measurement range nor higher than its high limit. Consequently, one needs to be reasonably careful when selecting the measurement range for a meter. Variations that could occur in the value of the process variable to be measured have to be taken into consideration.

When the measurement range is stated, it is customary to replace the word *to* with an en dash and to state the units in abbreviated form. Thus 0 to 1000 kilopascals becomes 0–1000 kPag, and 0 to 200 degrees Celsius becomes 0–200°C. The convention in which *to* is replaced by an en dash is satisfactory except when one of the limit values is negative, which is possible in the case of pressure and temperature measurements. In this case, the en dash can be confused with a minus sign. Consequently, it is better practice to write out the word *to*. Example: a differential pressure range of −5 *to* +5 kPa.

Although measurement ranges often start at zero, this is not mandatory. For reasons that will be explained, the lower limit of the measurement range may have a value above or below zero. A temperature range of 100 to 200°C is quite possible and, for a particular application, could be a better choice than 0 to 200°C.

3.4 Range Span

The *range span*, or simply the *span*, of a meter is the difference between the upper and lower limit values of the measurement range. Since the upper limit value is normally greater algebraically than the lower limit value, the span is usually a positive number. For a temperature meter whose range is 100 to 300°C, the span will be 200°C. If the lower limit is zero, then the span will be the same as the upper limit value. Inasmuch as the span is a difference value, the units of the span are differential units.

At least three important performance criteria of the meter—measuring sensitivity, readability, and accuracy—all depend on the range span. This makes it a reasonably important quantity, as the forthcoming discussion will show.

3.5 Elevation and Suppression

Accuracy, sensitivity, and readability are all expressed as a percentage of the range span. The shorter the range span, the better these performance criteria will be. Measurement ranges for level meters and flowmeters inevitably start at zero, so there is little opportunity to shorten the range span by cutting off the bottom end of a level or flow range.

Pressure and temperature measurements are a different matter. If the normal value for a particular process temperature is 110°C, it is not essential to specify a measurement range of 0 to 200°C. The range might be made 50 to 150°C, which would shorten the span to 100°C and improve accuracy, sensitivity, and readability, all by a factor of 2.

Adjusting the measurement range so that 0 kPag or 0°C is not the lower limit is referred to as *suppressing* or *elevating* the range. If the lower limit of the measuring range is higher than 0 kPag or 0°C, as it would be if the range were 50 to 150°C, then the range is suppressed. If the lower limit of the measuring range is below 0 kPag or 0°C, as it would be for a pressure range of −10 to +10 kPa, then the range is elevated (see Fig. 3.1).

Once again, the terminology is not particularly well chosen. *Elevated zero* and *suppressed zero* would have been more easily understood terms.

The limitations on shortening the range span by suppressing or elevating the range are (1) ensuring that all the anticipated values of the measured variable will still lie between the lower and upper limits of the range and (2) ensuring that the particular meter selected for the measurement job is actually capable of the required degree of suppression or elevation. Some meters have only limited capability in this respect.

3.6 Measurement Sensitivity

No industrial meter possesses unlimited sensitivity. As the change in the process variable being measured becomes smaller and smaller, eventually a point is reached where the meter refuses to respond. The measuring sensitivity is the smallest change in the process variable to which the meter will respond by altering its output. It is usually expressed as a percentage of the range span.

Most industrial-grade meters have a measuring sensitivity on the order of 0.2 percent of the span. Thus, for a temperature meter whose range is 100 to 300°C, the sensitivity would be 0.2 percent of 200°C,

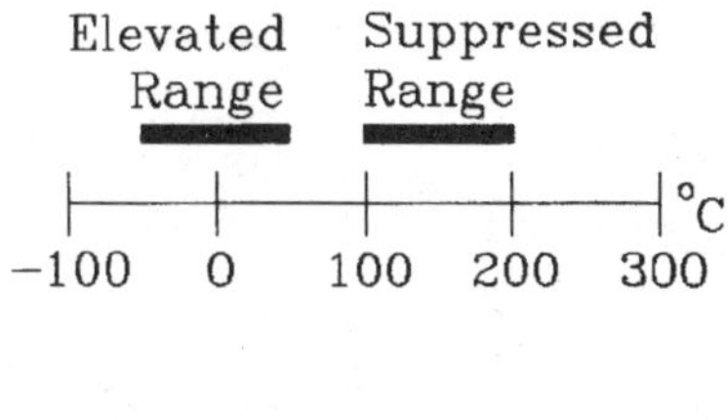

Figure 3.1 The distinction between an elevated range and a suppressed range.

or 0.4°C. This means that if the change in the measured temperature were less than 0.4°C, then the meter would not respond.

Notice that if the range could be shortened to 150 to 250°C, the measurement sensitivity would be improved to 0.2°C. Measurement sensitivity is important for automatic control. If the controller's measuring system does not react to changes in the controlled variable, then the controller will not generate any control action when in fact it may be needed.

3.7 Readability

Readability is often a concern with analog meters. When the pointer is in between two graduations, what is the correct value? There is, of course, a limit to the number of graduations that can be marked on a given scale or chart, usually about 100. A meter that has a range of 0 to 300°C would not normally have a scale with 100 divisions, since each division would then represent 3°C. The acceptable values for scale divisions are 1, 2, or 5 units or some factor of 10 times these values. A 0-to-300°C meter would likely have 60 divisions on the scale, with each division representing 5°C.

If the measurement range could be shortened to 100 to 200°C, then 100 scale divisions could be used, and each division would be 1°C, which would be an improvement in readability of five times. This improvement would be due partly to the shorter range span and partly to the fact that 1°C divisions could be used instead of 5°C divisions.

Reading errors are not a general problem with digital displays, which is one reason they are often preferred for process measurements. In fact, the opposite problem may prevail. The readability of digital displays is frequently pushed far beyond the accuracy of the measurement.

One should never be deceived into believing that there is a correlation between readability and accuracy. Any instrument can be made more readable by simply expanding its scale and putting in more graduations or more digits. This does nothing to improve its accuracy. An "honest" meter is one in which the readability is comparable with the accuracy. The fuel gauge in an automobile is usually graduated at the 25 percent points. As such, it is a pretty good example of an honest instrument, since its probable accuracy is also to within about 25 percent.

In the era preceding consoles, when process instruments were mounted on instrument panels, the one instrument in which all process operators seemed to have the utmost confidence was the multipoint temperature indicator. The long circular scale, which revolved when one of the temperature point switches was pressed, and the readability that was obtained through having a large number of graduations generated great confidence in the accuracy of the instrument. Was this confidence justified?

The sensors that were connected into the multipoint temperature indicator were thermocouples. Thus the indicator was not measuring temperature but small emfs. Each emf had to be converted to a temperature reading using a correlation between thermocouple output and temperature. The correlation used in North America is the one produced by the U.S. National Institute of Standards and Technology.

A typical multipoint temperature indicator in an oil refinery might have a range of 0 to 1200°F, with scale divisions at every 2°F. This means that the indicator could be read to 1°F, which is great readability over a 1200°F range. Is the accuracy of the temperature measurement this good also?

As an instrument for measuring emf, the accuracy of the temperature indicator is good; the error is probably no worse than 0.2 percent of the span, or within 2.4°F. There remains, however, the matter of how well any one thermocouple will behave in accordance with the NIST temperature/emf correlation.

Manufacturers who are in the business of making thermocouples, and who do their work carefully under controlled conditions, do not advertise their thermocouples as being any better than within 5°F of the values in the NIST correlation for any specified emf output. Consequently, even with a brand new well-made thermocouple and the temperature indicator in good working order, the measurement error in any temperature reading could be more than 7°F.

While this is a classic case of a mismatch between readability and accuracy, the fact that the old multipoint temperature indicator has been superseded by a console display has not improved the situation. The sensors used are still thermocouples. The correlation used is still the NIST correlation. Think about that the next time you see a temperature displayed to two decimal places on the screen of a console.

3.8 Accuracy

It is not generally appreciated that measurement accuracy exists in two forms, static and dynamic. Both types of accuracy are important in process measurement and control, though in different ways. Static accuracy is generally required in accounting situations such as billing, materials balancing, and process optimizing. Dynamic accuracy is important in automatic control, since control performance depends on the speed with which components react.

3.9 Static Accuracy

Static accuracy is a statement of how closely the reading of a meter matches the true value of the process variable being measured under static or steady-state conditions. It is an output-versus-input characteristic, the input being the true value of the measured variable, while the output is the reading of the meter. Another practical way of viewing this is that static accuracy is what is determined by a calibration. Most statements of accuracy that appear in sales brochures refer to static accuracy.

Notice that time is not a factor in determining static accuracy. When the value of the measured variable changes, the meter is allowed all the time it needs to assume its new reading. Static accuracy is usually expressed in terms of the error that can be expected. The potential error may be stated in units of the process variable being measured, but more often it is expressed as a percentage of the range span.

Accuracy statements are frequently loosely made; for example, "Orifice meters are 3 percent accurate." There are at least two things wrong with this statement. The first is that any meter that is "3 percent accurate" could technically be 97 percent in error. The second is more insidious, in that the implication is that no matter what the reading of the meter may be, it should be presumed to be out by 3 percent. This is not realistic. The calibration adjustments in any quality industrial meter are such that the meter can be calibrated for zero error at at least two points in its range. The correct way to state the accuracy of the meter is, "Orifice meters are accurate to within 3 percent."

Depending on the instrument, the error may be a percentage of the meter reading or a percentage of the range span. The specifications for the meter should be checked closely on this point. A temperature meter that is accurate to within 1 percent and has a range of 100 to 300°C would have a potential error of 1°C at the bottom of its range, increasing to 3°C at the top of the range, if the error were based on the meter reading. If the error were based on the range span, the potential error would be 2°C all across the measurement range.

If the error depends on the range span, as it often does, then the potential error can be reduced, and the accuracy improved, by shortening the range span.

3.10 True Value

An error can only be detected by comparing the reading of the meter with the true value of the measured variable. This raises the interesting question of how one knows what the true value really is. In many cases it can only be approximated. The best that can be done is to make the same measurement with a calibration standard that is known to be an order of magnitude more accurate than the plant meter being proved.

It is remarkable how often the accuracy of a meter is questioned, or even maligned, when the so-called true value against which its reading was compared was obtained from a second instrument or measurement technique that had no better or even poorer inherent accuracy than the plant meter. Most well-made calibration standards are accurate enough that they can be considered to provide the true value (close enough). They could not, however, be used in plant applications for a variety of reasons. They may be too expensive, not rugged enough, not electrically safe for a hazardous area, or not capable of being weatherproofed.

3.11 Accuracy Statements for Instruments

In some organizations, statements of instrument accuracy are made on a specific basis. To be able to interpret any statements of accuracy that are made, it is necessary to understand the basis.

One system that is fairly common involves calculating a statistical number known as the *standard deviation.* The reliability of accuracy values determined by this method improves as the number of calibration checkpoints increases. Therefore, as many readings as possible should be taken. A hundred readings would not be excessive. When all the readings have been recorded, there will be two columns of figures, one column listing the theoretical or "true" values and the other listing the observed instrument readings. Then the procedure is as follows:

1. For each reading, calculate the observed error, which is the difference between the observed value and the true value. The sign of the error, positive or negative, is not required, only the absolute value. Designate the error values as E_1, E_2, E_3, etc.
2. Determine the arithmetic average of all of the errors, that is,

$$\frac{E_1 + E_2 + E_3 + \cdots + E_n}{n}$$

where n is the number of readings. This quantity is called the *average deviation* and is designated D_a.

3. Determine the *standard deviation,* which is the quantity

$$\sqrt{\frac{\Sigma(E - D_a)^2}{(n - 1)}}$$

where Σ is the summation operator. This expression also can be written

$$\sqrt{\frac{(E_1 - D_a)^2 + (E_2 - D_a)^2 + (E_3 - D_a)^2 + \cdots + (E_n - D_a)^2}{n - 1}}$$

4. It is now necessary to discard any readings for which the error is so large that it should be attributed to experimental or observer's error rather than to the instrument. The readings to be discarded are those whose observed errors are greater than three times the standard deviation calculated in step 3.
5. Recalculate the standard deviation using only the valid readings.

The significance of the standard deviation determined in step 5 is that of all the readings taken, or which will subsequently be taken, 68 percent of the readings will lie within the bracket true value plus or minus the standard deviation, 95 percent will lie within the bracket true value plus or minus twice the standard deviation, and 99 percent will lie within the bracket true value plus or minus three times the standard deviation.

Stated in another way, if the standard deviation can be determined for an instrument, then it can be expected that 99 percent of the time the instrument readings will be within three times the standard deviation of the true value, 95 percent of the time they will be within twice the standard deviation of the true value, and 68 percent of the time they will be within one standard deviation of the true value. In other words, there will exist a certain *level of confidence.*

When a statement of accuracy is made for an instrument, it is important to know the confidence level that should be applied to the statement. If, for example, the statement is made at the 99 percent confidence level, it will mean that the potential error should be considered to be three times the standard deviation value determined from the instrument's calibration data.

Although this method has a certain appeal for people who like to dabble in statistical analysis, there are some weaknesses in using it as a basis for accuracy statements involving process instruments. First of all, the reliability of the standard deviation value depends on a very large number of checkpoints being taken. Second, it would be possible for an instrument to have a different standard deviation figure for each calibration point in its range.

For these reasons, there may well be better ways to make a statement of accuracy. For example, making a simple statement of the worst error that can be expected anywhere within a specified sector of the measuring range of the instrument is one alternative approach.

3.12 Dynamic Accuracy

Dynamic accuracy refers to the ability of the meter to follow changes in the value of the measured variable. In a dynamic accuracy test, the meter is given a test input that changes in a predetermined manner

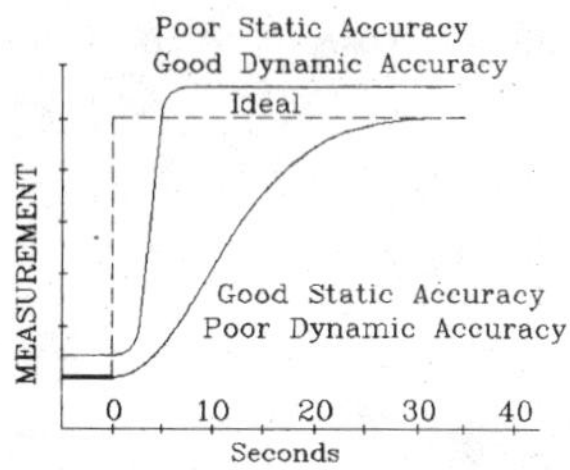

Figure 3.2 The distinction between static and dynamic accuracy.

with time. A step change from one fixed value to another fixed value is a popular form of test input for a dynamic test. From the time of the start of the test, the input and the meter output (reading) are both recorded on the same time base, as illustrated in Fig. 3.2.

A meter that has good dynamic accuracy will follow the input change without significant delay. If dynamic accuracy is poor, it will take an extended period of time for the output to catch up to the new value of the input. Notice that during the time the output is catching up, there is a considerable difference between the output and input values. This difference is dynamic error. Any difference between what the output is and what it should be after the output has ceased to change is static error.

The dynamic accuracy of a meter is an important consideration in automatic control, since the speed with which each of the components of a control system responds to an input change has a significant effect on how well the control system performs. The dynamic accuracy of temperature meters is often poor because the temperature sensor is installed inside a metal socket, or thermowell. Heat transfer through the wall of the thermowell and through the fluid films that coat the wall takes time. Due to the fact that the temperature sensor is really measuring the temperature of itself, there can be a delay of a minute or more before the temperature of the sensor changes to the new value of the process temperature that it is supposed to be measuring.

3.13 Reproducibility

Reproducibility is the ability of an instrument to deliver the same output, or reading, every time it receives the same input. As such, reproducibility is equivalent to precision. The output-input relationship for a meter may not be linear, and the readings of the instrument may not be particularly accurate, but if it has good reproducibility, a meter can be relied on to produce the same reading each time it measures the same value of the process variable.

As such, reproducibility is a reasonably important quality of any measuring device. A meter that gave different readings at different times when it was measuring the same value of the process variable would not be very useful.

3.14 Selectivity

An important and often overlooked feature of process measurement instruments is selectivity. *Selectivity* is the ability of the meter to respond only to changes in the variable that it is supposed to be measuring while ignoring all other influences. Information on selectivity is not often included in sales brochures. The user generally finds out about selectivity problems after he or she puts the instrument into service.

The first force balance DP cell, produced in the 1950s, was designed to measure differential pressure and provide a proportional pneumatic output. It was found that it would change its output, right enough, in response to changes in measured differential pressure. However, it also changed its output with changes in static pressure level, ambient temperature, air supply pressure, vibration, orientation, pull of the piping on the meter body, and the tightness of the cover bolts. It did not have very good selectivity.

An interesting observation is that one of the reasons why the cost of measurement transmitters with electronic output keeps increasing is the insistence of users that the selectivity of electronic transmitters be comparable with that which they had come to expect from pneumatic transmitters. One of the more frequent selectivity problems associated with electronic transmitters has been their tendency to be upset by signals from portable radios.

3.15 Reliability

Reliability implies the ability of the instrument to keep on working, at the desired level of performance, even under adverse operating or environmental conditions and with a minimum of attention. To some extent, the reliability of the instrument will depend on the care with which it installed. For an instrument to be successful in service, two basic requirements must be met. The instrument selected must be the right one for the job, and it must be used correctly.

Reliability is important because an instrument that needs excessive maintenance to keep it functioning will eventually become more expensive to own than some other, better instrument that might have cost more in the first place. The right way to use any instrument often has to be learned. This is why many instrument maintenance groups prefer to stick with a brand of instruments they are used to. Brands with which they are not familiar usually turn out to be less reliable, for a time anyway.

3.16 Deflection and Null Balance

Process meters can function on either of two principles: deflection or null balance. The deflection principle is the simplest; thus meters based on this principle are usually less expensive. Null-balance meters are capable of better accuracy, but they are more complex and consequently cost more money.

A meter using the deflection principle will have a sensor that can respond to liquid level, pressure, flow, temperature (L,P,F,T), or whatever process variable is to be measured and an output that will be either a display of the measurement or a pneumatic or electronic measurement signal.

The block diagram for a meter that functions on the deflection principle is shown in Fig. 3.3. Between the sensor and the output there will be a transducer, which may be pneumatic, electronic, or mechanical. In the general sense, a *transducer* is a device that receives an input in one form and converts it to an output of a different form. The correlation between the output and the input will be predictable. For just about every type of sensor used in industrial meters, the raw output of the sensor is not suitable for creating a display or for producing a signal that can be used to transmit the measurement. A transducer will be required to take the output of the sensor and convert it into a meter reading or a signal with the standard 3- to 15-psig or 4- to 20-mA range.

In a pressure gauge, the sensor is a Bourdon tube that produces motion of its free tip in response to applied pressure. This motion cannot, however, be read as values of pressure. What is needed is movement of a pointer across a scale. Thus a transducer is required to take the motion of the Bourdon tube and convert it into pointer motion. The transducer in this case is often a mechanical rack and pinion movement.

Meters that function on the deflection principle have two inherent problems. First, if there is any change in the characteristics of the transducer, caused by temperature, aging, or whatever, it will affect the meter reading or the output signal.

The second problem is that the power required to produce the meter output, whether a display or a signal, has to come from the sensor. Depending on the variable being measured, the sensor may be able to provide the necessary power. In many cases, however, the sensor has little capability to generate output power. Consequently, the measurement must be made without drawing power from the sensor.

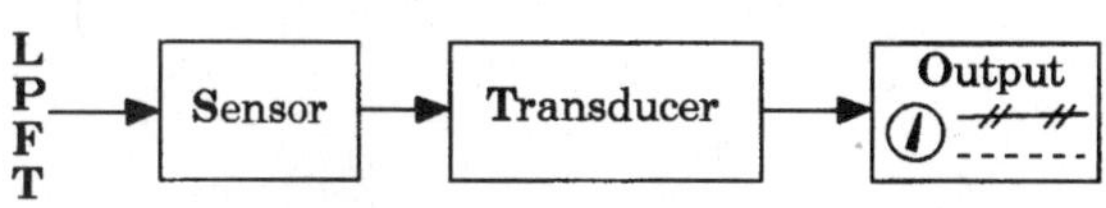

Figure 3.3 The deflection principle.

If there is any significant power drain, the sensor will act like a battery from which too much current is being drawn. Its output signal level, on which the measurement depends, will begin to fall off. For either or both of these reasons, it may be prudent, or even obligatory, to select an instrument that functions on the null-balance principle, even though the null-balance instrument will likely be more costly and its servicing more involved.

When faced with the choice of a deflection-type or a null-balance meter, one has to take into account the characteristics of the sensor, the overall performance required versus what the meter can do in service, and the cost. Ordinary pressure gauges and dial thermometers function on the deflection principle, and most are capable of an accuracy within 1 percent. This is certainly good enough for nearly all local reading pressure and temperature measurement jobs, so there would be no obvious need to specify a null-balance meter for this service.

A null-balance instrument also has a transducer, but the instrument functions in such a way that the accuracy of the meter is independent of changes in the transducer's characteristics. The transducer must remain operable; otherwise, the instrument will not function, but it is not essential that its characteristics remain strictly constant.

Figure 3.4 is a block diagram of a meter that uses the null-balance principle. The null-balance unit has a null detector that compares either the motions, forces, or emfs created by two opposing inputs. Input 1 comes from the measurement sensor. Input 2 is the output of the transducer. If there is any difference between the motion, force, or emf effects of these two inputs, the unbalance in the null detector produces a new signal that drives the transducer output to whatever new output level is necessary to achieve the balance, or null. The output of the transducer is also the electronic or pneumatic signal required to create a display or a pneumatic or electronic signal with the correct value.

Notice that there is only one possible output that can produce the null balance against the input from the sensor. Even though the characteristics of the transducer may change, the null-balance unit has to produce the output that matches the input received from the sensor. An error is still possible, however, if the characteristics of the sensor were to change. This problem would require a recalibration of the transmitter.

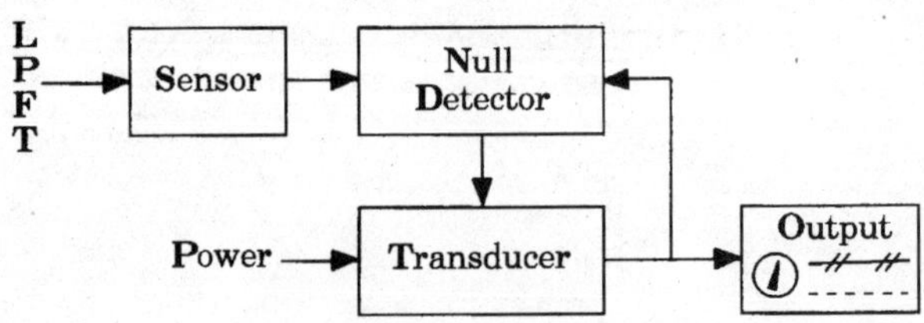

Figure 3.4 The null-balance principle.

Notice also that the power required to create a display or a measurement signal comes from the transducer, which has its own power source, not from the sensor. The sensor has to produce only that power required to unbalance the null detector. Even then, power is taken from the sensor only when the null detector is in an unbalanced condition. When it is in balance, no sensor power is drawn.

While in the majority of industrial null-balance instruments the null is achieved automatically as part of the instrument's function, it is possible to have a null-balance instrument for which the null is established manually. An example is an equal arm balance used for weighing objects. The object to be weighed is placed on one side of the balance, and weights with known values are placed, manually, on the other side until the balance is in its balanced, or null, state. The Wheatstone bridge, used sometimes in temperature measurements, is another example of manual null balance.

In the past, there have been process transmitters that were designed on the deflection principle and which performed with satisfactory accuracy on many applications. For most process transmitters, however, there is a general requirement for maximum accuracy, so null balance is specified, and the additional cost is paid. In fact, while most users would accept a deflection-type meter if they needed a pressure gauge or a dial thermometer, they would be unlikely to buy a process transmitter that did not function on the null-balance principle. Consequently, it is improbable that any instrument manufacturer will offer a deflection-type transmitter today.

3.17 "Smart" Measuring Instruments

The 1980s have seen the appearance of a new breed of process meters that have come to be called *"smart" instruments.* This is another of the many examples of poorly chosen names for process instruments. There is nothing particularly clever about smart meters. However, they do have capabilities above and beyond those of their predecessors, and these capabilities should be understood.

The additional capability has been made possible by the development of microelectronics and the inclusion of these remarkable components in process measurement transmitters. This means that the transmitter comes equipped with a small onboard computer that generally gives it the ability to do two things: modify its output to compensate for error-creating effects and be interrogated from the receiving end of the loop.

The peculiar capabilities of "smart" measuring instruments include

- The ability to transmit process measurements using a digital signal, which is inherently a more accurate method, rather than using an analog signal. The principal obstacle, however, is the lack of agreement as to what the digital format should be. Eventually, this will be resolved.

- All industrial measuring instruments contain components such as bellows, diaphragms, and linkages that exhibit nonlinear behavior or whose behavior can be altered by changes in temperature, pressure, vibration, power supply, or other external effects. In other cases, nonlinear effects are contributed by the laws of nature, as they are in orifice meters. The strategy until now has been to use more and more expensive components and materials to minimize these effects.

  Since "smart" instruments actually can be provided with computing power, more readily available components and materials can be used. The computing section can generate the appropriate equal and opposite effects required to compensate out the built-in nonlinearities and produce an output that is really linear with input all across the measurement range.

- In addition to transmitting information, "smart" transmitters can "listen." A practical benefit of this occurs in the prestartup checkout stage. From the operations center, a technician can hook onto a pair of wires and ask the transmitter at the other end what its identification number is. This obviates the requirement for having a second person at the transmitter location to verify that the transmitter is correctly wired in and considerably reduces the time required to complete the checkout of the many connections involved in a new plant installation.

- A "smart" transmitter can have its span and suppression altered easily through appropriate reprogramming commands rather than having to resort to mechanical calibration adjustments. This is a feature that is overrated. While on the surface it appears to have benefits, one should remember that since the *real* error remains unchanged, cutting the span in half by magnifying a section of the measurement range will double the *percent* error within the new span.

  In addition, for flow transmitters, the ability to alter the calibration from the receiving end of the transmission system does *not* provide the capability to do the one maintenance check that is required most often, namely, zeroing the meter. This can be done only by opening and closing the manifold valves at the transmitter.

Chapter

4

Flow of Fluids in Conduits

The flow rate of fluids in pipes and other conduits may become complicated because of the numerous ways in which the flow rate can be specified. Flow rates may be stated as

- The mass flow rate.
- The volume flow rate at line conditions.
- The volume flow rate at standard conditions.
- The mole flow rate.
- The average velocity of the stream.

4.1 Mass Flow Rate

The mass flow rate of any stream is basic, owing to the fact that the laws of nature prohibit the stream from accumulating or losing mass at any point. Therefore, the mass flow rate is unique to the stream. Its value does not change, irrespective of any changes in stream temperature or pressure or any change in the size or shape of the conduit.

Since all the other flow quantities can be computed from the mass flow rate, it makes sense to determine the mass flow at the outset and work from this quantity to whatever other values are required. Typical units are kilograms per second (kg/s) and pounds per hour (pph).

4.2 Volume Flow Rate at Line Conditions

The basic relation connecting mass, volume, and density is

$$\text{Mass} = \text{volume} \times \text{density}$$

If the derivative with respect to time is taken on both sides of this relation, it becomes

$$\text{Mass rate} = \text{volume rate} \times \text{density}$$

Rearranging this,

$$\text{Volume flow rate} = \frac{\text{mass flow rate}}{\text{stream density}}$$

In this case, the density must be the density of the fluid as it flows through the conduit, that is, the density at the prevailing temperature for liquids and at the prevailing temperature and pressure for gases or vapors. Typical units for volume flow rates are cubic meters or liters per second (m^3/s or L/s), gallons per minute (gpm), and cubic feet per hour (cfh, sometimes written acfh, for actual cfh).

4.3 Volume Flow Rate at Standard Conditions

The volume flow rate at standard conditions means that volume flow rate which would exist, at the same mass flow rate, if the stream were at standard conditions of temperature and pressure (even though it is not). Although it can be argued that the volume flow at standard conditions is a fictitious quantity, it is widely used in the petroleum and petrochemicals industries and consequently needs to be understood.

In these industries, the standard conditions of temperature and pressure are 15°C and 101.325 kPaa. Originally, standard conditions were 60°F and 14.696 psia (760 mmHg). The currently used standard conditions, brought on by the switch to the SI system, are virtually identical to the original ones: 15°C is 59°F, and 101.325 kPaa is 760 mmHg. Thus density tables for hydrocarbons developed prior to the introduction of the SI system can continue to be used without any significant error.

The volume flow rate at standard conditions can be computed as

$$\text{Volume flow rate at standard conditions} = \frac{\text{mass flow rate}}{\text{fluid density at standard conditions}}$$

For liquid flow, standard temperature only applies. For gases and vapors, both standard temperature and standard pressure apply.

Before making a calculation of the volume flow rate at "standard" conditions, it is a good idea to verify what the standard temperature value is considered to be. Some scientists and academics tend to use

0°C as the standard temperature rather than 15°C. Occasionally, a standard temperature of 4°C also comes up.

Notice that once the fluid has been identified and the standard temperature and pressure have been specified, the density of the fluid at standard conditions will be a constant for the stream. Therefore, the volume flow rate at standard conditions is really the mass flow rate divided by a constant. In other words, it is just another way of expressing the mass flow rate. Typical units are standard cubic meters per second (sm^3/s) and standard cubic feet per hour (scfh).

4.4 Mole Flow Rate

The mole flow rate is determined by

$$\text{Mole flow rate} = \frac{\text{mass flow rate}}{\text{molecular weight of the fluid}}$$

The point to be watched here is that if the mass flow rate is in kilograms per second, then the mole flow will be in kilomoles per second (kmol/s). Typical units are moles per second (mol/s) and kilomoles per hour (kmol/h).

4.5 Average Velocity

The volume flow rate, the average stream velocity, and the cross-sectional area of the conduit are connected by the relation

$$\text{Volume flow} = \text{conduit area} \times \text{average stream velocity}$$

In this case, the volume flow rate is the rate at the prevailing stream temperature and pressure. It is incorrect to use the volume flow rate at standard conditions. If the conduit is a round pipe or a tube, the area may be read from the manufacturer's tables, or it can be calculated from the internal diameter.

When a fluid flows through any conduit, its velocity will not be the same at all points across the cross section. The velocity is generally greatest at the center (on the axis). It decreases toward the inside wall of the conduit. Right at the wall, there is a film of fluid that does not move at all.

The *average* velocity of the stream can be calculated from the relation

$$\text{Average velocity} = \frac{\text{volume flow rate}}{\text{conduit area}}$$

Typical units are meters per second (m/s) and feet per second (fps).

4.6 Turbulence

Turbulence in a flowing stream, or the lack of it, can have a significant effect on what happens inside the conduit. For example, the degree of turbulence will determine whether the pressure drop across a restriction placed in the stream will be proportional to flow rate or to flow rate squared. This fact was first appreciated by a British engineer, Osborne Reynolds (1842–1912), who concluded that turbulence was an important enough phenomenon that it was desirable to be able to assign a number to the degree of turbulence in a fluid flow system.

From his own observations, Reynolds concluded that the degree of turbulence of a fluid flowing in a conduit with a round cross section (a pipe) would be affected by four separate factors. Turbulence would increase if either the stream velocity, the diameter of the pipe, or the density of the fluid were increased. However, turbulence would decrease if the viscosity of the fluid increased. Consequently, the extent of turbulence could be stated quantitatively by a number that was based on

$$\frac{\text{Velocity} \times \text{pipe diameter} \times \text{fluid density}}{\text{Fluid viscosity}}$$

Interestingly enough, if basic units are assigned to each of the four quantities, the ratio turns out to be dimensionless. This ratio has come to be called the *Reynolds number* for the stream and is often designated Nr.

High Reynolds numbers indicate a high degree of turbulence. As the Reynolds number decreases, at a value of about 2000, the turbulence virtually disappears, and the stream pattern becomes streamline, or laminar. For any Reynolds number higher than 4000, the stream will be turbulent. The region between 2000 and 4000 is a transition region in which the stream changes from streamline to turbulent flow.

Although the Reynolds number formula has several variations, a practical form is

$$\text{Nr} = \frac{1273W}{D\mu}$$

where W = mass flow rate (kg/s)
D = pipe internal diameter (m)
μ = absolute viscosity of the fluid at line conditions (mPa•s)

4.7 Flow Measurement

Flow measurement is a branch of technology in which people have made careers. It has a great many aspects. There are more ways to measure flow rate than there are ways to make any other industrial measurement.

Before evaluating particular techniques for measuring flow, there is some relevant background. First of all, flow can be, and is, measured on two different bases. It can be measured as *rate of flow* in a conduit, that is, the number of kilograms, cubic meters, or liters that are flowing past a point in the conduit *in one unit of time.* Flow rate is often used for plant feed and product streams. Or flow can be measured on a *cumulative* basis, that is, the number of kilograms, cubic meters, or liters that have gone through the meter since the last time the meter reading was taken. The meter may or may not be set back to zero after each run through the meter. The gasoline pump at a service station and the water meter in the home are examples of cumulative flow measurement. Obviously, the design of the meter must fit the requirement for either flow rate or cumulative flow measurement.

The second point, already mentioned, is that flow can be measured on a mass, mole, or volume basis. Mass flow rate is measured in kilograms per second. It is the most fundamental way to measure flow rate, yet apart from steam flow measurements, it seems to be used very little. Its virtue is that a kilogram is always a kilogram, no matter what its temperature or pressure may be. Mole flow rate is simply a variation of mass flow, since a mole of a substance is the number of grams of the substance equal to its molecular weight.

Volume flow can be measured either in volumes at line temperature and pressure or in equivalent volumes at standard temperature and pressure (15°C and 101.325 kPaa). The volume flow rate is obtained by dividing the mass flow rate by the appropriate density value. Note that for liquids only the temperature can affect the density, whereas the density of steam and gases will be affected by both pressure and temperature.

Finally, an important fact is that although flowmeters may be designed to read out in mass or volume flow units, many flow meters are really sensing the velocity of the stream. It is a known fact that the stream velocity will not be uniform over the cross section of a pipe. The maximum velocity will be somewhere near the axis (at the center). The velocity will become progressively less as the wall of the conduit is approached.

Accordingly, a key consideration is: Does the meter measure the average velocity across the conduit's entire cross section, does it measure the average velocity across one diameter of the conduit, or does it measure the stream velocity only at the particular spot at which the flow sensor is located? If either of the last two situations prevails, then determining the rate of flow by multiplying the measured velocity by the cross-sectional area of the conduit is a questionable procedure.

4.8 Custody Transfer

The term *custody-transfer metering* applies to flowmeters used for billing purposes. In other words, money changes hands based on the readings of a custody-transfer meter. Consequently, custody-transfer meters must be capable of high accuracy over an extended period of time. The accuracy must be verified periodically through a calibration check. In some industries this is called "proving the meter."

Custody-transfer meters for liquid service are almost exclusively cumulative flowmeters. Positive-displacement meters and turbine meters are the most frequent choices because of their inherent high accuracy. Even though the meter may have a calibration adjustment, the usual purpose of proving the meter is not to recalibrate the meter so that it registers exactly accurate volumes. Over a period of time, instrument service personnel have found that continually altering the calibration adjustment causes the mechanism to develop play, or looseness. When this happens the meter tends to produce inconsistent readings.

Instead, the purpose of the proving operation is to determine the current meter factor for the meter. The *meter factor* is defined by the following relation.

$$\text{Registration} \times \text{meter factor} = \text{true volume passed}$$

or

$$\text{Meter factor} = \frac{\text{true volume}}{\text{meter registration}}$$

Generally, the calibration of the meter is not adjusted unless the meter factor is outside the limits 1.000 $\pm$0.005.

4.9 Proving Requirements

The procedure for proving custody-transfer liquid meters conforms with certain rules, the more important of which are

1. Proving runs must be continued until two consecutive runs yield meter factors that do not differ by more than 0.0005. The average of these two consecutive meter factors is taken to be the current factor for the meter.
2. If the current meter factor differs by more than 0.0025 from the meter factor determined during the previous proving session, then proving runs must continue until three consecutive runs are made for which the meter factors do not differ by more than 0.0005.

4.10 Proving Techniques for Custody-Transfer Meters

Having set out the conditions for an acceptable meter proving, it is only proper that there be some discussion of the methods that can be used to achieve the required results. Bearing in mind that meter proving refers to custody-transfer liquid meters, two general methods can be used. In both cases, a short run through the meter is made at the design flow rate, and the liquid that passes through the meter during the run is collected and accurately measured. The registration of the meter over the course of the run is compared with this measured volume.

The difference between the two methods lies in the way in which the volume of liquid that actually went through the meter during the proving run is accurately determined. The most popular method is volumetric proving, in which the liquid that passed through the meter is measured in a proving tank whose volume is known accurately, possibly as a result of certification by some agency with official status.

The second method is gravimetric proving, in which the liquid run through the meter is collected in a weigh tank on a scale. The mass of the collected liquid is then determined, rather than the volume. Mass is always a more fundamental basis on which to work than volume. Once again, a kilogram or a pound is always a kilogram or a pound, irrespective of changes in temperature or applied pressure. The gravimetric method eliminates some of the errors inherent in volumetric proving, as will be explained. However, the fact remains that for proving a given meter or set of meters, a volumetric system will inevitably come in at a lower price.

Volumetric provers have the general appearance shown in Fig. 4.1. There is a wide main body with a seraphin neck at the top and the bottom. The top section has a level gauge with a reference mark. When the tank is filled up to this mark, it contains the certified volume of liquid. To obtain the factor for a meter, liquid is run through the meter into the proving tank until the liquid is up to the mark on

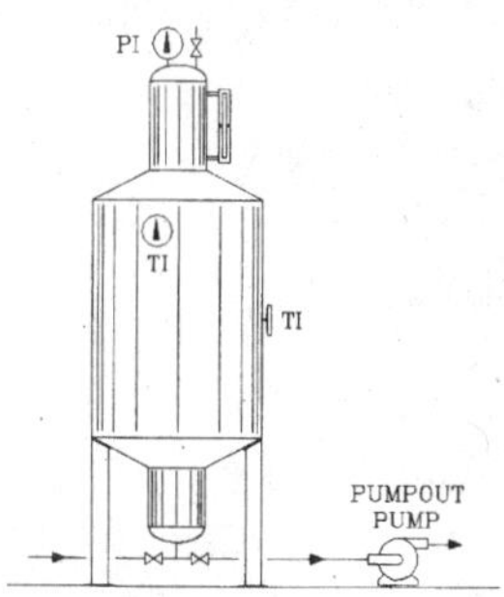

Figure 4.1 A proving tank for custody-transfer liquid meters.

the gauge. Then the registration of the meter is compared with the known volume of the proving tank.

This sounds straightforward enough, but there are complications. First of all, the meter registration can change with flow rate, so the meter should be proved at the rate of flow at which it will usually operate. Obviously, this rate of flow will not prevail when the proving run is started from zero flow. It also will not prevail at the end of the run when the tank is being topped off to the reference mark either.

Custody-transfer meters often will underregister at low flow rates. To minimize this potential error, it is necessary to have a long enough run at the design rate that any error incurred at the start and the end of the run, due to the low flow rate, will not have any significant impact on the overall accuracy. Consequently, the proving tank should be large enough to accommodate a proving run of 2 minutes minimum.

After a proving run is over, it is necessary to pump the tank out in preparation for the next run. If the liquid product has been degraded in the process of proving the meter, it will have to be pumped back to a rerun tank for reprocessing. Thus the personnel time required to prove a meter may not be the only cost involved.

Proving a custody-transfer meter by running liquid through it and collecting the liquid metered in a calibrated proving tank seems like a simple enough task. However, it often becomes laborious and frustrating because of the multitude of correction factors that are possible. It is not unusual to have to apply all the corrections below.

A correction is required if

- The calibrated volume of the proving tank was certified at one temperature, but the tank is at a different temperature when the meter is being proved.
- The volume of the proving tank was determined at atmospheric pressure but the liquid is held in the tank at line pressure to avoid loss of high-vapor-pressure components.
- The liquid passed through the meter at one temperature but is held in the proving tank at a different temperature. If the temperatures are different, then the volume registered by the meter and the volume collected in the tank must both be corrected to equivalent volumes at 15°C (or 60°F).
- The liquid goes through the meter at one pressure but is held in the tank at a lower pressure.

One might be inclined to question the last of these sources of error on the grounds that liquids are incompressible. The slight decrease in volume of a liquid as the pressure is increased can be ignored in many situations, but not in all situations. Recalling that meter factors from successive runs are required to agree within 0.0005, which is 5 parts in 10,000, it is quite possible that in this case the compressibility of the liquid should be taken into account.

As if these factors, in combination, were not enough, there is also the question of how much the volume of the prover tank is changed due to the liquid that is still wetting the walls of the tank from the previous proving run. For this potential error, there is no predictable correction when proving by volume. Gravimetric provers do not have this problem, which is one reason why they may be used for proving meters that handle heavy liquids such as bunker fuel oil.

Aside from the problems already outlined, proving custody-transfer meters using a volumetric tank can have "gremlins" that cannot even be accounted for. Instrument technicians have told me that if they have a custody-transfer meter to prove, they have to wait for a cloudy day. They found from experience that when the sun is shining on the proving tank, they could never get two consecutive readings that were within the required tolerance.

4.11 Pipe Provers

No one needs to be told that in the last 50 years developments in process instrumentation have been truly remarkable. Asked to name the outstanding examples, most people would come up with digital computer-based systems or sophisticated measurement techniques. Yet if one were to use as the criterion the number of problems that were solved without incurring new problems, one would have to give serious consideration to the calibrated pipe prover. Compared with the trials and tribulations associated with proving custody-transfer flowmeters using a proving tank, doing the job with a calibrated pipe prover is a breeze.

The essentials of a pipe prover are sketched in Fig. 4.2. As the name implies, the calibrated vessel is a length of pipe, but not just any pipe. Pipe used to build a pipe prover is carefully selected for straightness and roundness, after which the internal walls are honed and polished.

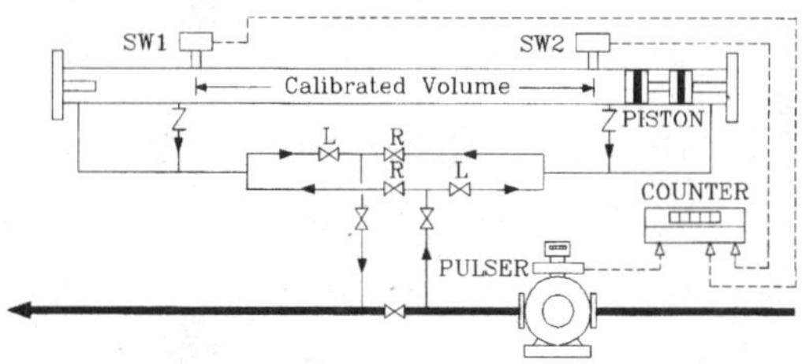

Figure 4.2 A calibrated pipe prover.

Inside the pipe is a piston that moves back and forth in the pipe during the calibration runs. The piston is fitted with cups so that leakage of the proving liquid past the piston is minimal. The weight of the piston is adjusted also so that the density of the piston is as close as possible to that of the proving liquid. Then, as the piston moves along the pipe, it does not wear on the bottom more than at any other place.

At either end of the pipe prover is a switch, shown as SW1 and SW2 in the figure. These switches detect the passage of the piston as it moves through the pipe. Mechanical switches and capacitance-type switches have both been used successfully, although other types may have possibilities. Precise detection is important, since the volume against which the meter is proved is the volume in the pipe between the two switches. This volume must be determined accurately, and, as in the case of the tank prover, the volume may have to be certified by some government or legal agency.

The four valves, two labeled L and two labeled R, are used to change the direction of flow through the prover. When the L valves are open and the R valves are closed, the flow through the prover is from right to left. Similarly, when the R valves are open and the L valves are closed, the flow is from the left to the right. As the liquid moves down the pipe, it pushes the piston ahead of it. One pass of the piston down the pipe is one proving run. Since the L and R valves all must operate together, that is, two are opening while the other two are closing, it is usual to have them motorized so that all four operate at the push of one button.

The three valves below the L and R valves in the figure are used to bring the pipe prover in and out of service. When the prover is operational, it is actually a part of the pipeline in which the meter is installed. The stop at either end of the prover provides the space necessary for the liquid to get behind the piston and start it moving down the pipe. The second branch, which contains the check valve, is necessary so that when the piston reaches the end of its travel, it will not block the line.

For the pipe proving method to work, the meter must be equipped with an electrical pulse generator on the drive that runs the meter register. The pulse generator breaks down the registration of the meter into finer increments. A typical pulse generator might generate 100 pulses for each digit which turns up in the least digit position (the right end digit) of the register. The output of the pulse generator goes to an electronic pulse counter.

A proving run is started by swinging the four L and R valves. As the piston passes the first detector switch, the switch trips, which starts the counter. When the piston trips the second switch, the counter is stopped. The registration of the counter and the calibrated volume of the prover are the data that determine the meter factor.

Only one correction is required—the temperature correction for the volume of the prover. There are no temperature or pressure corrections for the liquid because the liquid is at the same temperature and pressure in the prover as it is in the meter. And there is no slop to get rid of between proving runs. The biggest plus, however, is the savings of time and frustration in getting the job done.

All these virtues must be accompanied by some drawbacks. Actually there are two. First of all, pipe provers tend to be quite long; consequently, they require considerable space. Second, the prover itself, the seven manifold valves (four motorized), and the associated piping, all of the same size as the prover, plus the electronics that are required, total up to a relatively large sum of money. Still, when one chooses a calibrated pipe prover, one is going first class, and it has been proven to be worth it.

In an effort to reduce the requirement for space, hairpin-shaped pipe provers that bend back on themselves have been designed. While this is a gain as far as real estate is concerned, the loss is that the half-circle bend in the pipe makes it impossible to use a cylindrical piston. The piston has to be spherical instead. It is more difficult to get a positive seal between the piston and the pipe wall when the piston is a sphere, and leakage past the piston usually results in inconsistent readings.

In the pipe prover shown in Fig. 4.2, the piston travels in both directions. In other words, it is bidirectional. A bidirectional prover requires the four expensive L and R valves in order to change the direction of flow. Unidirectional pipe provers have been developed that do not need these four valves, thus eliminating their cost and the cost of the piping. The price that is paid for this, however, is that after each run the prover has to be opened and the piston removed and transferred back to the starting end. To do this, the prover has to be depressurized and emptied. Thus the savings in cost is nullified by the additional time and work required to do a proving.

Referring again to the prover in Fig. 4.2, in addition to the center section that contains the calibrated volume, the end sections must be long enough for the piston to be parked. More important, they must be of sufficient length that the piston can accelerate up to its normal velocity before it passes the first detector switch. Another requirement is that the body of the prover, as well as the associated piping, be insulated to ensure that the temperature of the prover stabilizes at that of the liquid in the pipeline.

Practical design guidelines for a calibrated pipe prover include

- The diameter of the pipe prover should be such that the velocity of the piston at the design proving rate will be not less than 0.06 m/s and not be greater than 0.90 m/s. If the velocity is too low, the piston may not move smoothly, and there may be leakage by the piston. If the velocity is too high, the operation of the detector switches may not be consistent.

- The calibrated volume of the prover between the detector switches should not be less than 0.5 percent of the volume passed by the meter per hour at the design proving rate. In addition, the volume between the switches must be great enough that an error of one pulse in the pulse count will amount to no more than 0.05 percent of the total pulses counted over a proving run. This amounts to saying that the pulse count for one proving should not be less than 2000. In any case, the distance between the switches should not be less than 5 m.
- The length of each end zone should be $vt/4$, where v is the design velocity of the piston between the switches (m/s), and t is the time required for each of the L and R valves to move full travel (s). Add to this distance, in meters, the length of the piston (usually 1 pipe diameter) and the length of the end stop (assume 0.30 m).

For some idea of what these guidelines would mean, a 6-in positive-displacement meter has a design flow rating of about 750 U.S. gpm, or 47.3 l/s. To establish the piston velocity in the required range, the prover and all the associated piping would have to be 12 in. The calibrated length of the prover, between the detector switches, would be 12 m. Each of the end zones should be 3 m long, so the overall length of the prover would be 18 m.

4.12 Custody-Transfer Gas Meters

Every household that has a gas furnace or other gas appliances will have a positive-displacement type of gas meter in the basement. These meters are relatively small, however, and proving them consists of replacing them at intervals with a new or reconditioned meter.

At the other end of the spectrum, there are industrial users of natural gas who buy it in large quantities, through large pipelines, and at fairly high pressures. They may not be acquiring natural gas in order to burn it either. Natural gas is one of the two raw materials from which hydrogen is made. There may be other chemical or physical reactions that require natural gas as one of the ingredients. Hence, whether one is buying or selling gas, the quantities purchased may be large enough to justify a full-fledged custody-transfer gas metering station.

Under these circumstances, displacement meters may not be capable of handling the pressures or the flow rates. Orifice meters have been used for custody-transfer purposes, although in time there may be more and more stations that use other types, such as vortex shedding meters. Since the potential accuracy of these gas meters is not as good as that of positive-displacement liquid meters, it is customary to have two meters in the station, both metering the same stream. If the meters do not check each other closely enough, an investigation is made to locate the meter that is not performing as it should.

Figure 4.3 shows one metering station piping arrangement that has been used. The meters shown in this figure are orifice meters, although vortex shedding meters also might be used.

Gas enters the station at the top left. The valves that have been blacked in are normally closed. Thus gas flows through the first measuring element ME1, back along the center pipe, and then on through measuring element ME2.

The entire meter run is assembled carefully using quality components. There is about 30 diameters of straight pipe ahead of both flow elements to ensure that the gas approaches the element with only random turbulence. Sometimes straightening vanes are used as well.

The piping and valving arrangement shown in Fig. 4.3 allows either of the meter runs to be bypassed so that it can be taken out of service for maintenance. The meter runs are located a foot or so above grade so that they are accessible but also so that if differential pressure meters are required, as in the case of orifice meters, they can be located above the meter runs. When this is done, the impulse lines run upward from the orifice pressure taps to the differential pressure (DP) meter. The result is that any liquid that attempts to get into the impulse lines will drain back down to the meter run, and the impulse lines will remain dry.

The argument as to whether two meters are sufficient to ensure a valid measurement is an old one. If the meters do not read the same, which one is in error? Nonetheless, in metering stations of this type, installing a third metering run is generally considered to be overkill, and too costly. Any significant difference in the readings of the two meters is taken to be a signal that something is not right. This is a better position to be in than having only a single meter whose readings have to be accepted at face value.

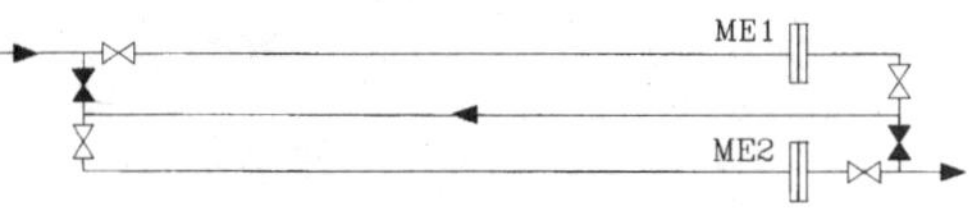

Figure 4.3 A meter run installation for custody-transfer gas meters.

Chapter

5

Orifice Flowmeters

During the design phase of the Syncrude plant, I was searching for a flowmeter that could measure the flow of bitumen froth at a fairly high rate with reasonable accuracy. I ended up with a list of 25 different flowmeters, along with the reasons why none of them would work on this application. This merely serves to point out that the variety of ways in which flow rate can be measured is greater than that for any other process measurement.

When one looks at the flow plans for oil refining and petrochemical plants, about half of all the variables that are measured and controlled are flow rates. One of the more important accounting jobs in a plant is the determination of losses. This is usually done by making a material balance around the plant. The measurements that go into a material balance are almost entirely flow measurements.

Surely, if the elaborate, expensive, on-line data-processing systems that are being installed today are to produce anything other than useless output, then something will need to be done to ensure that the flow measurements that are a big part of their input are decently accurate.

If a potential author were to put his or her mind to it, he or she could turn out not just a single book but a set of volumes on methods of making flow measurements. That will certainly not be attempted here. What will be discussed are some features of one flowmeter only—the orifice meter. This is not because the orifice meter is a good flowmeter, because it isn't. It is because the tradition of using orifice meters is so entrenched in many process industries that such meters are likely to be around for some time yet, even though better flowmeters are available.

Students in the Instrumentation Engineering Technology Program at the Northern Alberta Institute of Technology sometimes ask the question, "When are orifice meters going to be obsolete?" The answer the instructors give is, "They are obsolete now. The problem is to get people to stop using them."

At one time the Engineering Department of Imperial Oil was on the receiving end of a rash of complaints from the various refineries about errors in flow measurements using orifice meters. Material balances were not closing within the 5 percent tolerance that was considered to be acceptable error. When we were asked what accuracy one should expect to get with an orifice meter, we gave the standard answer, "within 2 percent for liquid flow, and within 3 or 4 percent for steam or gas, depending mainly on how well the properties of the stream, particularly its density, were known." We were told that the accuracy the refineries were getting was nowhere near this good. As a result, I was directed to make the rounds of the refineries to find out what was going wrong.

After looking at a number of installations, some facts of life became apparent. One was that for any orifice meter installation, there are four distinct areas that need attention. These are

- *The orifice meter run.* The meter run consists of the two sections of straight pipe upstream and downstream of the orifice, the orifice flanges, and the orifice plate.
- *The meter manifolding.* This is the small piping that transmits the pressures before and after the orifice to the high- and low-pressure connections on the differential pressure meter.
- *The differential pressure meter.*
- *The correlation between differential pressure,* which is what is actually measured, *and flow rate,* which is the measurement desired.

It is generally understood that the orifice plate, as an obstruction in the flow path, causes a transference of pressure energy to velocity energy. The difference between the pressures at the points of normal and higher velocity can be correlated with the velocity of the stream, which can then be correlated with rate of flow. The obstruction does not have to be a round hole in a flat plate centered on the axis of the pipe. A beer can would probably work. The point is, What would the differential pressure readings so obtained mean in terms of the rate of flow?

In the majority of cases, and these include the orifice plate, the transfer of energy from one form to another is not 100 percent efficient. There are some energy losses. Therefore, the correlation between differential pressure and flow rate cannot be obtained by sitting down at one's desk with a computer and starting from first principles. The correlation must be established by actual laboratory tests.

Another interesting and perhaps significant point was that when the accuracy of an orifice meter was in question, a peculiar scenario would be acted out. Operations, or accounting, or whoever needed the measurement, would take the complaint about the accuracy of the meter to instrument maintenance. After all, it is the responsibility of that department to look after the instruments. In response, an instrument serviceperson would remove the differential pressure (DP) meter, take it back to the shop, and recalibrate it, usually finding in the process that there was nothing much wrong with it. He or she would then reinstall the meter and report that he or she had fulfilled his or her obligations. Unfortunately, the next check a day so later would reveal that the inaccuracy was still there.

This situation was not particularly conducive to good relations between the instrument maintenance group and the people for whom they were supposed to be providing service. It did, however, lead to the realization that all problems with errors using orifice meters may not necessarily be caused by the calibration of the DP meter.

As a matter of fact, after looking at numerous orifice meter installations and their performance, a conclusion that was unmistakable emerged. By far the biggest contributor to excessive orifice meter errors was *substandard orifice meter runs.*

5.1 Orifice Meter Runs

The point has been made that the correlation between the differential pressure reading created by the orifice plate and the flow rate can only be established by laboratory experiment. Therefore, it follows that if the published laboratory-determined correlations are to be used for a field installation, then *the field installation must duplicate the conditions that prevailed in the laboratory when the correlations were determined.* It is curious, and somewhat pathetic, that this obvious requirement is so often ignored. Specifically, this means that

- Orifice meter runs should not be made from pipe smaller than 2 in in size. Certainly, there will be disagreement on this point. Orifice meter runs have been made up in 1½-in and even smaller pipe. The incentive behind this is not obvious, since 2-in pipe and flanges are not that much more expensive. How important is it to get the best accuracy possible?
- The stream must approach the orifice in a condition of random turbulence. It must not swirl. This means that there must be a certain allowance of straight pipe before and after the orifice to give the stream time and space to assume the random condition. The straight-run requirements can be found in any of a number of

instrumentation texts and handbooks. The point is, in many installations, the necessary straight pipe, especially on the upstream side of the orifice, is simply not provided.

One of the more common alibis for not providing enough straight pipe ahead of the orifice is that there is not enough room inside a structure to do it. The meter run will run up against a wall. An irritating question that often goes along with this situation is, "If we give you only 10 diameters instead of 15, how much error will it cause?"

People who ask this question are usually aware that it cannot be answered specifically. The only answer is a general one. "If we are to have a successful meter installation, there are certain rules that have been found to be obligatory. You are bending the rules. You are consequently asking for metering errors, and at this stage we cannot predict how large they will be."

If the space required to install the correct amount of straight pipe ahead of the orifice is not available, then the straight-run requirement can be reduced by installing straightening vanes at the entrance to the meter run. This alternative is valid but, for unknown reasons, is not used as much as it should be.

- Pipe should be in good shape. This means that it should be straight and round and clean and smooth on the inside. Pipe that has been bounced around on a flat-bed truck and then rolled off onto the ground will not necessarily meet these requirements. If the internal diameter were to be calipered at several diameters in the same plane, the pipe might very well turn out to be oval. Being smooth means that there should be no rust, scale, dirt, weld spatter, or other deposits on the inside. A half-hearted attempt to rectify any of these conditions with a hand-held grinder in the general region just ahead of the orifice is not a satisfactory solution.

- The correlation for flow versus differential pressure will show that it includes the inside diameter of the pipe to the second power, in other words, the internal diameter squared. This means that the figure used for the internal pipe diameter must be correct, at least to within 1 part in 1000. The measurement of the internal pipe diameter should be made and recorded for every orifice meter run before the meter run is fastened into place. The problem is that the necessity for doing this often goes unrecognized, so the meter run is welded into the line without the measurement being taken.

  The recourse is to get out the pipe manufacturer's tables and look up the internal diameter. What one sees in these tables is most reassuring. The internal diameter of 4-in schedule 40 pipe, for example, is given as 4.026 in, well within the requirement for resolution that should be met. We ignore at our peril, however, the question of how closely any one length of 4-in pipe made by this or any other pipe manufacturer will have an internal diameter of exactly 4.026 in. Another moot question is whether the pipe started out with a diameter of 4.026 in but was changed in the process of grinding it to make it clean and smooth on the inside.

- Assuming that the pipe is good in the first place, nothing should be done to it that will harm the internal roundness or alter the inside diameter. The culprit here is indiscriminate welding—welding to fasten brackets or fixtures for installing DP transmitters, instrument housings, seal pots, or whatever, and welding to install the orifice flanges. Welding for these purposes can produce boils on the inside of the pipe or, in the case of the flanges, a collar on the inside of the pipe. The end result is that the flow pattern is distorted in an unpredictable manner, or the figure used for the internal diameter of the pipe becomes questionable.

  If fixtures must be attached to the meter run, they should be held on with U-clamps, not by welding. As for flanges, the simple fact is that the cheapest type of flanges is the worst, and the most expensive type is the best. Slip-on flanges are the least expensive. They slip on over the end of the pipe and then are fastened on by welding all around the back of the flange. The heat applied usually causes the inside wall of the pipe to sag inward, with inevitable consequences to the pipe inside diameter and the flow pattern.

  The most expensive flanges are weld neck flanges. They are butted to the end of the pipe and welded on with a butt weld. The result is that about 1 diameter of the flow path right next to the orifice is not pipe, but the machined opening in the flange. This usually can be relied on to be uniformly round, smooth, and of a predictable diameter. However, what contractor will use weld neck orifice flanges if the client's specifications, or lack of them, permit the use of slip-on flanges?

- The orifice diameter ratio, that is, the ratio of the diameter of the orifice to the inside diameter of the meter run pipe, must be within the specified acceptable limits. Once again, authorities on this subject may object to the values given here. The fact remains, however, that if optimal accuracy is to be ensured, then the orifice diameter ratio should not be less than 0.15 or greater than 0.70.

  In far too many cases, orifice plates with an orifice diameter ratio of 0.80 and even greater are used in order to increase the range of the meter. The technically correct alternative, which is to increase the size of the meter run one pipe size, is, of course, considered to be too costly. The result is that a questionable orifice diameter ratio is used, usually with a detrimental effect on the accuracy of the meter.

- Finally, there is the orifice plate itself. The correlations for orifice meters are based on a flat plate with a smooth, polished upstream face, a smooth, circular orifice, and a sharp corner at the entrance to the orifice. This sharp corner is most important. If the corner is eroded off, then the plate will still be a differential producer, except that now it is no longer an orifice, it is a nozzle. Nozzles have their own peculiar correlations.

It would be better for all concerned if orifice meters would fail completely once the orifice has started to wear at its upstream edge. The trouble is, they do not. They go on producing differential pressures, and people go on using the differential pressure readings and converting them to flow readings using the correlations for a round, sharp-edged orifice. At a meeting of the Edmonton Section of the Society for Measurement and Control, a representative of Dow Chemical reported that the company had found that after only 1 month of service in ethylene gas (not even liquid), the sharp upstream corner of a brand new orifice plate had eroded to the point where the meter calibration was significantly changed.

If plants that use orifice meters were really serious about the importance of accuracy, then they would replace used orifice plates with new ones at least once a year. In how many cases have orifice plates been in service since the plant was built? What will be the effect of this on the validity of the flow information being fed into the big new data processor?

There is a practical answer to the problem of eliminating errors that are caused by deficiencies in orifice meter runs. This is do not field-fabricate orifice meter runs. Buy factory-made orifice meter runs from a company that is in that line of business and knows how to make them properly. If this policy is followed, one can be assured that the pipe will be round and clean, and orifice flanges will be welded on square. The internal roundness of the pipe will be ensured by turning the pipe on a lathe, not by hand grinding. Finally, the real inside diameter of the pipe will be measured and stamped on the outside of the meter run.

In the larger sizes, factory-purchased meter runs may be more expensive. However, this is a cost that is paid only once. Flowmeters that cannot measure flow rate with decent accuracy generally have their own ways of running up costs, and these go on indefinitely. It is a curious fact of life that whereas we seem to be able to justify millions of dollars to install a computer-based supervisory control system with its colorful graphics on monitors, we cannot find a few extra dollars to do a proper job of installing an orifice meter.

5.2 Meter Manifolding

The term *manifold* refers to the small piping that transmits the pressures on either side of the orifice to the high- and low-pressure connections on the differential pressure (DP) meter. The manifold is made up of ½-in pipe or tubing along with ½-in valves or specialty valve assemblies designed for flowmeter installations. The two runs of pipe or tubing that transmit the pressures are called the *impulse lines.*

An accepted and generally safe practice is to mount the DP meter below the orifice for liquid flow measurements and above the orifice for gas flow measurements. This ensures in the liquid flow case that bubbles of gas do not get trapped in the DP meter or the manifolding and in the gas flow case that liquid does not collect in the DP meter or the manifolding. For steam flowmeters, the best method is to mount the DP meter below the orifice, with condensate pots at the orifice flanges to provide a reservoir of condensate.

In any case, it is most important that the slope of the impulse lines be maintained so that there are no humps or valleys anywhere along their length. In addition, it is bad practice to parallel two DP meters (e.g., a transmitter and a local recorder) on the same set of orifice taps. The increase in volume encourages fluid flow in and out of the impulse lines as the differential pressure varies with flow rate. This, in turn, can lead to pressure measurement errors, especially when metering a compressible fluid. If it is necessary to run two DP meters from the same orifice meter run, then the second set of pressure taps, on the opposite side of the orifice flanges, should be used for the second meter.

There is an embarrassing trap that unwary persons can fall into. Sometimes a brand new orifice meter installation is turned on for the first time, and it produces no readings at all, even though the hookup is correct and all of the valves are open. What has probably happened is that through an oversight or lack of know-how, the pressure taps in the orifice flanges were not drilled through. The flange manufacturer sets the holes and threads them at the outside edge of the flange but may not drill the last inch or so. The user is supposed to do this.

A further concern is to what extent the meter manifold can affect the accuracy of an orifice meter. This leads to the consideration of a well-established argument regarding whether the DP meter should be grade-mounted or line-mounted.

5.3 Elevation Differences

Whenever the orifice meter run is at one height and the DP meter is at another, there is an elevation difference that has the potential for error. For any liquid flow orifice meter in which the orifice meter run is above the DP meter (a common enough arrangement), the pressures that exist at the connections to the DP meter are not the same pressures that are present on either side of the orifice. On both the high- and the low-pressure sides, the pressure at the DP meter is the pressure at the meter run connection plus the pressure created by the head of liquid in the impulse line. This additional pressure is proportional to the difference in elevation between the pressure connections at the orifice and at the DP meter and the density of the liquid in the impulse line.

It would appear that since the difference in elevation is the same on both the high- and low-pressure sides, there should not be any problem. However, even though both impulse lines contain the same process liquid, the possibility of a difference in density exists if the impulse lines are not at the same temperature. This could be caused by proximity of one of the impulse lines to some hot object or uneven application of heat tracing used for the purpose of winterizing the impulse lines. Thus the DP meter would not be measuring the same differential pressure as is being created across the orifice plate.

Before this subject gets entirely out of hand, the magnitude of the potential error should be examined. In an oil refinery, the orifice meter might be up in a pipe rack, with the DP meter at grade, so that the change in elevation over the length of the impulse lines is 3 m. Assume that the liquid being metered is an average hydrocarbon with a specific gravity of 0.80 at 15°C and that due to local conditions one impulse line happens to be at a temperature of 15°C while the other is at 20°C, a difference of 5°C.

A hydrocarbon liquid with a specific gravity of 0.80 at 15°C would have a specific gravity of 0.796 at 20°C. Using the relation

$$P\,(\text{kPa}) = 0.0980 \times sg \times h$$

the difference in the pressures created by the heads of liquid in the two impulse lines turns out to be 0.118 kPa. If the DP meter had a range of 25 kPa (the equivalent of 100 inH_2O), the error would be 0.47 percent of the DP meter range.

This discovery in itself is not entirely relevant because it is the error in the measurement of flow rate that is of concern, not the error in measuring differential pressure. In a head-type flowmeter, flow rate is proportional to the square root of the differential pressure. If we designate F for flow rate and h for differential pressure, then we can write

$$F = C \times h^{1/2}$$

If we are dealing in percentage values, h is 100 percent when F equals 100 percent. Substituting these values in the preceding relation gives a value of 10 for C. Therefore,

$$F = 10h^{1/2}$$

We now need to apply a limited amount of differential calculus to this relation. The result is

$$\frac{\Delta F}{\Delta h} = 10 \times \frac{1}{2} \times h^{-1/2} = \frac{5}{h^{1/2}} = \frac{50}{F}$$

What this relation is saying is that the percentage error in flow rate created by a 1 percent error in measured differential pressure will vary depending on where the flow rate is within the range of the flowmeter. At 100 percent of flow range, $\Delta F/\Delta h$ will be 0.5:1. In other words, an error of 1 percent in the measurement of the differential pressure produces a 0.5 percent error in the flow rate. At 50 percent of flow range, $\Delta F/\Delta h$ is 1:1. At 20 percent of flow range, $\Delta F/\Delta h$ is 2.5:1.

For the example, the percentage error in flow measurement created by a 0.118 kPa error in differential pressure, which is a 0.47 percent error in a 25 kPa DP range, would be 0.24 percent at 100 percent of flow range, 0.47 percent at 50 percent of flow range, and 1.2 percent at 20 percent of flow range. The percentage error figures here are percentages of the flowmeter range. The same amount of flow error, expressed as a percentage of the *actual* flow rate, would become higher and higher as the flow rate decreased within the range of the meter.

The objective of this example was to provide an insight into the magnitude of the error that can be created by a difference in the temperatures of the two impulse lines. Other aspects of this subject are discussed in the grade-mounted versus line-mounted section of Chap. 7.

5.4 The DP Meter

Not much needs to be said about the differential pressure (DP) meter, since, after all, it is the part of the orifice meter that seems to get the attention when complaints about metering errors start coming in. In order to do its part of the job adequately, the DP meter needs to have its calibration checked at reasonable intervals, as determined by experience.

It goes without saying that the quality of the DP meter should be good in the first place. The meter should be capable of being calibrated, to within at least 0.5 percent, right across its range. Many years ago, the mercury manometer was the workhorse for measuring differential pressures created by orifice plates and other differential producers. Although the mercury manometer has now been all but replaced by aneroid units and DP cells, it nevertheless had two sterling qualities. It seemed to be easily calibratable to within the desired limits of error, and it would hold that calibration for months on end.

Today, however, mercury is considered to be a hazardous material. This would probably prove to be true if it were actually ingested. The claim that it is hazardous on any other basis is hard to justify, since its vapor pressure at room temperature is just about zero. A mercury manometer might still be a good choice for steam flow measurements.

In any event, it is important to realize that the accuracy specification is not the only one that is important. Perhaps equally relevant is the matter of selectivity. There are, unfortunately, DP cell transmitters that react to changes in such factors as ambient temperature, vibration, and for electronics, signals generated by portable radios. Good DP transmitters may cost more money, and if they do, it is more likely that they have superior selectivity than superior accuracy.

When the DP measurement is transmitted, there will inevitably be some error in converting measured differential pressure into a signal and then converting the signal back to a reading. The error will be worse in the case of analog signals (3 to 15 psig or 4 to 20 mA) than for digital transmission and could be as much as 1 to 2 percent.

What this means is that if optimal accuracy is to be obtained, the measured differential pressure should be converted directly into a reading, using a local indicator or recorder, rather than transmitting it.

5.5 The Correlation

Since differential pressure is what is being measured but flow rate is what is required, there must be a correlation between the two variables. The next step is the appropriate calculations, since no matter whose correlation is used, it will contain a number of factors.

The correlation that will be used here will be that developed by L. K. Spink, a flowmeter engineer with considerable credibility, who spent his career in the employ of the Foxboro Company. He has to his credit a text entitled *Principles and Practice of Flow Meter Engineering*. This text is now out of print, which is regrettable, since it had the rare quality of being understandable.

Actual calculations will not be done here, so there should be no hardship involved if the SI form of the correlation is used:

$$W = 1.110 \times S \times D^2 \times Fa \times Fr \times Y \times (\rho \times h)^{1/2}$$

where W is the mass flow rate (kg/s). 1.110 is a numerical factor that takes care of the units used. S is the efficiency factor for the orifice meter. It is required because the conversion of pressure energy to velocity energy when using an orifice plate is not 100 percent efficient. S is a function of the ratio of the diameter of the orifice to the inside diameter of the meter run pipe. This ratio is known as the *orifice diameter ratio,* and is usually identified by the Greek letter ß. D is the internal diameter of the meter run pipe (m). Fa is a temperature correction factor, made necessary because the orifice will be bored out to the specified diameter at room temperature but will be in service at a different temperature. Fa is a function of the fluid temperature and the material used to make the orifice plate (usually type 316 stainless steel). Fr is a correction factor for the Reynolds number of the metered stream. Fr is a function of the Reynolds number, as well as the orifice diameter ratio ß and the pipe size of the meter run. Y is an expansion factor for compressible fluids, that is, gases, vapors, and steam. For liquids, the value of Y is 1.00. The expansion factor is required because as the compressible fluid jets through the orifice, its pressure drops, with a resulting decrease in the density of the fluid. This has to be taken into account in the correlation. The expansion factor Y is a function of four factors: the ratio of the differential pressure at the orifice to the absolute line pressure, the orifice diameter ratio ß, and whether the line pressure is measured upstream or downstream of the orifice. It also depends on the fluid itself, specifically, its specific heat ratio Cp/Cv. ρ is the density of the process fluid being metered, at line temperature and pressure (kg/m^3). And finally, h is the measured differential pressure (Pa).

Most orifice meter calculations these days are done by a computer so that, presumably, the calculations should not pose any problem. What should be remembered, however, is that any high school student can take an equation out of a textbook and make a computer program from it. What is important is that the principal virtue of using a computer rather than a calculator is that the computer saves time. Little time is saved by using a computer calculation if the person who requires the result has to spend considerable time and energy searching for values for S, D, Fa, Fr, Y, and ρ.

A proper question to be asked of anyone who is promoting a computer program for orifice meter calculations is, "Exactly what input data do we have to provide?" A good computer program should be able to do the following:

- Determine the value of D, given the pipe material and schedule.
- Determine the value of ß, given the orifice diameter.
- Determine the value of S, using the value of ß.
- Determine the value of Fa, given the line temperature and the material of the orifice plate.

- Determine the value of the Reynolds number and *Fr*, using the value of *D*, the pipe size, and the orifice diameter ratio ß and given the viscosity of the fluid at line temperature. It would be even better if the program contained correlations for the viscosities of some of the more common fluids with temperature.
- Determine the value of the expansion factor *Y*, from the value of ß and given the line pressure, the nature of the fluid (specifically its ratio of specific heats), and information on whether the line pressure is measured upstream or downstream of the orifice.
- Determine the value of the density ρ of the process fluid, given the name of the fluid and its temperature, and for compressible fluids its pressure, at line conditions.

In other words, a good computer program will contain not just the equations that are peculiar to the orifice meter but also pipe tables and a considerable amount of physical data pertaining to the fluids that may be metered.

5.6 The Expansion Factor *Y*

In the correlation for orifice meters, the expansion factor *Y* applies to compressible fluids only. Fortunately, it has a value of 1.00 in the case of liquids.

The expansion factor needs special consideration. It is a bad actor, inasmuch as *Y* is a function of the differential pressure at the orifice, which will change with flow rate. This means that as the flow rate changes, the value of *Y* is changing, and the flowmeter is actually changing its calibration. Complications such as this are not a happy prospect.

There is no way that the problem of the variable calibration can be circumvented. It has to accepted as a fact of life. The only recourse is to do the best that can be done to limit the extent of the change. With proper design, it can be be limited to 1 percent or less.

Figure 5.1 contains some graphs that show how the value of the expansion factor *Y* varies with those factors which affect it. These graphs reveal important fine points and are worth some study. The values of *Y* are plotted on a base of the ratio of differential pressure to absolute line pressure. The SI values kilopascals per kilopascals absolute (kPa/kPaa) are used.

This particular set of graphs applies to gases or vapors that have a Cp/Cv ratio of 1.3. This would include steam and light hydrocarbon vapors. The graphs in Fig. 5.1 also apply to orifice meters that have the pressure connections in the flange, usually referred to as *flange taps.* Orifice meter installations involving vapors and gases with dif-

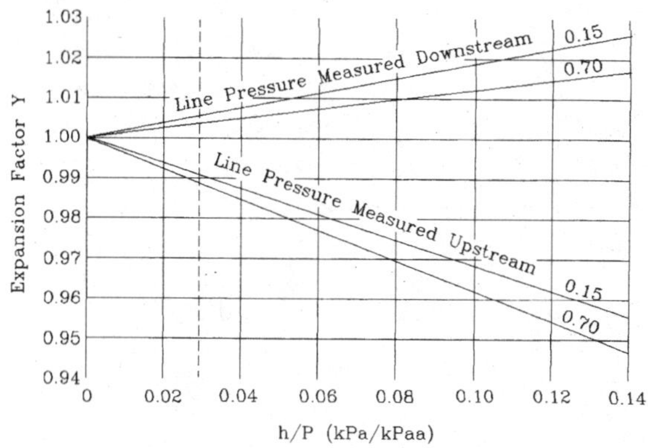

Figure 5.1 Variation of the expansion factor Y with h/p ratio.

ferent Cp/Cv values, different pressure tap locations, or the units for the differential pressure to line pressure ratio in inches of water per pounds per square inch absolute will have similar straight-line graphs but with different values for Y.

The graphs are straight lines, in pairs, one pair for line pressure measured downstream of the orifice and the other pair for line pressure measured upstream. Each pair of lines shows how the value of Y is affected by the value of the orifice diameter ratio ß. The two lines are drawn at ß = 0.15 and ß = 0.70, which are the practical limits for the value of ß. Graphs for intermediate values of ß will be straight lines and will fall somewhere between these two lines.

Perhaps the first consideration should be the significance of where the line pressure is measured. For a compressible fluid, any change in line pressure will affect the density of the fluid, which is one of the terms in the general equation. The density of the vapor varies directly with the absolute line pressure. However, the density term is under the square root sign, so fluid density is not as influential as some of the other factors in the equation.

Nevertheless, if there is any significant variation in line pressure, the pressure value, at the moment, will need to be taken into account for the sake of accuracy. In some gas metering installations, it is customary to have a second measurement, the absolute line pressure, alongside the primary measurement, which is the differential pressure. The flow calculation then becomes

$$W = C_1 \times \sqrt{P \times h}$$

where P is the absolute line pressure, and C_1 is a constant, usually called the *meter factor*.

It is obvious from Fig. 5.1 that the variations in Y are only half as bad when the line pressure is measured downstream as they are when the line pressure is measured upstream. This makes a pretty good case for measuring the line pressure on the downstream side of the orifice. On the other hand, getting a credible flow measurement becomes simpler if the line pressure is essentially constant rather than varying, in which case it would not be necessary to measure the line pressure at all. If the pressure is constant and known, then the density of the fluid is not affected by pressure changes, and the density value can be included in the meter factor. The relationship then becomes

$$W = C_2 \times \sqrt{h}$$

However, when an obstruction such as an orifice plate is placed in a pipe, the differential pressure is created by lowering the downstream pressure in varying degrees, not by changing the upstream pressure. Therefore, if the pressure is presumed to be constant, it is the value of the upstream pressure that should be used when the gas or vapor density is determined.

If the objective is to limit the variations in the expansion factor Y to 1 percent or less, then it means that in Fig. 5.1 variations in Y must be limited to those segments of the graphs which lie to the left of the vertical broken line. This conclusion is based on the graphs for line pressure measured upstream, since they represent the worst case.

The broken line occurs where the value of the h/P ratio is about 0.03, or roughly 1/30. Consequently, in a good design, the numerical value of the differential pressure range selected for the meter, in kilopascals, should not be greater than 1/30 the numerical value of the absolute line pressure, in kilopascals absolute. In the fps system, the equivalent statement is that the value of the differential pressure range, in inches of water, should not exceed the value of the line pressure, in pounds per square inch absolute.

Of all the orifice meter design rules that are destined to be broken or bent, this rule is the one that is abused most frequently. In many cases, designers do not understand orifice meters well enough to realize the consequences of what they are doing. In other cases, the rule is broken because to conform with it would (what else?) cost more money.

For example, low-pressure steam at 50 psig (65 psia) is being piped to a plant through a 12-in line. The rule, based on limiting variations in the expansion factor Y to 1 percent, says that the differential pressure range selected for the meter should not be greater than 65 inH_2O. However, it turns out that 65 inH_2O is not a great enough differential pressure range to obtain the flow range desired.

The correct course of action is to increase the size of the meter run to 14 or perhaps 16 in so that the required flow range can be obtained with a differential pressure range of less than 65 inH_2O. However, this would require a meter run that is more expensive than a pipe-size meter run, 12 in. Thus, to obtain the required flow range, a differential pressure range of 100 or 200 inH_2O is specified instead. The result is that variations in Y, the expansion factor, may be as great as 3 percent over the range of the meter. This either goes unrecognized or is shrugged off.

5.7 Successful Installation

Users of orifice meters who are concerned about accuracy are obliged to remember that a successful orifice meter installation means attention to four areas: the orifice meter run, the meter manifolding, the differential pressure meter, and the correlation between measured differential pressure and flow rate. Of these four, the leading contributor to excessive errors is invariably meter runs that are in poor condition.

5.8 Vortex Shedding Flowmeters

The vortex shedding flowmeter is one flowmeter that shows promise as a replacement for the orifice meter. The feature that has traditionally kept the orifice meter entrenched in the petroleum and petrochemical industries is that it conforms with the requirement that every process measuring instrument must be capable of being maintained without shutting down the operating equipment on which it is installed. Orifice meter maintenance has consisted for the most part of calibrating the differential pressure meter. The impression that this constitutes proper maintenance is a fallacy, since it ignores the fact that the orifice plate is wearing and, because of this, the calibration of the meter is changing.

The positive features of the vortex shedding meter are

- Its output is *linear* with the *average* velocity of the stream.
- Its calibration is less sensitive to changes in fluid density or viscosity than is the calibration of an orifice meter.
- Its calibration is less affected by erosion of the bluff body by the flowing fluid than the calibration of an orifice meter would be affected by the equivalent amount of wear on the orifice plate.
- It has no impulse lines that contain process liquid. The connections from the sensing element in the pipe to the transmitter are wires. This virtually eliminates winterizing problems and costs.
- The meter can be zeroed simply by unplugging the sensor cable.

5.8.1 Operating principle

A bluff (flat-faced) object placed in a moving fluid stream produces a wake consisting of a train of vortices that are shed from one side of the bluff object and then the other in an alternating pattern. This effect can be seen in the fluttering of a flag (the pole is the bluff object).

The frequency of shedding of the vortices from the bluff body is described by the equation

$$f = \mathrm{St}v/w$$

where f = frequency of shedding (Hz)
v = stream velocity (m/s)
w = width of the bluff body (m)
St = Strouhal number (a dimensionless parameter)

This equation shows that there is a linear relationship between the frequency of shedding and the average velocity of the flowing stream. It also shows that the shedding frequency depends on the width of the bluff body. However, it does not appear to depend on whether or not the corners of the bluff body facing the stream are sharp or rounded. If this is true, then erosion of the corners of the bluff body will not affect the calibration of the meter. Erosion of the sides of the bluff body, which would change the effective width, is another matter.

Tests have shown that the parameter St remains constant over a range of Reynolds numbers from 10,000 to 1,000,000 for the flowing stream.

The major obstacle in the development of the vortex shedding (VS) meter has been measurement of the shedding frequency. Early versions of the VS meter had the frequency detectors mounted on the bluff body. Thermistors and piezoelectric crystals were two types of detectors that were tried. This restricted applications of the VS meter to those in which the fluid temperature and pressure did not exceed the limits that could be tolerated by the frequency detectors.

Fortunately, later models have improved on this situation greatly. The shedding frequency can now be detected by sensitive strain gauges or piezoelectric sensors mounted on the bluff body extension bar, which is outside the process stream. Thus the working pressure is now virtually unlimited, and the temperature limits have been increased to the point where a VS meter can be used to measure steam flow at 400°C or cryogenics at −40°C.

5.8.2 Size and accuracy

Currently, vortex shedding meters are available in sizes from ½ to 8 in. Typical accuracy, with a Reynolds number not less than 10,000, is within 1 percent of flow rate for liquids and within 1.5 percent for steam or gases.

5.8.3 Vortex shedding meter problems

Any of the following can cause or contribute to poor accuracy from a vortex shedding meter:

1. *Low velocity.* Vortices that are shed at Reynolds numbers below 10,000 are weak and irregular. Most VS meters are virtually useless at these low Reynolds numbers because they are unable to detect positively the pressure pulsations.
2. *Distorted flow profile.* An irregular velocity profile, or swirling, will interfere with detection of the vortices. A proper meter run is required. Typically, it should be the same size as the meter, with a minimum of 10 diameters of straight pipe upstream and 5 straight diameters downstream.

   Slip-on, screwed, or other types of flanges that locally distort the inside diameter of the meter run must *not* be used if accuracy is to be ensured. The transition from the meter run to the body of the meter must be smooth. Weld neck flanges are the best, even though they may be the most expensive.

   The flow profile also can be distorted if thick gaskets or gaskets with an inside diameter that is too small are used. Finally, special care is required to ensure that the meter body is lined up true with the meter run. The alignment should be checked using gauges, not by eyeballing.
3. *Pipeline "noise."* Interference caused by mechanical vibration of the pipe makes the detection of weak vortices even more difficult. This is a common cause of VS meter malfunction and excessive error.
4. *Pressure drop.* The bluff body takes up a sizable fraction of the cross-sectional area of the pipe. Accordingly, the possibility of significant pressure drop exists. The manufacturer of the VS meter can estimate the pressure drop for a given application.

Chapter 6

Plant Stream Analyzers

6.1 A Case Study

The extensive publicity that has been given to on-line computers in process control notwithstanding, the contribution to improved plant operations made by plant stream analyzers should not be discounted. There have been many petroleum refinery and chemical plant applications in which a properly applied plant stream analyzer has provided a monetary return even greater than that which has been obtained through the use of process computers.

Most articles written about plant stream analyzers have tended to emphasize the tricky analytical measurements they have been able to make and how much money was saved through their use. Not enough space has been dedicated to the effort that had to be made to get them to do the job that was expected of them. As a means of presenting a broader view of plant stream analyzer application, let us consider the following fictitious situation.

Imagine that you and I are process superintendents and that we are responsible, among other things, for the operation of a plant that makes a rather expensive chemical. The essential specification for this chemical product is that the percentage of a certain impurity that it contains in its finished form is not to be greater than a specified limit. The plant is equipped in the conventional manner with meters and automatic controls. The product stream is provided with a flowmeter that tells the plant operator how much product is being made at any given moment.

Product quality is, of course, equally important as quantity. Unfortunately, the quality can only be determined by sending samples of the product to the Inspection Laboratory for manual analysis. Traditionally, about 3 to 4 hours elapse between the time that a sample is taken and sent to the laboratory and the time that the process operator receives the report. The disadvantages of this situation are obvious. First of all, the operator must base his or her judgment on information that is 4 hours old. Second, in addition to the fact that the information is out-of-date, it also represents only a single point in time. The operator cannot really tell whether the impurity in the product is increasing, decreasing, or remaining steady.

In the past, this situation has led to off-specification product being run into the final product tank. The entire tank of product was then ruined and had to be rerun. Of course, when this happened, the responsible operators were reprimanded, with the result that the incidences of off-specification product were substantially reduced.

However, later on it was discovered that the costs of operating the plant had risen abruptly. We seemed to be making product at a decreased rate, while at the same time we were using more fuel, power, and other utilities. Besides this, in one section of the plant a rather expensive catalyst is used. Whereas the life of this catalyst was normally considered to be 6 months, this had now decreased to 4 to 5 months. Changes in operating procedures were apparently behind all of this, but why had the procedures been altered?

We eventually figured out that our operators were deliberately making a product that was better than the specification called for. Actually, the product was acceptable if it contained not more than 0.5 percent of the impurity. On checking the product reports over the last few weeks, we found that the impurity figures were running 0.2 percent at most, and there were many occasions when there was no impurity at all. Obviously, since they lacked good information on the quality of the product as it came off the plant, the operators were running the plant better than it should have been run in order to stay out of trouble. This explained the increasing operating costs.

One day we were discussing our problem over coffee. We agreed that what we really needed was an instrument that would tell the operators what the percentage of the key impurity in the final product was at all times. Taking samples once or twice a shift and having them analyzed by the inspection laboratory was not working out. By coincidence, the same day you happened to be glancing through a technical magazine, and you noticed a short article about a continuous plant stream analyzer that made an analysis very much like the one we needed to make. The resemblance was so striking that we decided to investigate this development further. To this end, we called our purchasing department and asked them to try to contact the company that manufactured the analyzer. In their customary efficient manner, purchasing arranged to have a representative of the analyzer manufacturer call on us on Monday of the following week.

Our conversation with the sales engineer was enlightening. He confirmed that his company did make an instrument that would measure the amount of impurity in our product on a continuous basis. The standard version of the instrument cost $30,000, but it was the company's practice not to make a formal quotation until its engineering department had been provided with a complete report on the composition of the stream that was to be analyzed. It happened that we had such a report in our files, so we made a copy and gave it to the sales engineer. Armed with this information, he departed with the promise that he would prepare a quotation and send it to us as soon as possible. It seemed to us that $30,000 was a great deal of money. The most we had ever paid before for one instrument was $1800. However, we decided to reserve judgment until we saw his formal quotation.

A few days later, the quotation arrived in the mail. The opening paragraphs contained comments to the effect that the stream we wanted to analyze was in a special category so that the standard version of the the analyzer could not be used. Nonetheless, the company would provide a modified version that could make the analysis for the sum of $38,000. We were naturally not pleased with this development, but as we read on, things got worse. There were specific installation requirements that would have to be met, including the requirement for an elaborate sampling system, along with proper cold weather protection for the analyzer. When we consulted the estimating group about these requirements, their opinion was that the requirements would add an additional $13,500 to the cost. Thus, at that point, the cost of the analyzer installation had gone from the original $30,000 to $51,500.

Our immediate reaction was to file the quotation and forget about it. It was an idea that had looked good at first but turned out to be out of our reach financially. However, fate intervened. A few weeks later, following some trouble at the plant that could have been avoided if a product analyzer had been available, we decided to look into it again. True, $51,500 was a lot of money, but perhaps we should have made an estimate of how much we could save if we had the analyzer. Once we got started, it was less difficult than we would have thought to come up with a figure. This is how it added up:

1. Each time a tank of finished product was spoiled because it did not meet the impurity specification, it cost us $8000 to rerun the off-spec material. Last year we had to run three tanks over again. It would be reasonable to suppose that if the operators had had the benefit of a product stream analyzer, then the number of tanks of product that had to be rerun might be reduced from three to one. This would save $16,000 per year.

2. We were able to determine that if we could raise the current average impurity figure from 0.2 to 0.4 percent, then the value of the increase in production, plus the lower plant operating costs, due to increased catalyst life and reduction in utilities charges, would be close to $11,000 per year.
3. At the particular plant where we work, the cost of maintaining the inspection laboratory is charged against the process units that use it. The inspection laboratory charge for the product purity test we require is $20. At this time, our operating procedures require the senior operator to order at least one product inspection during each shift, and at the operator's option, a second inspection can be called for. As a result, product inspections are averaging close to five per day. We reasoned that the product analyzer should cut this number to three per day, which would represent a saving of $40 per day, or $14,000 per year.

The return from these three items totaled $41,000 per year, and this we could apparently get from an investment of $51,500. In other words, the product analyzer, on paper anyway, would pay for itself in 15 months. Generally speaking, in our company it was considered that a 3-year payout was a good investment. On this basis, the return on the product analyzer application should be excellent despite its relatively high cost.

We consequently made a formal application for a $51,500 addition to the current year's budget, accompanied by the economics. It required only one review by management for the project to be approved. The only comment made by the review committee was, "Why don't we get more investment items like this one?"

At this point things looked promising. However, we were not so naive as to believe that our success with this new, elaborate instrument would be automatic. Even though we had been granted the funds to buy it, we had certain reservations that would have to be cleared up. This suggested that a second discussion with the sales engineer from the analyzer company was in order.

The first thing we had to be reassured about was that the analyzer could actually make the measurement that we wanted it to make. So far we had only a general confirmation that it could do it, based on a single report on the composition of our product stream.

The sales engineer agreed that this was hardly an adequate basis on which to proceed. Therefore, what we should now do is send a 20-l sample of our product to their head operation, where it would be run through a plant stream analyzer of the type on which they had quoted. We would then get a formal report one way or another as to whether the analyzer the company was proposing was really capable of making the analysis we wanted.

Our second reservation had to do with servicing. We have a competent instrument maintenance department, but our technicians had never had to deal with an instrument that was this complex. Sooner or later the analyzer would require servicing. Who was going to provide it?

The sales engineer had an answer for this also. If we had a person in our instrument maintenance group who was considered to be a good learner and who had received training in electronics, then we could send him or her to the factory, where he or she would be given a 1-week comprehensive course in all the techniques required to maintain the analyzer properly. The course was free of charge; the only costs would be the person's travel and living expenses. We happened to know that there was such a person in instrument maintenance. His name was Arnold Goodman.

Accordingly, two important questions were cleared up. We thanked the sales engineer for his assistance, and the next day we arranged to have a 20-l sample from our product stream forwarded to his head office by the quickest route. About one week later we received a report to the effect that the analyzer had been tested on our product, and it had made the analysis we required without any problem. The record it had made during the test was attached, and we had to admit it looked impressive. There now seemed to be no reason to delay placing the order for the analyzer any longer, and we instructed our purchasing agent to proceed.

Shortly thereafter, arrangements were made for Arnie Goodman to take the training course on the maintenance of the analyzer. He departed full of enthusiasm. When the course was duly completed, he reported back to us, radiating confidence and carrying two rather large binders that appeared to contain operating and maintenance manuals.

While we were waiting for the analyzer to be delivered, we frequently talked about our analyzer application. One day we came up with a disturbing thought. What if, despite our best intentions, the analyzer did not win the acceptance of the people who were supposed to use it: the plant operators. Most of the operators on this plant had worked there since the plant was built 12 years ago. They had been through many startups, shutdowns (scheduled and otherwise), and upsets of one type or another. They felt that they knew the plant's every idiosyncrasy. Now we were introducing a new state-of-the-art item of instrumentation, with the message that with the assistance of the information it would provide them, they would be able to run the plant better than ever before.

Human nature being what it was, some of the operators might accept this, but others might feel that it was a reflection on their performance. If this were the case, then they would resent the analyzer or at least be indifferent to it, and they would ignore it. Under the circumstances, we realized that it would be a bad mistake on our part if the operators had no indication that the analyzer was coming on board until they saw it being carried into the control center and installed in the new opening that had been cut in the instrument panel for it.

We decided, consequently, that our operating personnel should be kept fully informed of developments. We would do this by giving informal talks at the control center, in which the operators would be brought up-to-date on the plan, with all the details of what the analyzer could do and with special emphasis on the benefits which they themselves would derive from it.

By a democratic flip of a coin, we divided up the morning, afternoon, and night shifts for presenting our message. When the orientation program was complete, we compared notes back at the office. In one respect, our experience was consistent—the reaction we got from the operators was inconclusive. As they listened to the presentation, their faces were blank. At the conclusion, when we asked if there were any questions, there were none.

Later on, we received somewhat better feedback by way of two of the instrument technicians who were assigned to the plant and who were on more intimate terms with the operators than we were. The consensus was that most of the operators were skeptical that the analyzer would work; however, they were prepared to reserve judgment until they saw what happened.

At this point we stood back and reviewed our position. First of all, we had established that there was real economic incentive to having the on-line product analyzer. Second, by sending a sample of our product to the analyzer manufacturer, we had verified that the analyzer could make the analysis we needed. Third, we had a capable person from our instrument maintenance group trained in the servicing techniques that would have to be carried out to ensure that the analyzer would have a satisfactory service factor. Finally, we had orientated the operating personnel to what we were hoping to achieve by bringing in this ultranew instrument.

Having done all of this, every aspect seemed to be covered. What could we have overlooked? What could possibly go wrong?

Eventually, the plant stream analyzer arrived, and in due course it was installed. It took 5 weeks longer than we anticipated because just about every trade in the plant seemed to be involved in the installation, and none of the trades could do their job until someone else had finished theirs. For some reason, the mechanical department seemed indifferent to our requests for priority on the services of the people required to install the analyzer.

In any event, the monumental day arrived at last. Arnie Goodman, who had been following the installation work, was elected to turn the analyzer on. Of course, you and I were both present in the control center to witness the event. We checked with Arnie that everything was in order. He confirmed that it was. "OK," we said, "start it up."

With that, Arnie proceeded out to the shelter where the analyzer was located. He opened the sample valve to start the sample flowing through the analyzer. He established the required flow rate and turned on the power on the analyzer. Then he came back to the instrument panel and threw the switch on the recorder to the ON position. After a brief warmup period, the pen began to rise on the chart. It continued to rise through midscale until it had gone all the way to the top of the chart, where it stopped.

You and I, Arnie, and the operators who were watching knew that this was not a very good reading of the percentage impurity in our product. Consequently, everyone looked expectantly at Arnie Goodman. Arnie stepped forward, muttering something about sensitivity. He opened the door of the recorder, swung back the large carriage that contained the chart-drive assembly, and with a screwdriver turned a small screw way at the back of the instrument. Nothing happened. He turned the screw a little farther, but the pen remained at the top of the chart. After this he closed up the recorder at the front and went around to the back of the panel. In a moment he emerged again and peered at the pen, but it was still at the top of the chart.

By this time we could see that Arnie was getting a bit flustered, so we made some offhand remark like, "Probably just a minor adjustment," and we told Arnie that we were going back to the office and would check back with him some time after lunch.

When we arrived back at the control center a couple of hours later, we found, to our dismay, that the analyzer was dark, and there was a sign hanging on it which said, "Out of Service." Our technician was not working on the analyzer either. He was, in fact, sitting at a table in a corner of the control center reading from one of his large binders of instructions. He said that he had not found the trouble, but he hoped to do so shortly.

Unfortunately, 2 days later Arnie still had not been able to identify the problem and correct it. At that time, we asked him how things were going, and he had to admit that he was stumped. We found this somewhat aggravating, especially since we had gone to the expense of having Arnold Goodman trained to service the analyzer.

Eventually, however, the truth dawned on us. During the week he was at the training school, Arnie was taught how to do routine maintenance and repairs. This included locating and replacing bad components and the general servicing procedures necessary to keep the analyzer in good operating shape. Nonetheless, in the short time that he was there, Arnie could not possibly have acquired the knowledge needed to troubleshoot the whole application. This kind of know-how comes only after years of experience. Not only would a person have to understand in detail the workings of the analyzer, he or she would have to appreciate the peculiarities of our plant application as well.

Out of desperation, we decided to contact the sales engineer again and tell him about the trouble we were in. He suggested that we call his company's service group at the head office and ask them to send out a service engineer. In response to our plea that the situation was urgent, a service engineer was able to come the following week. He stayed 3 days, during which time he made several modifications and at one point phoned the factory for some parts, which were duly flown in. These he installed, whereupon everyone was elated to see the analyzer present a reading that was apparently accurate.

At last we had the results we had been hoping for. We both shook the hand of the service engineer and thanked him for his trouble. Our enthusiasm for the analyzer application attained such heights that it was hardly dampened when, a week or so later, we received an invoice for $2,600 to cover the time and expenses of the service engineer.

For a while, there were no further problems. The product analyzer became part of the general operation of the plant. As far as we could tell, the anticipated benefits were being obtained. Then one day, while you were in the control center on a routine inspection, you noticed an operator come into the building with a full sample container in his hand. You then heard the operator phone the inspection laboratory and request that the sample be picked up. Your immediate reaction was that with the analyzer in service, there should be no reason for taking samples, so you asked the operator why this was being done.

The operator's answer was that they had been obliged to take samples because the analyzer was not working. You then asked whether they had attempted to contact Arnie Goodman. The operator said that they had done this, but that he had been out on an important job in Area C and was unable to come and fix it for them.

When you heard this you got into your truck and drove over to Area C to locate Arnie Goodman and bring him back to the control center to deal with our problem. Arnie seemed a bit upset about all of this. However, when he finally arrived at our control center, it required only about half an hour of effort on his part to correct the problem and get the analyzer on stream again.

Talking to Arnie later, we asked him not to allow this to happen again. Despite our pointed request, there were two other similar occurrences in the following few weeks. The analyzer developed a problem, the operators reported it, but the analyzer remained out of service for several shifts because Arnie Goodman was not around to fix it. When he was finally located and prevailed upon to come to the control center, he cleared up the problem in a reasonably short time. Unfortunately, until he did his job, the "Out of Service" tag was prominently displayed on the front of the analyzer.

This was another problem that we had not anticipated. When we mulled it over together, certain truths became evident. For the most part, Arnie Goodman had acquired the knowledge to service the analyzer when it needed it, but we had neglected to consider whether he was going to have the required time. After all, we had picked Arnie for the job because we thought he was one of the more capable instrument service technicians.

Since he really was better qualified, his services were in demand all around the plant. Furthermore, when he had a number of jobs competing for his time and he had to set priorities, he was guided by two pretty basic human traits.

- It is natural to undertake first the job with which one is most familiar and which, subconsciously, one feels most competent to handle. Despite the fact that he had had the benefit of the training course, Arnie sensed that this did not exactly make him an expert on the plant stream analyzer. The startup incident did nothing to build his confidence either. Consequently, other jobs that he knew how to handle tended to get priority over problems with the analyzer.
- If he were obliged to evaluate the urgency of two jobs that he had to do, Arnie was inclined to take into account that the plant had run for 12 years without any plant stream analyzer, so it should be able to survive without the analyzer for another day or so until he could get around to fixing it.

The question of the priority which the analyzer should have on Arnie's time was cleared up only after we had a serious discussion with Arnie and his supervisor. We emphasized, among other things, how much it meant to us in dollars to have the analyzer in operation.

In the end, we reached an understanding whereby Arnie Goodman was to make uninterrupted operation of the analyzer his number one priority.

A month or so later, we found out, unexpectedly, that some process operators were once again sending samples to the Inspection Laboratory. On these occasions Arnie Goodman was not at fault. He had met his commitment to give priority to maintaining the analyzer in operation. In fact, the in-service time of the analyzer over the past 6 weeks had been excellent.

After questioning the operators, however, we learned that the daily laboratory inspections, which were still allowed as a routine check, had occasionally given a different impurity number than the analyzer itself had recorded. This had happened often enough that the operators had begun to notice some disagreement between the analyzer and the laboratory readings. After a time, the accuracy of the plant stream analyzer had become suspect. The result was that the operators were reverting back to their old practice of obtaining their product quality information from samples sent to the inspection laboratory for manual analysis.

This was a really annoying development. We notified the operators that the next time they noticed a disagreement between the measurements obtained from the plant stream analyzer and from the inspection laboratory, they were to report it to us. Having come this far, we were not prepared to give up on the analyzer because it produced different values than the laboratory.

One day we decided to try an experiment. We went out to the plant and drew a sample ourselves. We then split the sample into two parts. One part we sent to the laboratory for analysis right away. As soon as we received their report, we sent the other part of the sample to the laboratory. When we compared the two analyses made on the same sample, we found that the impurity values differed by 6 percent. Furthermore, the reading of the analyzer at the time we drew the sample was midway between the two values which the laboratory came up with.

This was, to us anyway, convincing evidence that the analyzer, when well maintained by Arnie Goodman, was capable of more consistent accuracy than the traditional method of grabbing samples and sending them off to the inspection laboratory for analysis. This should be the case, if for no other reason than that the plant stream analyzer had a much superior sampling system that prevented the lighter components from escaping and which did not allow any extraneous substance such as air to contaminate the sample.

Much to our disappointment, however, this line of argument made little impression on our operating people. The conclusion was disturbing. Although there were some exceptions, of course, it seemed that most of the operators really did not want the analyzer to show them the real value of the impurity. What they wanted was an analyzer that gave them the same value that the laboratory gave them. Only under such circumstances would they believe that the analyzer was really useful.

We had proved to our satisfaction that there was a fair amount of inconsistency in the technique involving manual sampling and analysis by the laboratory. The readings that were ultimately obtained would be influenced by how carefully the samples were taken and handled and by the techniques and equipment which the laboratory used. Consequently, it would be all but impossible to calibrate the analyzer so that it did what the operators thought it should do.

As if this were not enough, we came up against still another problem that we had not anticipated. Originally we thought we had confirmed the operability of the analyzer by sending a sample of our product to the manufacturer so that the analyzer would be designed to function on the same material it would be analyzing when it was actually in service. On this basis we had assumed that the analyzer was applied correctly.

The point that had eluded us, however, was the fact that while the sample that we submitted was representative of what would be in our product stream *most* of the time, it was not representative of what was there *all* the time. As a matter of fact, the feed stream for our plant came from another process unit immediately ahead of it in the overall processing sequence. In this other plant the feed stream was treated with a fairly strong chemical additive. Occasionally, an upset occurred in the upstream plant that allowed some residual treating chemical to be present in the feed stream to our plant. Since our plant was not designed to remove this chemical, it eventually showed up in our product.

If the chemical got into the sampling system of the analyzer and into the analyzer itself, it laid down a grayish-white muck that quickly put the analyzer out of service. Each time this happened, it took Arnie Goodman the best part of 2 days to dismantle the sample system and the analyzer and clean them up. On one occasion a sensitive part of the analyzer was damaged, and it took 3 weeks to obtain a replacement, even though we asked for the best possible delivery. While all this was going on, the ominous "Out of Service" tag hung on the front of the analyzer. Even though the analyzer itself was not at fault, the appearance of the "Out of Service" sign did nothing for its acceptance by the operating staff.

And as if this were not enough, by this time another sinister influence was beginning to make itself felt. In many plants, the attitude of the operating personnel is that no matter how great the potential of any new operating equipment may be, they are prepared to wait only so long for the results they were told it would achieve. After this psychological deadline has passed, they will assume the new device or system is useless, and they will write it off. No matter how it performs after that, it will not be accepted. We had to admit that we were beginning to sense this attitude in some of the operators in our plant.

Like suddenly waking up from a dream, or maybe a nightmare, our story is interrupted at this point without revealing whether we were able to get our plant stream analyzer application to work or, if we did, whether we were able to get our operators to use it as we wanted them to. It is worth noting, however, that despite all the problems that we encountered, the credibility of the economics was never in doubt. If the application did prove to be successful, then the analyzer must have made a great deal of money for our plant and for the company. This is typical of the majority of plant stream analyzer applications.

It has been said that if users could achieve one successful plant stream analyzer application out of every three they attempted, then they would end up ahead of the game economically. There is considerable evidence to support this statement. The payout is that good. Even though on-stream analysis applications may be fraught with unexpected problems, still each application, successful or otherwise, will increase the knowledge of the user. Thus the chances of success should improve with each successive attempt.

6.2 What Did We Learn From All of This?

1. No matter how attractive the projected return from an investment in state-of-the-art instrumentation may be, the instrument or system will not earn $1 if you cannot get it to work.
2. When you have convinced yourself that you have all the potential problems covered, that's when there is sure to be something else that you have not thought of.
3. The sample that you send to the instrument manufacturer so that the performance of the instrument can be verified will be representative of what is in your plant stream most of the time but not all the time.
4. If the plant has run for some years without the new instrument or system, it will be a difficult job to convince the operating staff that the plant really needs it.
5. The operating staff may not want the new plant stream analyzer to give them a true reading. They may want it to read the same as the analyses, good or bad, they are getting back from the inspection laboratory.
6. The human factor has a monumental bearing on the ultimate success of any new process measurement or control system venture. If there are enough people who are indifferent to whether the instrument or system works or not or who may actually be happy if it fails, then it probably will.

6.3 PSA Maintenance

Sometimes we forget that a successful instrument application requires not only obtaining the correct hardware but also using it properly. Proper usage has two elements: correct installation and adequate maintenance.

When any plant gets its first on-stream analyzer, the question of who will be responsible for keeping it in good operating condition is usually answered, "The instrument department, who else?" The implication is that the job of doing the analyzer maintenance should logically be added to the workload of the mechanical department instrument maintenance group. After all, it is an instrument, isn't it? In other words, the group who has been trained to service pneumatic and electronic transmitters and controllers and control valves also should provide the service for the plant stream analyzer.

As we learned from our experience in the analyzer story, analyzer maintenance is not to be taken lightly. During the 1950s, one of the major Canadian oil companies attempted to recognize the requirement for analyzer maintenance during the project evaluation stage by including the cost of maintenance in the economic analysis. The rule of thumb at that time was that, on the average, a competent analyzer serviceperson could take care of 10 plant stream analyzers. Thus $\frac{1}{10}$ of the serviceperson's wages should be included in the original cost estimate. This very credible strategy broke down, however, because when the tenth analyzer was bought, another analyzer serviceperson should have been hired, but this never happened.

After several plant stream analyzers have been installed and have failed to live up to the expectations originally held for them, someone will come up with the suggestion that perhaps a special analyzer service group, made up of better-qualified instrument service personnel, should be formed within the instrument department. While the impression is given that the analyzer service team would be an elite group, the persons selected to be a part of it may very well consider it to be otherwise and may try to avoid being a part of it, for these reasons:

- They will have to spend considerable time studying the maintenance techniques for these complicated instruments, and for this onerous task, they will probably get nothing extra in their paychecks. They could draw the same money by staying where they are, calibrating transmitters and tuning controllers, which they already know how to do.
- If an automatic controller acts up and causes process problems, someone will be called out to fix it no matter what the time of day

or night. For this service, overtime will be paid. On the other hand, if a plant stream analyzer quits, well, the plant ran without it for years; therefore, the problem can wait until tomorrow morning when the analyzer technicians show up for work. In the meantime, the plant operator can get the information he or she wants by taking a sample and sending it to the laboratory. Result: no overtime.

The crux of the matter is that for the most part, adding the care of plant stream analyzers to the workload of the existing instrument department, simply because the analyzer is "an instrument," usually ends up being an exercise in failure. Companies that set up a special analyzer service group are on the right track, but the group should not be attached to, or a part of, the regular instrument maintenance department. Where, then, should it be in the organizational structure?

The logical and best place for the analyzer service group is within the inspection laboratory. There are three formidable arguments for this:

1. Technicians who service the analysis instruments that are the property of the inspection laboratory are familiar with the principles on which these analyzers work. Therefore, they will not have any appreciable difficulty learning how the plant versions of the same analyzers function. The main difference in most cases is in the operation of the sampling system.
2. The inspection laboratory should have a vested interest in ensuring that the plant stream analyzers work reliably, because if they do not, the operators will collect samples and send them to the laboratory for manual analysis instead. This will increase the workload in the laboratory.
3. By placing the responsibility for the performance of plant stream analyzers in the inspection laboratory, it puts an end to any interdepartmental argument over who is right, the instrument department and their on-line analyzer or the inspection laboratory and their manual analyses.

6.4 Sampling Systems for PSAs

The number of plant stream analyzer (PSA) installations in which the sensing element is located right inside the process pipe or vessel are relatively few. In most cases, a plant stream analyzer will require a sampling system that will continuously draw off a sample of the fluid to be analyzed and pipe it to the analyzer at the correct conditions of pressure, temperature, flow rate, and cleanliness, along with other possible factors. When all the hardware necessary to meet the requirements for sample conditioning are added up, some sample systems begin to look like a small process unit.

The assertion has been made, and proved out by many users, that when a plant stream analyzer application fails and has to be abandoned, by far the most frequent cause is failure to obtain a satisfactory sample. In view of this, the design and construction of a PSA sampling system merit careful consideration. Entire textbooks have been written on this subject. People have devoted careers to analyzer sampling systems alone. Nonetheless, in this text, the matter will be left, with only a single question being asked.

The principal cause of on-line analyzer failure is known to be failure to obtain a consistently reliable sample. Why then, when an analyzer installation costing $50,000 to $100,000 is being undertaken, doesn't the procedure call for building the sampling system first, and verifying it as workable, before the analyzer itself is placed on order?

Chapter

7

Installing Process Instruments

Successful instrument applications go beyond selection of the right hardware or system for the job. Instruments also must be used properly. A large part of proper instrument use is correct installation, a consideration that is often taken for granted.

This chapter will not be taken up with pages of instrument installation diagrams. Diagrams intended to illustrate the correct installation method for various types of instruments can be found in the evergrowing collection of International Society for Measurement and Control (ISA) standards and recommended practices. The American Petroleum Institute (API) Standard 1500 also has a complete set of instrument installation diagrams. Consequently, diagrams will not be included in this chapter unless they serve to illustrate some important principle.

It is a fact of life, and an unfortunate one, that in certain continuous processing plants such as oil refineries there are unwritten but nevertheless inviolable rules that have a great deal of influence on how instruments may or may not be installed. The first of these rules is that when any instrument is installed on process piping or equipment, the installation must be made in such a way that the instrument can be removed for repair or replacement without shutting any part of the process down.

The second is that accessibility has a high priority when installing any process instrument. Instruments that are hard to reach or which require more than one serviceperson to handle their servicing or removal will usually end up getting less than the amount of maintenance required to keep the instrument functioning at its optimal level.

The subject of instrument installation methods can become argumentative, especially when there is a substantial cost difference to be

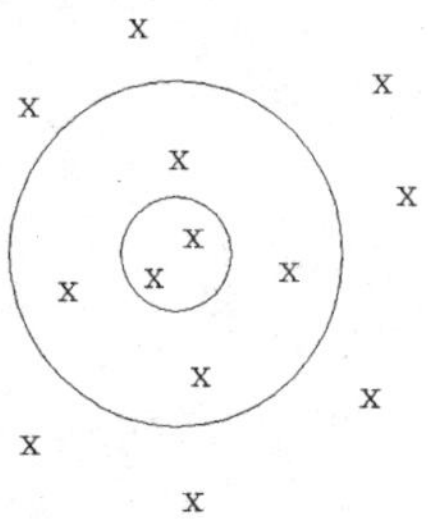

Figure 7.1 An instrument installation may or may not prove to be successful depending on circumstances.

justified if the better method is to gain approval. Contractors may justify their promoting a cheaper method by saying that they have used it successfully somewhere else. What the owner instrument specialist has to keep in mind are the areas in Fig. 7.1. This diagram is intended to illustrate, by the small x's, the number of ways in which an instrument might be installed, one x per method. The x's inside of the inner circle represent those installation methods which are guaranteed to succeed. The x's outside of the outer circle represent those ways which have no hope of working.

Unfortunately, there is an intermediate zone. The x's there represent methods that will work at some times but not others, or in some locations but not others. Whenever the assertion is made that a method has been used successfully before, be on guard that it is not a method in the intermediate region. Not all installation practices that work fine on the Gulf Coast will also work in an Alberta tar sands plant.

7.1 Piping Standards and Codes

Most plants that need extensive piping will likely have adopted a piping code that will state how heavy the pipe and pipe fittings must be for specified situations. These situations will be identified by the nature of the fluid handled and its pressure and temperature. Since piping has to be connected up, the specifications also will state what kinds of connections are to be used. By implication, the specifications also will tie down the type of connections to be provided on process vessels, since the piping, which is covered by the code, must mate up with the connections on the vessel.

It is understood that what the piping code calls for is the minimum acceptable for safe operation. Using pipe and fittings that are heavier than required by the code is not a problem, except in the sense that money may be spent unnecessarily.

Whenever the requirement is to measure a process pressure, a small connection has to made on the vessel or piping at the proper location. This has to be connected, through small piping, to the pressure-measuring instrument so that the instrument senses the process pressure. At this point we should define two terms. The small pipe or tube that transmits the pressure from the process connection to the instrument is called a *pressure lead line,* or an *impulse line.* For our purposes, we will call it an *impulse line.* The impulse line plus any valves or other fittings that may be required for isolating the instrument, calibrating it, or otherwise servicing it are called the *instrument manifold.*

Over a period of years, specialty instrument valves have been developed for making it easier to assemble instrument manifolds. Most of these specialty assemblies have distinct advantages. As an example, for a differential pressure measurement of the type found in an orifice meter, a specialty assembly that contains the two shutoff valves and the bypass valve is available. Unfortunately, the use of these specialty assemblies sometimes raises objections. There is no question that specialty instrument manifold valves are useful, in that they reduce the amount of labor required to assemble the manifold, plus the fact that they result in an overall neater job. They also have the required pressure rating for the application. The problem is, however, that very few plant piping codes recognize these specialty assemblies, and many plant piping people do not like to use valves which they cannot find in their code book.

Is there a solution for this problem? There must be, because many plants use the specialty instrument valves and get the benefits, even though their plant piping code makes no mention of them. The agreement that permits this is that where the pressure connection is made on the process vessel or piping, an isolating valve that complies with the piping code must be installed right against the vessel or pipe. Thus, if any problem develops in the instrument manifold, it will be beyond the isolating valve, which can be used to shut down the entire instrument manifold along with the instrument.

7.2 Process Instrument Connections

When measurement instruments and control valves are to be installed in process piping or in or on process vessels, the question of the best type of connections has to be decided. Three types of connections for plant instruments are in general use, namely, threaded, flanged, and welded.

When an instrument or valve is welded into place, the connection is virtually permanent. Removing the instrument for repair work then becomes awkward; it is necessary to actually cut the instrument out.

Reinstalling it after repairs are made will require welding, possibly preceded by restoring the beveled ends through some process of grinding or machining. Consequently, welded connections tend to be used only in those cases where zero leakage for the duration of an operating run must be ensured. High-pressure steam and boiler feedwater are examples, as are hydrogen and other chemicals that would be hazardous if they were allowed to leak, even in small amounts.

Many plants have a policy of using threaded connections for small (1 in and less) pipes and fittings and flanged connections for larger pipes and fittings. If the instrument or valve has threaded connections, then it follows that in order to make it removable for servicing, unions will be needed. Whether union connections are good enough to be used in industrial plant installations is open to question. They are really in the gray area of installations that work on some applications or in some locations but not in others. Since they have metal-to-metal mating surfaces, they can be difficult to make leakproof. If they do leak, the tendency is to pull them tighter together. This eventually makes them difficult to take apart, when it becomes necessary to do so to remove the instrument or valve.

Experience seems to have shown that the overall best type of connection for instrument components is a flanged connection. The mating flanges are pressed against a resilient gasket and can be pulled up tight enough to prevent leakage in just about all cases. Even so, it is still relatively easy to break the connection and remove the instrument, though it may have been in service for a year or more.

For this reason, some plants standardize on flanged connections for control valves and for certain instruments, even in small sizes. There is a possible complication in the case of control valves, since some manufacturers do not supply flanged ends on valve bodies smaller than 1 in. The solution to this problem is to specify a 1-in flanged body with a reduced-size inner valve.

7.3 Instrument Connection Sizes

In small sizes, 1 in and less, the usual method of making an instrument connection on a process pipe or vessel is to drill a hole and weld in a pipe coupling. In larger sizes, the practice is to weld in a flanged pipe nipple.

One would think that the only criterion for specifying the size of an instrument connection on a process vessel or piping would be to match it to the size of the connection on the instrument or valve. While this is certainly a consideration, there is somewhat more to it. There is a frequently a risk involved in specifying instrument connections that are smaller than ¾ in. Fires have been caused in oil refineries because a vibrating pressure gauge installed on a ½-in connection eventually broke off the connection and flammable hydrocarbon spewed out. Bear in mind, too, that it is not unusual for workers to climb on instrument connections if they have no other easier way to reach something they want to get at.

The smallest recommended instrument connection is therefore ¾ in. It should be made with at least a 3000-psi-rated coupling in order to ensure that there is enough metal in it to give it adequate physical strength, plus some additional allowance for corrosion. If the plant is handling a dangerous or valuable fluid, then strengthening the connection by gusseting it should be considered. The only exception to the ¾-in minimum recommendation is the pressure connection size for orifice flanges, which is conventionally ½-in internal pipe thread.

7.4 Pipe Versus Tubing for Impulse Lines

It often happens that a pressure-measuring instrument should not be installed right on the instrument connection at the pipe or vessel simply because it would not be convenient to read it or to work on it at that particular location. It would be better to have the instrument installed at ground level (grade) or at some convenient platform on a structure. In order to do its job, however, the instrument will have to be connected to the process with an impulse line.

Prior to 1950, it was conventional to make impulse lines from ½-in pipe and fittings. At every point where a change in direction was required, it was necessary to install a threaded elbow. Long runs required threaded couplings to join lengths of pipe. The cutting, threading, and fitting were time-consuming. Furthermore, the potential for a leak existed at every threaded connection. Finally, commercial steel and iron pipe is not always clean on the inside. It may contain dirt or scale, which can lead to instrument problems.

Hence the present trend is to use ½-in tubing. A section of tubing of any desired length can be pulled off a reel, so there will be no intermediate joints. The connections made using high-quality tube fittings are generally more reliable than those made by screwing a threaded pipe end into the socket of a pipe fitting. If a bend is required, it can be formed easily, without additional fittings, using a tubing bender. All this minimizes the problem of leaks.

Tubing can be obtained in a variety of alloys, such as stainless steel, so that it will be internally clean and corrosion-resistant. Where connections have to be made, the tubing connectors that are available now are of excellent quality.

The ideal situation with respect to tubing fittings would be to have one brand only in use throughout the plant. This would minimize the number of fitting types that have to be kept in stock and the number of different wrenches required to do them up or undo them. It seems that this goal is impossible to meet, however. Usually, the contractor who builds the original plant, or an expansion to it, never orders enough fittings initially to finish the entire project, since he or she does not want to have any left over. Toward the end of the job, the contractor suddenly needs more fittings, and the original supplier cannot provide them on the desired delivery. Consequently, the contractor orders a different brand of fittings from another vendor who can meet the delivery requirements.

Aside from this, some instrumentation components, control valves, for example, come from the factory with accessories such as I/P converters or valve positioners mounted and hooked up. The instrument supplier never seems to use the same brand of fittings as the contractor.

Perhaps this is not really all that great a problem. The important thing is that the fittings used are of good quality. The criteria of quality are that the connection does not require much in the way of manual skill to make it up and that the connection is leakproof when it is tightened. The old compression fittings with the round ferule, and flared fittings, do not meet these criteria. One easy test for the quality of a fitting is to see how snugly the two parts fit together with the nut completely loosened off.

7.5 Orifice Meter Valve Manifolds

In addition to the two shutoff valves at the pressure connections in the orifice flanges, an orifice meter requires valves at the differential pressure (DP) meter itself so that the DP meter can be turned on an off properly and, perhaps more important, so that the meter can be checked for an accurate zero reading. Figure 7.2 shows the valving arrangement for both a three-valve and a five-valve manifold.

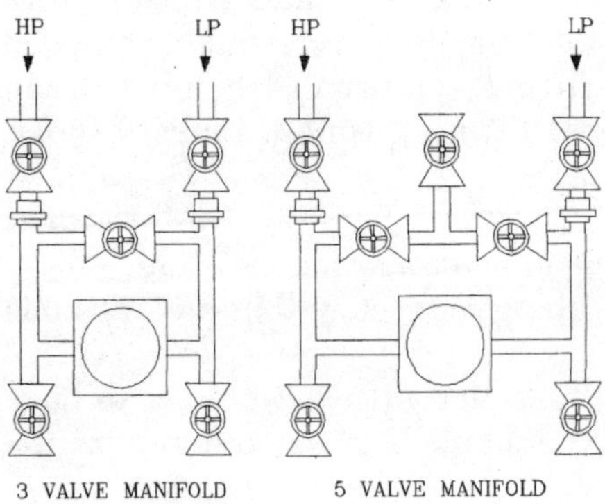

Figure 7.2 Orifice meter manifolds.

The two valves at the bottom of the impulse lines do not count when describing the manifold as a three-valve or a five-valve manifold. These valves used to be incorporated into orifice meter manifolds some years back so that the impulse lines could be blown down. This presumably cleared them of any sediment that had accumulated in the lines. Blowdown valves are rarely used anymore, probably because they created more problems than they solved.

Allowing process liquid to spew out through an open valve is potentially dangerous. Allowing the process liquid to build up on the ground under the meter also may be hazardous. Beyond that, when the cold liquid in the impulse lines is replaced by hot fluid from the process pipe, the heat could reach and damage the DP meter or possibly any electric tracing installed to keep the impulse lines warm in the winter time. Self-limiting electric tracing, which is sometimes used for instrument winterizing, is supposed to provide heat, not receive it. Blowing down the impulse lines of a steam flowmeter is a proven method of ruining self-limiting electric tracing.

The two shutoff valves at the top of both manifolds make it possible to isolate the meter from the process without having to climb up to the shutoff valves at the orifice flanges to do the same thing. The union connection just underneath each of the shutoff valves allows the meter to be removed for calibration or repair.

The bypass valve serves two purposes. It permits the DP meter to be brought on-line without shocking the meter, and it also allows the meter to be zero-checked. In service, the DP meter will be measuring a small differential pressure, perhaps 15 kPa or 2 psi, at a pressure level that could be 7000 kPag or 1000 psig. If there were no bypass valve and the DP meter was turned on by opening the shutoff valves one at a time, then when the first valve was opened there would be full line pressure on one side of the meter and zero on the other. This would be enough to damage some DP meters or at least put them off calibration.

Today, good-quality DP meters and transmitters have built-in protection against ill effects due to having full line pressure applied to one side of the meter. Even so, it is obviously best for the DP meter if the correct startup sequence is used.

For the three-valve manifold, starting with all three valves closed, the correct startup procedure is to open the bypass valve, open the low-pressure (LP) valve, close the bypass valve, and open the high-pressure (HP) valve. When the DP meter is taken off line, the order in which the valves are closed is not important, as long as one remembers that all three valves should not be open at the same time. If this is allowed to happen, there will be flow through the impulse lines from the high-pressure tap to the low-pressure tap at the orifice flanges.

Zeroing the meter requires only that the HP and LP shutoff valves be closed and then the bypass valve be opened. This creates zero differential pressure at the meter, which should then read zero or give an output that is equivalent to zero, such as 3 psig or 4 mA.

For most orifice meters, the three-valve manifold is all that is needed. The only concern is the possibility that the bypass valve might leak. Should this happen, there will be flow through the impulse lines, and the differential pressure which the DP meter sees will be less than the differential pressure at the orifice taps.

The five-valve manifold is used where zero leakage through the bypass line must be ensured. This is achieved by having two bypass valves instead of one, with a vent valve that remains open in between them. Then there is no possibility that anything can leak from the high-pressure side to the low-pressure side of the meter.

Figure 7.2 shows the two types of manifold valve assemblies as they would be assembled from individual valves and pipe fittings. This can still be done, but many users of orifice flowmeters have found that it is easier, and better, to purchase the factory-made three- and five-valve assemblies. They do the job just as well, are more compact, have fewer leakage points, and generally reduce the amount of labor required to complete the orifice meter hookup.

7.6 Grade-Mounted Versus Line-Mounted

This section is devoted to a subject on which there are decidedly conflicting views. Before the advent of the DP cell, differential pressures created by head-type flowmeters (orifice meters being the most common) were measured with mercury manometers or large aneroid manometers. The size and mass of these DP measuring instruments were such that they had to be mounted on a sturdy fixture. The result was that the only practical way to mount the whole instrument was on a stand at ground level (grade) or on a convenient platform of a nearby structure. The pressures on either side of the primary flow element had to be piped down to the DP meter through impulse lines.

This arrangement had the potential for error, since the pressures that were present at the connections to the DP meter were not the same pressures that existed on either side of the primary element. In both cases, the high- and low-pressure sides, the pressure at the DP meter was the pressure at the primary element plus the pressure created by the head of liquid in the impulse line. This additional pressure is proportional to the difference in the elevation between the two sets of connections and the density of the liquid in the impulse line.

It is obvious that the difference in elevation is the same for both impulse lines, so this should not be a factor. However, even though both impulse lines contain the same process liquid, the possibility of a difference in density exists if the impulse lines are not at the same temperature. Thus the DP meter would not be measuring the actual differential pressure being created across the primary element. The magnitude of the error that might be expected as a result is investigated in Chap. 5.

In many cases this error will not be great enough to require any special treatment of the meter installation. Nevertheless, there is a school of thought that holds that any potential error should be avoided if it is possible to do so. The state-of-the-art DP cell transmitter, owing to its compact size and relatively low mass, can be mounted right on the meter run so that the impulse lines from the connections on the pipe or flange to the connections on the DP meter are horizontal. The difference in elevation is zero. Thus the DP meter will be measuring the true pressures on either side of the primary element, irrespective of the condition of the fluid in the impulse lines. There could actually be liquid in one impulse line and vapor in the other, and no error in measured differential pressure would result. This is called a *line-mounted installation.*

There can be no doubt that the line-mounted method is theoretically the correct one to use. In some companies, the specifications for installing orifice meters reflect this fact. What proponents of line mounting tend to forget, however, is that a vital ingredient of reliable instrument performance is conscientious routine maintenance. Furthermore, the extent to which the required routine maintenance is performed on any instrument is directly related to the accessibility of the instrument. Instruments that are grade-mounted are readily accessible. Instruments that are line-mounted require ladders or other means to reach them. They also may need two servicepeople, rather than one, to do the job. Thus there are accessibility problems that will inevitably influence the quality of the maintenance the instrument gets.

Line mounting is an installation method used for no other reason than to get around a potential error created by a difference in the temperatures of the two impulse lines. It also has been shown that this error, in some cases, will not be worth worrying about. Taking all the aspects into consideration, therefore, leads to the conclusion that the practical choice will be to grade mount the DP meter for the sake of the benefits associated with accessibility and take whatever precautions are available to ensure that there is no appreciable difference in temperature between the impulse lines. One technique that might achieve this end would be to run the impulse lines side by side (touching) and insulate them together.

Another possibility is the arrangement shown in Fig. 7.3. This installation can be used if the pipe that carries the fluid being metered

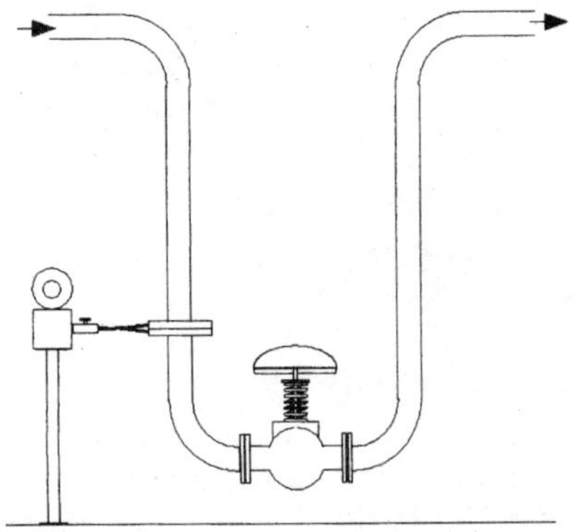

Figure 7.3 Accessibility and horizontal impulse lines in the same installation.

has to be looped down from a pipe rack to grade so that a control valve can be installed. In Fig. 7.3 the control valve block valves and the bypass have not been shown in order to simplify the sketch.

Installing the meter in this way achieves both grade mounting and horizontal impulse lines at the same time. A possible objection to this method is that the meter run will be vertical, and there is a belief in some circles that this is unacceptable. Although tradition is strongly in favor of having the meter run horizontal, just what error will be created if the meter run is vertical instead is not entirely clear.

While this arrangement has the appearance of having the best of both worlds, accuracy and accessibility, it does have its limitations. The pipes in the pipe rack will be about 4 m above grade, and the bypass around the control valve might be 1 m above grade. This leaves about 3 m in which to install an orifice meter run. When the fitting at the head end of the meter run is an elbow, about 14 diameters of straight pipe are required ahead of the orifice and 4 diameters after the orifice, or a total of 18 pipe diameters in all. If this length of straight pipe is to be installed in a distance of 3 m, the maximum size that the meter run can be is 6 in.

When the meter run is vertical, as in Fig. 7.3, one of the pressure taps on the orifice flanges is 2 in higher than the other. A spirited debate can sometimes be started with the question, "Does this mean that the calibration of the DP meter should be biased by the pressure created by a 2-in column of line liquid in order to compensate for the difference in the height of the pressure taps?"

The correct answer is, "No." If the flow in the line stops completely, the differential pressure at the meter will be zero, since the extra 2 in of liquid in one of the impulse lines is balanced by the 2 in of liquid inside the pipe that acts on the other impulse line.

7.7 External Float Chambers

A consequence of the rule that requires all sensors to be removable on the run is the use of external float chambers for liquid level meters. A metal chamber is connected to the side of the process vessel by two connections, one at the top and the other at the bottom of the chamber. A shutoff valve is installed in each of the two connections. Liquid from the vessel will flow into the chamber through the bottom connection, while the upper connection acts as a pressure-balancing line. Presumably, the liquid will establish a level in the external chamber that will match the level in the vessel itself.

The float, or other level sensor, is located in the external chamber rather than inside of the vessel. If it were in the vessel and anything went wrong with it, it would be necessary to take the vessel out of service to repair the level sensor. This would be, as stated, unacceptable. If the sensor is in the external chamber and a problem develops, the external chamber can be isolated from the vessel by closing the shutoff valves, and the sensor can be removed. Consequently, one of the installation methods sketched in Fig. 7.4 is usually used.

The float-chamber installation has two potential problems. First, the laws of nature do not require that the level in the external chamber be the same as the level inside of the vessel. Nature requires only that at the lower (liquid) connection, the pressure created by the column of liquid inside the vessel be the same as the pressure created by the column of liquid in the external float chamber. The pressure developed by any head of liquid is determined by the product of the density of the liquid, the gravitational constant, and the height of the liquid column.

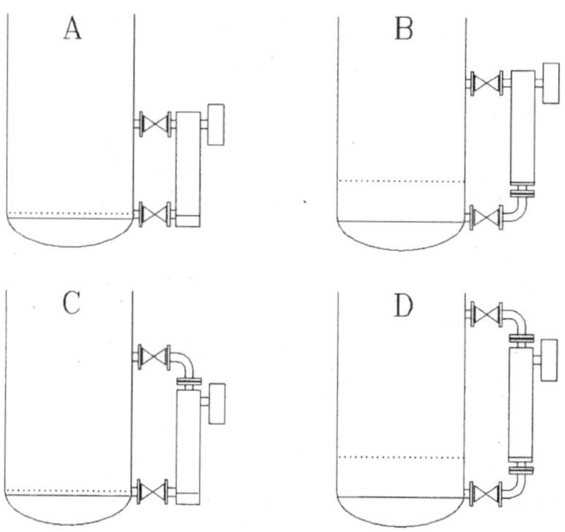

Figure 7.4 Possible external float-chamber installations for displacers.

The assumption is that since the same liquid is involved in both columns, the liquid density in both liquid columns will be the same. This ignores the effect of temperature on density. If the liquid in the external chamber is colder than the liquid in the vessel, which is quite possible, then the height of the liquid column in the external chamber will be less than the height of the liquid in the vessel, and the sensor will not be measuring the true level. This problem is accentuated when the level measured is an interface between two liquids.

The second problem is one of dynamics. In order for the liquid level to change in the external chamber in response to a change in level in the vessel, a certain amount of liquid must flow into or out of the external chamber, because the chamber has capacitance. This transfer of liquid into or out of the chamber must take place through the lower connection, which imposes a certain amount of resistance to flow. The combination of resistance and capacitance creates a fairly small but nevertheless measurable time constant, which may contribute to poorer control.

Figure 7.4 shows four different methods of installing a displacer-type level sensor. Is any one to be preferred over the others, and if so, which one is it?

First consider installation *A*, in which the two connections are both on the side of the float chamber. Offhand, this appears to be the neatest installation, since the float chamber can be bolted right on to the vessel connections without the need for any other fittings aside from the shutoff valves. The problem is, however, that the connections on the float chamber are welded on by the instrument manufacturer, whereas the connections on the vessel are placed there by the vessel manufacturer. What if they fail to match up by several millimeters? This would be quite possible owing to the tolerance that is allowed in spotting both sets of connections. Or what if the connections match correctly, but the connections on the vessel are not directly one above the other so that the float chamber could not be installed plumb. Although the side-and-side installation is occasionally used, it is not a good choice for a standard because it allows no leeway for compensating for incorrectly placed vessel connections.

This suggests that the choice of a standard for attaching the float chamber to the vessel should be made from one of the three other methods illustrated in Figure 7.4, that is, *B, C,* or *D*. There are at least three governing factors. First, the level-measurement range always starts at the bottom of the displacer, and it ends at the top of the displacer. Second, if a connection is on the *side* of the float chamber, it will be exactly opposite either the bottom or the top of the displacer. Third, the liquid in the vessel will be in the bottom of the vessel, and in many applications, the requirement is to measure the level down to as low a point in the vessel as is possible.

The sketches in Fig. 7.4 show that the lowest point on the vessel at which a horizontal connection can be placed is just above the weld line between the straight vertical side and the elliptical bottom of the vessel. In each sketch, the dotted line shows how far down in the vessel the level can be measured using that particular installation method.

If the bottom connection on the float chamber is on the side of the chamber, then the connection on the vessel will line up with the bottom of the displacer, and the measurement range will start from there. On the other hand, if the float-chamber lower connection is on the bottom of the chamber, the ell in the piping required to join it to the vessel will require about 6 in of vertical space. This would mean that the lowest level which the displacer could measure would be 6 in above the vessel connection. Thus, if the objective is to measure the vessel level down to the lowest possible point, it follows that the lower connection on the float chamber should be on the side of the chamber. Since the arrangement that had both float-chamber connections on the side has already been eliminated from contention, the remaining choice is method *C* in Fig. 7.4.

For these reasons, installation *C* is a good method on which to standardize when installing displacer-type level meters that have an external float chamber.

The importance of ensuring that the float chamber for a displacer-type level instrument is installed plumb is often underestimated. There is not much clearance between the displacer and the inside wall of the float chamber, and if the float chamber is not truly vertical, the displacer may rub on the float chamber, and inconsistent level readings will result. Incidentally, external float chambers that start off being vertical, as they should, do not necessarily remain that way. Over a period of time, process vessels may settle unevenly and tilt slightly. While this may not bother the vessel, it can upset a level instrument.

7.8 Thermowells

Plants in which thermocouples are the accepted method of measuring temperatures invariably install thermocouples in *thermowells,* sometimes called *sockets.* Once again, the objective is to allow the temperature-sensing element to be removed and, if necessary, replaced without shutting down the pipe or vessel in which the sensor is installed. In the case of a thermocouple, there is a second reason. A thermocouple is a relatively fragile device, and if it were installed bare, it would die quickly in most plant streams due to erosion, corrosion, or buffeting.

Owing to its low mass, a bare thermocouple responds quickly to changes in temperature, and in this respect it is unfortunate that it is necessary to encase it in a thermowell to ensure its life and to make it possible to service it. The point has been made already that a thermocouple does not measure the temperature of the process fluid that is in contact with the outside of the thermowell. The thermocouple measures the temperature of itself. Thus, if a thermocouple in a thermowell is to respond to a change in temperature of the process fluid, a certain amount of heat transfer must take place between the thermocouple and the process, and this heat must flow through the wall of the thermowell. Since the metal in the wall has capacitance for heat and imposes resistance to heat flow as well, there will be a time constant involved.

This is not all the bad news, however. Around the outside of the thermowell, and in contact with it, there will be a film of process fluid that is not moving. Against the inside wall of the thermowell there will be a film of air that is likewise immobile. The resistance of these two films to heat flow may well be considerably greater than the resistance of the metal in the thermowell.

The combined effects of all these resistances is that a thermocouple in a thermowell may take 20 times as long to respond to a temperature change as it would if the same thermocouple could be installed bare. This is not necessarily bad if the temperature is for indication only. If the temperature is controlled, however, the delay will increase the recovery time in a temperature control loop.

When thermocouples in thermowells are a fact of life in a plant, outlining all the temperature-measurement problems that can occur is not very helpful. The point is, What can be done to minimize these problems? There are at least two things:

- Use butt-welded thermocouples, not thermocouples formed by twisting the wires together. Then make sure that the thermocouple junction is pushed tight against the bottom of the well. There are spring-loaded thermocouple assemblies that will ensure this, and they are worth evaluating.
- Install thermowells in pipes so that the tip of the well is facing the stream, that is, so that the stream impinges against the tip. This helps to reduce the film effect. The best installation is to install the thermowell in an elbow so that the stream bears right on to the tip of the well. The next best method is to install the thermowell at a 45-degree angle projecting upstream. The worst installation, from the viewpoint of response time, and the one that is most common is to install the thermowell so that it projects inward at right angles to the pipe wall. These three methods are sketched in Fig. 7.5

Notice that in each case the better installation is more difficult to make, because it involves welding a coupling onto the curved surface

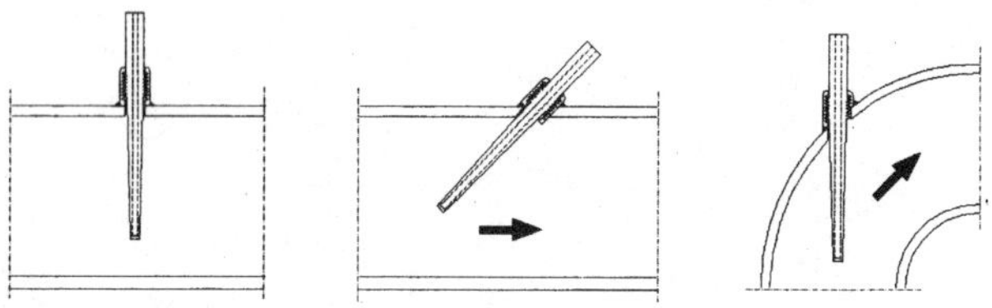

Figure 7.5 Methods of installing thermowells.

of an elbow or at an angle of 45 degrees to the outside surface of the pipe. This is just another of the numerous cases where instrument performance is sacrificed for the sake of easier installation or savings in cost.

It is also possible to obtain armored thermocouples that can be installed without using a thermowell. The same applies to using temperature elements other than thermocouples, such as filled system temperature sensors. A gas thermometer, for example, is potentially an accurate temperature-measuring device, and in addition, it is capable of shorter range spans than can be obtained with a thermocouple. It would seem logical to consider using these temperature sensors where the process fluid is not hazardous and where the pressure is low.

It goes without saying that if a temperature sensor is installed bare to improve the response time, then the installation must be marked "No Thermowell" in a very visible manner. A potential disaster would result if an instrument serviceperson were to inadvertently remove the temperature sensor when the line or vessel is in operation.

Chapter

8

Winterizing Process Instruments

In many plants, process instruments have to be located out of doors. They consequently require protection from the prevailing elements, of which the weather is the most likely to receive attention. However, the elements also will include construction and maintenance materials such as insulation and paint, along with dirt and other debris, that are swept off of structures, all of which can be detrimental to reliable instrument performance.

The term *winterizing* implies that the protection required is from cold weather, and for the most part, this will be the kind of protection discussed, since it is the priority problem in Canada and the northern United States. In other parts of the world, the need for instrument protection may be against high humidity, which can cause molds to grow inside instrument cases, blowing sand, or salt corrosion in regions that are close to the ocean.

In order to get the optimal service from any process instrument, two requirements have to be met. It is necessary to buy the right instrument for the job, and then it is necessary to use the instrument properly. Using the instrument properly means, among other things, that it must be installed properly. Adequate winterization for an instrument located outside is part of the installation requirement.

8.1 Basic Concerns

When instrument winterizing needs are being evaluated, two potential problems must be considered. These are

1. What can happen to the instrument itself when it is exposed to very low temperatures? Some electronic transmitters have a low ambient temperature limit of about −30°C. At these low tempera-

tures, the instrument may go off calibration until the temperature returns to a reasonable level, or it could be damaged.

2. What can happen to the liquid that fills the impulse lines between the process vessel or pipe and the meter and which are supposed to transmit the process pressure to the meter? This liquid also enters the body of the meter. It will be stagnant, and if it freezes, pressures will no longer be sensed by the meter, and it will quit. If ice forms, the meter also may be damaged.

   The same applies to the process liquid in a level-meter float chamber mounted external to a process vessel. It will eventually cool down to ambient temperature, and if it sets up or freezes, the vessel level can no longer be measured.

These are the basic concerns. The block diagram in Fig. 8.1 illustrates them. Notice that on the right side, maintaining fluidity of a viscous liquid is also included. This is not strictly a winterizing matter, because such liquids have to be kept hot all year round so they will retain their fluidity. However, since the hardware used to accomplish this is the same as that used for instrument cold weather protection, this concern is also included under winterizing.

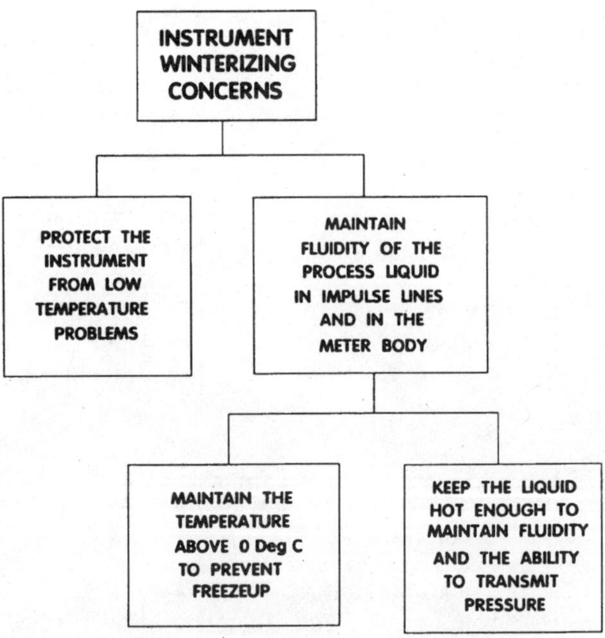

Figure 8.1 Purposes underlying instrument winterizing.

8.2 Order of Preference

Of the several ways of achieving instrument winter protection, some are definitely to be preferred over others. The order of preference is

1. Locate instruments indoors.
2. Use instruments that do not need winterizing.
3. Fill impulse lines and the meter body with a nonfreezing liquid seal so as to keep the process liquid or vapors out.
4. Heat trace and insulate impulse lines and float chambers to keep the contained process liquid at a high enough temperature that it will not freeze.
5. Run a dry gas purge down the impulse line to the process to keep process liquid or vapors out of the impulse line and the instrument body.
6. Use a nonfreezing liquid purge to keep process liquid out of impulse lines and the meter body.

These methods, along with the reasons for the prescribed order of preference, require explanation.

8.3 Locate Instruments Indoors

The whole subject of instrument winterizing presupposes that instruments have to be located outside, but do they in all cases? Every effort should be made to locate instruments inside a building that is maintained at a reasonable temperature. This not only obviates the need for winterizing, but it also creates a better environment for doing maintenance.

Adequate service and care are always a factor in successful instrument applications, and the extent to which this requirement is met is often influenced by the comfort factor where the work is to be done.

Furthermore, winterization can be expensive. Locating instruments where they do not need winterizing eliminates added installation costs and, in cases where energy is continuously required, operating costs as well.

Instruments that cannot be installed indoors should be provided with an instrument housing, even if it is not necessary to heat it. It is better for the housing to take the wear and tear from rain, snow, hail, paint, dirt, insulation, and whatever else. In most cases, a housing will improve instrument reliability and life. Accordingly, it is a good investment. There are competent manufacturers in the business of making instrument housings, so there is no need to fabricate them on site.

One should not expect any instrument that has a hinged door to be weatherproof, even though it is advertised to be so. The door gaskets will seal tightly for only a limited number of closures, after which rain and dirt will penetrate. Provide an instrument housing.

Generally speaking, an instrument whose cover bolts or screws down onto a gasketted surface can be considered weatherproof. However, *weatherproof* only means that undesirable stuff will not get into the inner works of the instrument. It does not mean that the instrument will be safe from low temperatures.

8.4 Use Instruments That Do Not Need Winterizing

It has been said already that winter protection is needed to prevent damage to instruments and/or to prevent freeze-up or set-up of the process liquid in impulse lines or in an external float chamber. A list of the measuring instruments that are vulnerable to the second of these two problems would be

- Displacer-type level meters
- Ball-float level meters and level alarm switches
- Level meters employing differential pressure measurement
- Head-type flowmeters, especially orifice meters

Locating instruments indoors, as a policy, obviously has its limitations. Therefore, as an alternative, why not use instruments that do not have pressure lead lines or float chambers and thus do not need to be winterized?

This is where the instrumentation systems designer runs up against a formidable obstacle, in the form of the unwritten plant rule. The rule states that measuring instruments must be installed in such a way that they can be removed for servicing without shutting down the process pipe or vessel to which they are connected. It is the general acceptance of this rule that makes it necessary to have float chambers and impulse lines, which in turn leads to a need to winterize these items.

From the purely winter protection point of view, the next best thing to locating an instrument indoors is to locate it right inside the process vessel or pipe. If this were permissible, then capacitance and sonic level meters could be used for level measurement. Rotameters, vortex shedding meters, turbine meters, and magnetic meters could all be used for flow measurement. Unfortunately, the selection of these instruments, which would not have to be winterized, runs contrary to the plant rule, unless they are installed with expensive bypasses around them.

The popular orifice flowmeter is often perceived as consisting only of the differential pressure meter, which can be removed for servicing without shutting down the pipe in which flow is being measured. The orifice plate, which creates the differential pressure, is felt to be immune to any harm. The fact of the matter is, however, that it is eroding away. People who believe that the justification for using orifice meters is that they can be serviced without shutting down any operating equipment are ignoring the orifice plate and deluding themselves.

8.5 Use a Liquid Seal

For any measuring instrument that has impulse lines, the process liquid or the condensate from process vapors eventually will collect in the impulse lines. The liquid will be stagnant, and its temperature will drop to ambient. Under the circumstances, it may freeze or set up and will then be unable to transmit pressure. To overcome this problem, the impulse lines must be filled with a different liquid that will retain the required degree of fluidity at all times.

Generally speaking, if the liquid in an impulse line is different from the process fluid, then the instrument is said to be *sealed.* The instrument is not considered to be sealed if the liquid in the impulse line is the same liquid as in the process pipe or vessel to which the impulse line is attached.

The principal benefit of sealing instruments, rather than using some other winterizing measure, is that this method does not depend on externally applied heat. This circumvents two possible problems. First, heat rarely comes free, so there is an operating cost. Second, any failure of the heat in cold weather means a failure of the instrument.

Seals actually come in two forms, and the onus is on instrumentation personnel to choose the better way for any given application.

- *Open seal.* In an open seal, the seal liquid comes in direct contact with the process fluid, usually in a seal pot.
- *Closed seal.* Also called an *integral seal,* this type of seal can be used with pressure and differential pressure instruments that have only a small change in meter body volume over their measurement range. Each pressure connection on the instrument has a length of capillary that ends in a flexible capsule or diaphragm. The capillary system is evacuated and then filled with a liquid that has a freezing point below the lowest expected ambient tempera-

ture. The diaphragm or capsule senses the pressure and transmits it to the seal liquid, which transmits it through the capillary to the measuring instrument. Thus the process fluid does not come in contact with the seal liquid. A differential pressure instrument will require two such sealed capillary systems, one for each measured pressure.

8.5.1 Open seals

The attraction of open seals is that they can be field assembled. Figure 8.2 shows two pressure measurement applications in which an open seal is used. Figure 8.2*a* is a process vessel on which a differential pressure (DP) meter is being used to measure liquid level. The impulse line from the vapor space down to the meter needs to be sealed off with a liquid seal; otherwise, the vapors from the vessel would eventually find their way into the impulse line and condense there. This would create a variable head of liquid in the impulse line, which the meter could not distinguish from a change in level in the vessel. The DP meter is actually responding to the differential pressure at its connections, and this differential pressure will be changed by a change in the liquid head in either of the impulse lines.

Since the DP meter should respond only to changes in the level in the vessel, the impulse line up to the vapor space must be filled right to the top so that the head of liquid on this side of the meter will be known and constant. The pressure created by this head of liquid must be biased out when the meter is calibrated as a level instrument. A DP meter in a liquid-level application usually has an elevated range.

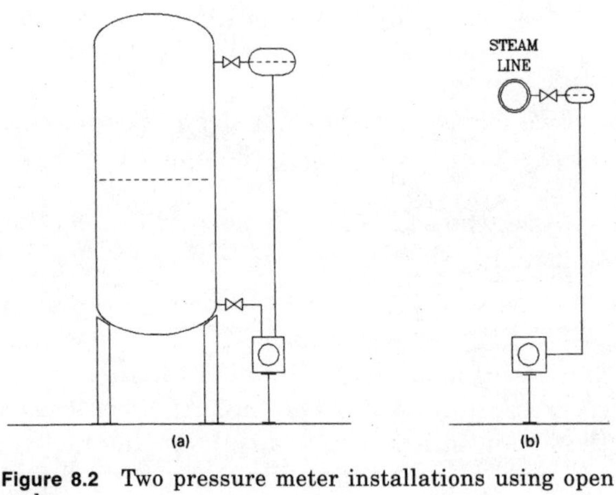

Figure 8.2 Two pressure meter installations using open seals.

The low volumetric displacement of modern force balance pressure and differential pressure transmitters notwithstanding, it is proper practice, when using an open seal, to install a seal pot at the process connection. The purpose is to provide a reservoir of seal liquid, which will minimize variations in the level of the seal when the pressure changes. This is important, since the pressure measured by the meter will be the process pressure plus the pressure created by the column of seal liquid. Any change in the height of the column of seal liquid will change the pressure created by the liquid column and cause a zero shift in the meter. A level-meter installation of the type shown in Fig. 8.2*a* should never be made without using a seal pot with adequate capacity. Typically, 6 in in diameter by 14 in long is a standard size.

There is no seal pot in the impulse line to the liquid space, since this impulse line is not sealed. Using a liquid seal against a process liquid rarely works in practice. Even two liquids that are supposed to be mutually immiscible will form an emulsion if they are left in contact for months at a time. When this happens, the height of the column of seal liquid in the impulse line, which figures in the calibration of the level meter, becomes indefinite.

The correct procedure for this impulse line is to dispense with the seal and allow the vessel liquid to run into the meter. If the vessel liquid is such that it would freeze up outside the vessel, then the impulse line must be heat traced, and the meter must have a heated housing.

Figure 8.2*b* shows a steam pressure meter. Steam is the only known vapor that conveniently provides its own seal liquid. Accordingly, the best technique for this meter is to let nature take its course and allow steam condensate to fill the impulse line and the meter body. Here again, a seal pot is necessary to establish a consistent head of liquid in the impulse line. For steam meters, however, the seal pot can be smaller than for the level-meter installation. Usually, 3 in in diameter by 9 in long is satisfactory.

For a steam meter installation there is never a concern that the seal liquid will disappear, nor that the head of seal liquid which is pressuring the meter, and must be calibrated out, will change. Nonetheless, there is the problem that steam condensate can freeze up. When the installation is made outdoors, as in Figure 8.2, the seal pot and the impulse line have to be heat traced, and the meter has to have a heated housing, since the condensate is also inside the meter body.

A steam meter is probably the only situation in which winterizing requires heat tracing in addition to having a seal. Usually, either sealing or heat tracing is used. However, if the meter is outdoors, seal pots, heat tracing and insulation on the impulse lines, and a heated meter housing are standard installation procedure for steam meters.

8.5.2 Seal liquids

There are three general requirements for any sealing liquid used in an open seal:

- It must not freeze or become viscous at any prevailing ambient temperature.
- It should be immiscible with the process fluid or its condensate.
- The rate of evaporation should be minimal at the prevailing temperature.

Inability to meet either or both of the last two requirements is often the reason that some form of winterizing, other than an open seal, is used.

A desirable, though not necessary, quality of a seal liquid is that it should not have a low surface tension. Liquids with low surface tension tend to leak out of the smallest crevice. Manifolding has to be dead tight in order to contain them.

Commonly used sealing liquids for open seals include light oils, mixtures of ethylene glycol (automobile) antifreeze and water, and mixtures of glycerine and water. The table in the Appendix gives the freezing points and specific gravities at 15°C for different mixes of glycol and water. If a glycerine and water mixture is used, it may be necessary to heat the water in order to get the glycerine to blend in completely.

Open seals have a drawback in that they require continual maintenance. Seal pots have to be checked regularly and topped up if the level is not up to the process connection. Having to repeat this chore is a real source of annoyance in instrument maintenance departments. Over time, the seal seems to disappear, probably through evaporation into the process line or vessel, possibly through leakage. If there are a considerable number of open seals in a plant, a technician could make a career of keeping them topped up with seal liquid.

8.5.3 Closed seals

Closed seals are made at the instrument supplier's plant, where they go by the name of *factory integral seals*. Integral seals can only be used with pressure instruments having a low volumetric displacement over their pressure range, owing to the limited capability of the capsule or diaphragm to flex. For accuracy, the process pressure must be freely transmitted to the meter through the integral seal. The pressure must not be partially used up in stressing the diaphragm. Force balance DP transmitters meet the low volumetric displacement requirement well.

While the closed capillary system filled with liquid will transmit pressure, it also can act as a filled-system thermometer. If, for a differential pressure measurement, the capillaries are of the same length and they are at the same temperature, the thermometer effect will be same in both seals, and it will cancel out.

DP cells with integral seals are expensive, and the seals cannot be repaired in the owner's shop. They have potentially more error due to the thermometer effect and the fact that the pressure has to be transmitted through a metal diaphragm. Consequently, they should be specified only for those applications which will make full use of their advantages. This generally means liquid-level measurement by differential pressure and flow measurement on liquids that are so viscous that the pressure lead lines and meter body would have to be maintained at a very high temperature in order to keep the liquid fluid enough to transmit pressure.

DP cells with factory integral seals have proven useful for measurement of liquid level by measuring differential pressure. They eliminate the disappearing seal problem associated with open seals. They also eliminate the winterizing requirement—heat tracing and insulating the impulse lines and a heated housing for the meter—because the seal fluid inside the capillary can take the low temperatures. Some DP cells with factory integral seals have a minimum differential pressure range of about 20 inH_2O, or 5 kPa, which may restrict their use.

Instrument maintenance people have been known to dislike instruments with integral seals owing to the difficulties with calibration and repair. In addition, the absence of any manifold valves makes it impossible to do a zero check at the meter.

8.5.4 Pressure-gauge diaphragm seals

There is one type of integral seal that is different from the others in that it can, and should, be field filled or refilled. This is the *diaphragm seal,* which is designed for use with pressure gauges. A sketch of a diaphragm seal is shown in Fig. 8.3. The pressure gauge threads into the top of the body above the diaphragm. The pressure-gauge Bourdon tube and the cavity above the diaphragm are pumped down with a vacuum pump and then filled with a seal liquid such as a light oil. All the air must be removed if the pressure gauge is to read accurately, because the flexing capability of the diaphragm is limited.

The space under the diaphragm is connected to the process. Any insulation on the pipe or vessel should be formed up around the connecting piping and the body of the diaphragm seal so that heat from the process will reach the diaphragm seal. The length of the connec-

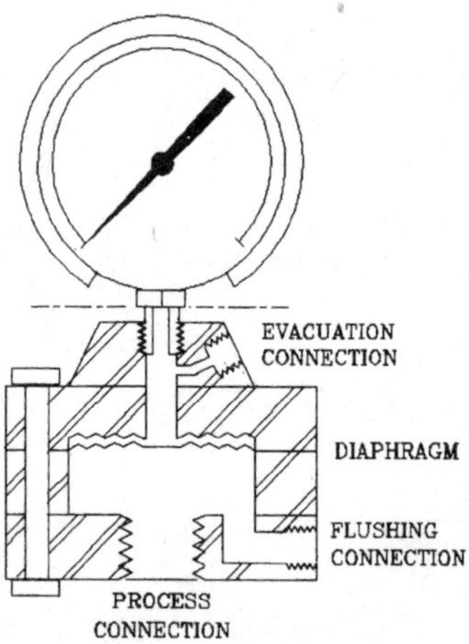

Figure 8.3 Cutaway of a pressure-gauge diaphragm seal.

tion between the diaphragm seal and the process must be the minimum possible.

Pressure-gauge diaphragm seals are particularly useful for liquids that are viscous at ambient temperature, as well as for pressure gauges on outdoor water and steam lines. These last two applications are annoyingly difficult to winterize. The diaphragm seal is the best solution. The alternative is to heat trace, insulate, and perhaps house the pressure gauge, at many times the cost of the gauge.

The main problem that afflicts diaphragm seals is, oddly enough, a human factor. In some plants that have large numbers of pressure gauges, instrument maintenance personnel have turned over the job of changing out pressure gauges to the plant operators. Unfortunately, since no one has taken the trouble to enlighten them, many operators do not know what the purpose of a diaphragm seal is. As a result, they may unscrew a questionable pressure gauge from a diaphragm seal, screw in a new gauge, and then wonder why the second gauge reads even farther out than the first one.

This is not the fault of the diaphragm seal, but it has contributed to their getting a bad reputation. The real problem is one of procedures and education.

8.6 Heat Tracing and Insulating

There are two situations in which it is necessary to draw the process fluid outside its piping or vessel in order to make the measurement required.

- To measure pressure or differential pressure, the fluid must fill the impulse lines.

- To establish a level of process liquid in an external float chamber for the purpose of measuring the liquid level in the vessel.

Once the fluid is outside the vessel or pipe, it will be stagnant and will be exposed to ambient temperatures. As such, it could freeze, or it could become so viscous that it would no longer establish a representative level or transmit pressure. This potential problem can only be prevented by applying heat to the impulse lines or float chamber. A tubular or strip-type heater, called a *tracer,* is attached to the impulse line or float chamber, and then they are both covered with suitable insulation.

Heat tracing without insulation would be useless, since all the heating effect would be lost. Consequently, from this point on, where heat tracing is mentioned, insulation is also implied.

Where pressure and differential pressure are being measured, there will be process fluid in the meter body as well as in the impulse lines. However, this fluid will be kept warm by a heater inside the meter housing. Heat tracing is intended for impulse lines, not for instrument housings, which should have their own heaters.

An important consideration in the use of heat tracing for instruments is the source of heat. For many years steam was used exclusively, and heat tracing automatically meant steam tracing. The tracer consists of ½-in copper tubing connected to a source of steam at one end and with a steam trap at the other. Steam as a source of tracing heat has some distinct advantages:

- Steam is safe. It cannot cause a fire in a plant handling flammable materials. In some locations, this consideration preempts all others, and who can argue?
- It can be easily field installed. After the instrument has been mounted, one only has to pull as much tubing as is required off a large reel, attach it to the impulse lines or float chamber, make the necessary end connections, and insulate over it.

While this second point is generally true, the correct application of instrument heat tracing does require some skill and local know-how. If the job is not done properly, it is possible to end up with too little heat or too much.

The use of steam for instrument tracing also has drawbacks that cannot be ignored:

- If the steam flow stops in winter, even for a short time, the condensate in the tracer will freeze up. It is usually an onerous task to get a frozen tracer back on line.

- Steam has a waste product, steam condensate, which has to be dealt with. Tracing steam traps that are allowed to discharge condensate for days on end in winter can create icy conditions that are hazardous.
- Steam can be too hot. Even low-pressure steam (150 kPag) has a temperature of 128°C. This would be hot enough to boil some lighter liquids inside an impulse line.
- Steam can be messy. Fittings that were tightened up at ambient temperature sometimes begin to leak when they are heated up to steam temperature. The leak will show up as a wet spot on the insulation covering the tracer. Repairs to the tracer and the insulation will be required to correct the situation.

In an attempt to get around the problems of steam tracing, instrument people began to investigate the possibility of using electric tracers instead. Electric tracing obviously would have none of the problems experienced with steam. The question was, Could electric tracers be made safe enough for use in plants that were rated hazardous because they were handling flammable materials?

In some plants, electric tracing is almost a dirty word. It has been tried in important applications, has failed at a bad time, and has caused considerable grief. Such experience seems to be fairly well limited to electric tracing on process pipes, where the performance obtained from electric tracing has been inconsistent. Poor experience with electric tracing on pipes should not be a reason for writing it off for instrumentation, where it has had a relatively good record.

There are actually two types of electric tracing, resistance and self-limiting. Both have been used for tracing instrument impulse lines and float chambers. They are not, however, interchangeable. Some discussion of their relative merits and application situations is needed.

8.6.1 Resistance tracing

The first installation of electric tracing for instruments was very likely done with resistance tracing, since self-limiting tracing had not been developed at that time. A resistance tracer consists of a pair of bus wires inside a copper jacket about ½ in in diameter. The separation of the bus wires from each other, and from the metal jacket, is maintained by filling the inside of the jacket with a densely packed insulating powder, such as magnesium oxide. Resistance heating cable is sometimes called *MI cable.*

Power is applied to the bus wires at one end of the tracer. At the other end of the tracer, the bus wires are spliced together. Thus heat is developed by the flow of electric current along the bus wires. The resistance drop along the wires converts electrical energy into heat.

The heat output of a resistance-type electric tracer depends on three factors:

- The size of the bus wires
- The length of the bus wires, that is, the length of the tracer
- The voltage applied to the tracer

Difficulties with regulating heat output, along with a ruling from provincial inspectors in Canada regarding the safe use of resistance tracers, made it awkward to use them. When approval was sought to use resistance electric tracers in plants processing flammable materials, the concern of officials was not so much with the tracer itself, which is structurally quite strong. Rather, it was with the connections between the tracer and the power source. The collective judgment of the inspectors, therefore, was that this connection should not be made in the field. The tracer should be cut to length and finished off with proper power leads at the factory, where the quality of the job could be controlled.

The outcome was that each tracer had to be factory fabricated. Obviously, no order for a factory-built electric tracer could be filled unless the length were specified. To do any tracing job, the owner or contractor had to either guess at the length of each tracer from an isometric sketch of the meter installation, or wait until the meter had actually been installed, and then make measurements. Both these tactics had problems: the first with tracers turning out to be the wrong length, and the second with the possibility of instruments having to wait to be winterized until the tracers were delivered. This could well be after the plant had started up and the instruments were in service.

There were other difficulties with resistance tracers as well. Heat-traced instruments will need more heat when the weather is really cold. Most do not need any heat at all if the temperature is above freezing. If the heat output of the tracer is set for the coldest conditions, then when the weather is merely chilly, the heat output will be too great. At best, energy will be wasted. At worst, the excessive amount of heat could boil the process liquid out of the impulse lines.

How, then, can the heat output of a factory-made resistance tracer be varied to suit the climate? The size of the wires or their length cannot be altered once the tracer has been built. This leaves only the possibilities of varying the applied voltage or turning the tracer on and off with a thermostat. To vary the voltage, one would have to install a power transformer with a number of taps on the secondary and then manually change the tracer power connections as the weather required. Not a very practical arrangement. Equally questionable is the use of a thermostat. The cost, should it have to be an explosion-proof model, is one problem, yet it is probably not as significant as the matter of where the temperature-sensing element should be located so that the thermostat can be depended on to do its job.

The use of electric tracing would probably have met a premature end if resistance tracing had been the only form available. Nevertheless, it is worth remembering that resistance electric tracing is rugged, and because of this fact, there are certain applications where it may be the best choice. Hot tracing impulse lines for steam meters located outdoors is one. The drawback is that to get the length of the tracer correct, one really has to completely install the meter first and then make the necessary measurements.

8.6.2 Self-limiting tracers

The point has been made that if resistance tracers were the only means available to electrically trace instruments, then in all likelihood the use of electric tracing would have been considered impractical and would have been discontinued. The situation was likely saved when a new development, in the form of self-limiting electric tracing, appeared at just the right moment.

A self-limiting tracer has two bus wires, as does a resistance tracer, but the heat is not developed by the flow of current along the bus wires. The bus wires are connected to the power source, but they are not spliced together at the opposite end of the tracer. The bus wires run parallel to each other, separated by a flat plastic core about 1 cm wide. The heat is developed by the flow of current between the bus wires and through the core.

The flow of current, hence the heat developed by the tracer, depends on the resistance of the core, which depends on the ambient temperature. As the ambient temperature decreases, so does the resistance of the core. This causes the current drawn by the tracer to increase, and it puts out more heat. The reverse occurs when the ambient temperature rises—hence the self-limiting effect. The tracer's heat output increases when the ambient temperature is low and decreases when the ambient temperature is high. This is done inherently without having to resort to the complication of a thermostat to turn the tracer on and off.

For instrument electric tracing, the self-limiting tracer looked as though it were almost the answer to a prayer. In service, self-limiting tracers have in fact fulfilled the hopes that were held for them. A self-limiting tracer puts out a specified number of watts per meter, and this output holds irrespective of the length of the tracer. Thus self-limiting tracing can be purchased by the reel, and the required length can be cut off after the instrument installation has been completed—

an advantage that formerly belonged exclusively to steam tracing. However, as always, there are some potential problems that must be taken into account.

First is the matter of electrical safety. The most vulnerable point is the junction of the tracer and the power supply lines. In order to satisfy the electrical inspectors on this point, a special fitting, or grommet, had to be devised by the manufacturer. After the splice has been made inside the body of the fitting, the fitting is filled with a sealing compound to render the splice weatherproof and electrically safe. Certification of the tracer must include the fitting.

It also has been found from field experience that self-limiting tracers are relatively fragile. They must be handled carefully in the installation stage; otherwise, a broken bus wire or a burnout due to an electrical short may result.

If insulation is applied over the tracer, it must be done by workers who realize what they are insulating. Driving staples at random through the cladding can cause shorts, either between the bus wires or, more likely, between one of the bus wires and the protective braided metal shield of the tracer, which will be grounded. The arc created by a short can cause the core of the tracer to catch fire, a development that does not impress people who work in plants that handle flammable materials.

Finally, a self-limiting electric tracer is designed to provide heat, not receive it. The ambient temperature that the core of the tracer can withstand is definitely limited. Self-limiting tracers on steam meters have been burned out when the meter impulse lines were blown down and the condensate in the lines was replaced by live steam at a temperature that was much higher than the tracer could tolerate.

Before self-limiting tracing is selected for a winterizing application, some investigation must be made into the ambient temperatures to which it will be subjected, even if only on a short-term basis. A burned out tracer means a frozen impulse line will occur when cold weather strikes.

8.6.3 Impulse tube bundles

One of the facts of life concerning the tracing and insulating of instruments is that there is a considerable amount of field labor involved. The work will probable require the pipe fitting, electrical, and insulating trades. The concern is not so much the cost of doing things this way, although cost is a factor. The real concern is how to ensure consistently high-quality work.

It was with this in mind that in the design phase of the Syncrude project, Syncrude engineering personnel, along with people representing the engineering contractor and the manufacturers of electrical cable and self-limiting tracers, developed the concept of the impulse tube bundle in 1975. The bundle was visualized as being made up of a ½-in stainless steel impulse tube, or tubes for a flowmeter, with a self-limiting tracer alongside each tube, a wrap of high-quality insulation around them, and finally, an outside jacket of flexible weatherproof plastic such as PVC.

The bundle would be shipped on large reels like electrical cable. The field work would consist of cutting off sections to the length required, forming and supporting it, and making the end piping and electrical connections. The requirement for field insulating would be practically eliminated, since the ends of the bundles, for the most part, would be inside heated enclosures.

Prototype sections of the tube bundle were made and tested in a cold chamber to confirm that the heat output from the self-limiting tracer was adequate to keep the temperature of water in the impulse line at no less than 5°C when the ambient temperature was −50°C. A second objective was to determine how long the insulation would sustain the temperature of the water in the impulse line after power was shut off to the tracer with the outside temperature at −50°C.

The results were encouraging, and based on them, the decision was made to use the impulse tube bundles throughout the Syncrude plant. This decision was not without some risk, since winter temperatures at the plant location often go down to −40°C. Nevertheless, another plant in the same area had trouble enough with steam tracing, so if the venture proved successful, and we believed that it would, there was much to be gained.

Over the first winter, the performance of the impulse tube bundles was generally good. The number of freeze-ups, due to tracer failure or other causes, amounted to about 5 percent of the installations. This was a high enough number that in the following spring we undertook an investigation of the failures to see if the cause, or causes, could be identified. The conclusion that resulted was very clear.

By far the leading cause of bundle failures was bad installation work. For example, the impulse tube and tracer, with the wrap of insulation and the final plastic jacket, made up a bundle with a diameter of about 5 cm. It was obvious from the outset that if the bundle had to be bent to make a change in direction, a certain minimum radius had to be adhered to. Sharp bending could squash the insulation or stretch or even snap the self-limiting tracer. Consequently, the manufacturer of the bundles made up and supplied two special bending jigs which, when used, would ensure safe, proper bends.

The problem was (and still is) that in a plant of this nature, the installation of the small piping and wiring, which includes most instrument hookup, inevitably comes toward the end of the job, when operations wants the contractor out so that they can get into production. This happened in the case of the Syncrude plant, and as a result, a host of electricians and fitters all descended on the installation of the tube bundles at the same time. There were not enough bending jigs to go around, and the electricians began to bend the bundles with the bending bars that they normally use for bending metal conduit. Since no one took the trouble to educate them, they were unaware of the damage they were wreaking on the tubing bundles.

In hindsight, a decision to use the impulse tube bundles on another project would be a good decision, because of the benefits in the form of reduced field costs and the better overall quality of the work, along with the several benefits of using electric tracing. However, this work should not be given to workers who are ignorant of the nature of the materials they are installing. A small, specialty group that is thoroughly cognizant of what impulse tube bundles are and what they do should get the job. Otherwise, expect trouble, and do not blame the tube bundles.

8.7 Glycol Systems

For performance, an excellent alternative to both steam and electric tracing is to circulate hot glycol through tracers. In addition to being completely safe and eliminating the problem of winter freeze-up, the temperature of the glycol can be easily controlled to the optimal value.

The problem with the glycol system is, of course, the initial cost. A glycol system requires both supply and return lines, so the amount of piping required is roughly double that required for steam. Heating and pumping equipment are required, along with whatever control devices are needed to ensure that the flow of glycol is distributed equally to all the branch lines that need it.

Energy consumption will be greater because, unlike electric tracing, there will be heat losses in the distribution system. In addition, hot glycol solution has a low surface tension, which means that it will leak out of the smallest crevice. Consequently, sections of pipe or tubing must be welded or soldered together or at least connected with high-quality fittings. All this notwithstanding, some plants have elected to use glycol systems in the belief that the advantages are greater than the drawbacks.

8.8 Instrument Purges

Instrument purges are sometimes used in the hope that they will protect instrument impulse lines from the problem of process liquid freezing or setting up in an impulse line. The theory behind instrument purges is that if a steady flow of purge gas or light liquid can be established down the impulse line from the instrument to the process connection, then this will prevent the process liquid from entering the impulse line while still permitting the process pressure to be transmitted accurately to the instrument.

In the list of winterizing options, purges are at the very bottom, which is where they rightfully belong. There are enough valid arguments against the use of purges that, with the possible exception of one or two special cases, they should not be considered. There will always be some better method of winter protection. The arguments against include

- The amount of small hardware and piping needed to set up an instrument purge makes the purge system complex and costly. Inasmuch as the flow of purge liquid or gas must be steady, a pressure and/or flow regulator, a rotameter, a needle valve, and a filter are all required for most installations. The combination of these items, all piped together, virtually constitutes a miniature process unit.
- In most cases, the best purge liquid or gas is at some location other than the location where the purge is required. This usually requires a long run of small piping, possibly winterized, from the source to the point of use.
- In some gas-purged installations, the desired gas is not available in the plant at the required pressure. Compressed gas from cylinders has to be used. This means that cylinders have to be changed regularly. The time and effort required to exchange cylinders are not all that is involved. It also means someone has to remember to order full cylinders and check to see that they are delivered in time.
- If the purge liquid or gas can be obtained from a plant process stream, it will invariably be a product stream, in other words, one of the more valuable materials in the plant. Assuming that the purge liquid is a light gas oil and a purge velocity in the impulse line of 1 cm/s is considered adequate, then the cost of the purge oil, for one connection, would be on the order of $20 per day.
- Aside from the cost of the purge gas or liquid, the purge will mix into the process liquid in the vessel, with the possible problem of contamination. A light gas oil purge for a pressure instrument on an asphalt tank sounds like a textbook application of a liquid purge. However, the gas oil will dilute the asphalt, usually considered to be an undesirable situation.

- The purge flow rate must be held within close limits. Too high a flow rate will waste the purge gas or liquid. It also will cause a backup of pressure in the impulse line, with the result that the instrument will measure a higher pressure than the pressure at the process connection. If the purge flow rate is too low, there is a risk that a pressure bump in the vessel will force a slug of process liquid into the impulse line, where it will set up.

These are the more obvious objections to instrument purges. All of these deficiencies could be lived with, however, if instrument purges could be relied on to do the job they were intended to do. They cannot. It takes just one purge failure, even for only a short term, to allow process liquid under pressure to back into an impulse line and seal it off, and it seems to be only a matter of time before this happens. In the light of current measurement technology, the way to measure the pressure of a heavy, viscous, and perhaps hot process material in a pipe or in a vessel is to use a factory integral seal.

8.9 Where an Instrument Purge Will Work

An application on which an instrument purge has been successful is so unusual that it stands out. The classic case is that of a gas-purged dip tube for measuring liquid level, as shown in Fig. 8.4. A stream of purge gas, often air, flows down the dip tube and bubbles up through the liquid. The pressure needed in the dip tube to cause the bubbles to emerge will be equal to the pressure created by the head of liquid above the bottom of the dip tube. Thus the pressure of a dirty or viscous liquid is converted into the same pressure in clean gas. By measuring this pressure, the level of the liquid can be inferred.

The dip tube level meter works best on open vessels that are at atmospheric pressure, although it is possible to use the method on vessels under pressure. Two purges are then required, one for the dip

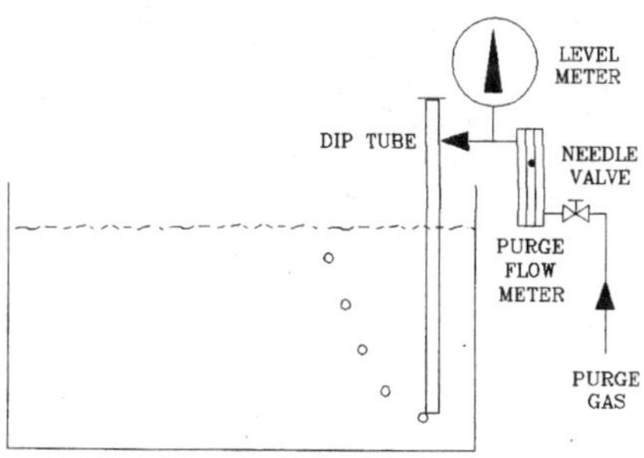

Figure 8.4 A dip tube is a special case of a gas purge.

tube in the liquid and the other for the vapor space above the liquid. The possibility of contaminating the vessel contents should always be considered.

8.10 Three Great Ways to Have a Failure

1. Arrange steam tracing so that the steam has to follow two parallel paths. A flowmeter has two impulse lines, so it requires two tracers. Split one tracer into two at the connection to the steam supply and then join them together again just ahead of the trap.

   What will happen is that the steam will take whichever path, of the two, that offers the least resistance. The steam flow in the other path will stop, and the condensate that has formed in that tracer will freeze. Splitting the tracer into two at the steam supply end is all right, but each of the tracers so formed must have its own trap.

2. Project the pipe pedestal through the bottom of the meter housing so that the meter can be mounted on the same pedestal. Assume that the law of thermodynamics, which says that if given a suitable conductor, heat will flow from a point of higher temperature to a point of lower temperature, will not apply in this case.

   When it gets really cold outside, the heat put out by the heater inside the meter housing will drain away down the pipe pedestal, and the meter will probably freeze up. If a pipe pedestal is used, the meter housing should sit on top of it. The meter should be on a separate stand inside the meter housing so that the escape route for heat is broken.

3. If a finned heater is to be used inside the meter housing and there is not enough room along any of the three walls to install it horizontally, then install it vertically. What's the difference?

   The difference is that for a finned heater to give off heat, air must flow between the fins. Unforced air moves by convection, which is vertically. The fins of the heater must be vertical; otherwise, there will be no flow between them, no transfer of heat, and the meter will freeze up even with the heater turned on.

Chapter

9

Alarm and Shutdown Systems for Operating Equipment

For some process or equipment monitoring situations it is not necessary to know the actual value of the process variable. The requirement is to provide an alert, sometimes combined with automatic action, if the value of the process variable passes a lower or upper safe limit. Process alarm systems perform this function.

Alarm systems live in the digital world. They suffer from the misconception, by designers, that since they are relatively simple and do not require any control theory, then it does not take much background to specify or design them. In fact, there are many aspects of alarm systems that should be taken into consideration if they are to do the job that the designer and the user want them to.

Alarm systems perform a valuable process function, yet they should not be specified indiscriminately. Too many alarms on an instrument panel or console inevitably leads to their being ignored. A weakness of today's state-of-the-art digital systems is that it is almost too easy to generate alarms.

9.1 Components

Process alarm systems generally consist of three components, as shown in Fig. 9.1. These components are

- An alarm switch unit with a measuring system that can respond to changes in the process variable being monitored. It compares the measured value of the process variable with an alarm point that has been calibrated or programmed into it. If the process variable exceeds the safe value, the switch is tripped. This produces a digi-

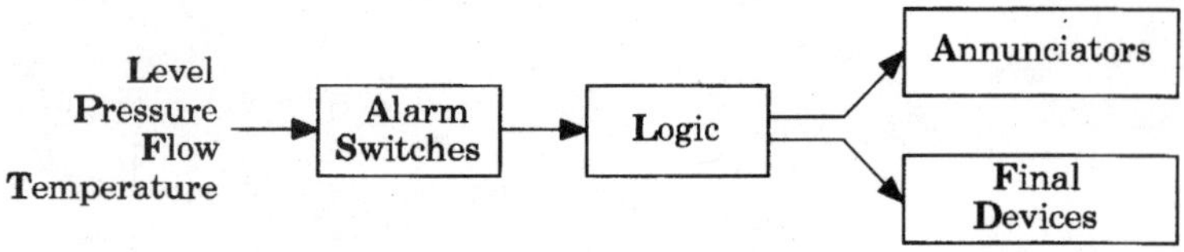

Figure 9.1 The alarm system block diagram.

tal output which correlates with On-Off, Safe-Unsafe, Ready–Not Ready, and so on.

- A logic component that receives the output signal from one or more alarm switches and then decides, based on preplanned logic, what output or outputs should be sent to an annunciator and any final devices that are to be automatically activated.
- An annunciator, possibly with a final device or devices, that will produce appropriate action in response to the outputs from the logic. Alarm annunciators can be flashing lights, illuminated plaques, messages on a video display terminal (VDT) screen, and other visible forms. Audible alarms are often used as well.

A final device causes an automatic shutdown of operating equipment. Common types of final devices include

- An electric relay acting as a second Stop or Start button in the control circuit of the starter for an electric motor.
- A solenoid-operated emergency valve. The maximum practical size is on the order of 1 in.
- A solenoid-operated pilot valve, which can load or vent air pressure from a spring/diaphragm actuator on an emergency valve. The pilot valve will usually be a ¼-in three-way type. The emergency valve can be almost any size, as long as the spring in the actuator is strong enough to close or open the valve, whichever is appropriate, when the air loading pressure is zero.
- A motorized valve, damper, or gate. These are not as practical as the first three above because they have no fail-safe position. Power is required to drive them to their safe status. If the emergency involves loss of power, then motorized devices cannot respond.

9.2 Alarm System Types

Process alarm systems come in five different types, which will be identified here. While for purposes of this text it is convenient to discuss alarm systems under these headings, the types themselves are not formally recognized as such in industry.

9.2.1 Type 1 alarms

These are not alarms in the strict sense of the word. They are really process or equipment status indicators. The status combinations may be On or Off, Open or Closed, Running or Stopped, Ready or Not Ready, or others of a two-state variety.

If this type of alarm is specified, there should be a positive indication of each state. A single light which is On for one state and Off for the other is hazardous. If the light is Off, does it indicate that safe status is not confirmed or that the bulb is burned out?

A possible color code for some status combinations is

White = On	Blue = Off
White = Open	Blue = Closed
Green = Running	Amber = Stopped

9.2.2 Type 2 alarms

A type 2 alarm system is one whose function is to call the attention of the plant operator to an abnormal reading of a meter to which he or she has access. The alarm will be initiated by a signal-line alarm switch, likely located in the operations center. The alarm switch will be tied in to the measurement signal going to the panel instrument in question. Figure 9.2 shows a type 2 alarm system. A level transmitter (LT) measures the vessel level and transmits a measurement signal to a level indicator, recorder, or controller.

This measurement signal is picked off and sent to an alarm switch (LSL) that activates an annunciator (LAL) if the unsafe low-level point is reached.

If signal lights are used for annunciation, a logical color coding is amber = Too Low and red = Too High. This avoids the use of green, which most people associate with a safe situation. Instrumentation people should bear in mind that the color-coding standards used by electrical people for their equipment may be different from the ones listed here and may take precedence. A red On light usually means

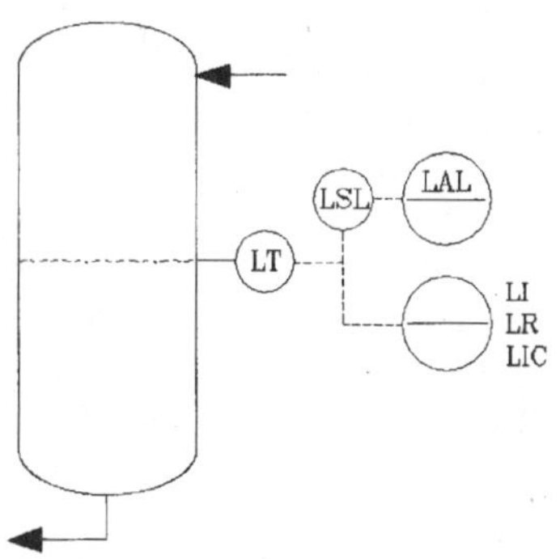

Figure 9.2 A typical type 2 alarm system.

electrical equipment is powered up and electrically hot but nevertheless operating normally.

A type 2 alarm system has a limitation that must be recognized, particularly since most alarms configured in distributed computer control systems are of this type. A type 2 alarm does not provide backup for a failure of the measurement to which it is linked. Unfortunately, designers and operators tend to specify this type of alarm because it is easy to implement, with the expectation that it does have backup capability.

9.2.3 Type 3 alarms

These are alarms which are specified *instead of* a meter for the process variable. The implication is that it is not necessary to know the actual value of the variable as long as it remains on the right side of a safe limit. If the process variable crosses the low or high safe limit, however, an alarm signal is sent. Most alarms on automobile instrument panels are of this type.

A type 3 alarm (Fig. 9.3) is the simplest form of the general configuration shown in Fig. 9.1. It requires an alarm switch operated directly by the appropriate process condition and located at the point of measurement. The on-off output of the alarm switch activates an annunciator directly. No other logic is needed.

Also included in this category are alarms that signal that an automatic shutdown system has been tripped. The shutdown alarm switch (SS) and shutdown alarm (SA) in Fig. 9.4 are an example. Note that this alarm does not initiate the trip. It simply signals that the trip has occurred. The alarm switch will be connected into the pneumatic or electrical loading line to the final device of the automatic shutdown system, not into the measurement signal line. The measuring element of the alarm switch will be designed to accept a 4- to 20-mA or a 3- to 15-psig input.

9.2.4 Type 4 alarms

These alarms serve as a backup in the event of a failure of an instrument that measures the same process variable (Fig. 9.3). The measuring instrument (LT) could be providing the measurement information for an automatic controller. It follows, then, that the measuring system for the alarm must be independent of the measuring system for the meter that the alarm is backing up. Specifically, this means that

- The alarm switch (LSL) must have its own measuring element. It must not be a signal-line alarm switch, as in type 2.

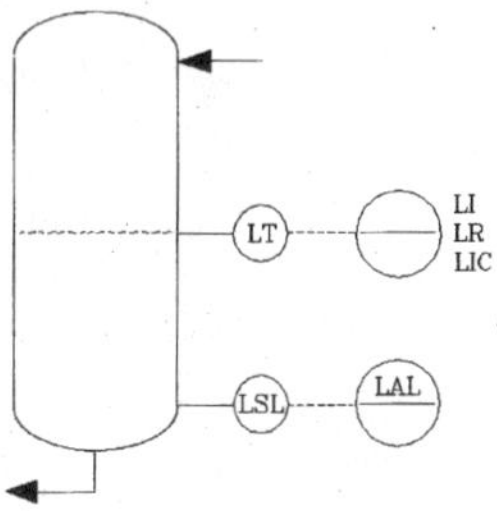

Figure 9.3 A type 3 alarm system.

- It must be operated directly by process conditions.
- Since a possible reason for a failure of the primary instrument is a plugged instrument connection, it must have its own connection on the vessel or piping.

9.3 Type 5 Automatic Shutdown Systems

The design of an alarm system that automatically causes things to happen is shown in Fig. 9.4. The low-pressure alarm switch (PSL) detects the low-pressure condition and sends a signal to the logic. The output of the logic sets off a low-pressure alarm (PAL). A second logic output is a signal to the final device to activate it, in this case a shutoff valve that closes. A second alarm (SS and SA) tells the operator that the shutdown signal has gone out to the final device.

A type 5 alarm system involves several factors whose significance should be understood. In many cases, automatic action means automatic shutdown. If a machine has lost its lube oil pressure or a reactor is running too hot, an automatic shutdown is called for. However, there are cases in which the appropriate action will be an automatic startup. Examples are startup of an emergency generator on loss of regular electric power or startup of a standby pump or fan if the regular unit goes down. Nonetheless, since the type of alarm system described under type 5 is so often referred to as an *automatic shutdown system,* this name will be used.

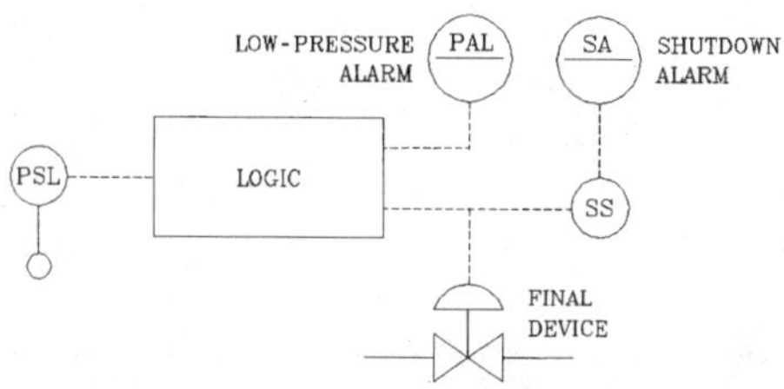

Figure 9.4 A type 5 alarm and shutdown system.

9.3.1 Design rules for automatic action systems

When any system that is to take automatic emergency action is designed, there are five important design rules that should be adhered to closely:

1. The system should be designed so that the plant operator can do manually what the system does automatically.
2. Automatic action, shutdown or startup, must be signaled. Automatic action must never be taken without alerting the plant operator to the fact that it has happened.
3. Automatic action should lock out. Equipment that has been shut down automatically should not automatically restart simply because the condition that caused the trip has gone back to normal. Once automatic action has occurred, it should remain in its shutdown status until the plant operator has satisfied himself or herself that the situation that caused the shutdown has been corrected. The operator can then, at his or her discretion, restart the equipment using the proper startup procedure.
4. If the system has more than one input (sometimes called a *permissive*) and any input causes the operating equipment to shut down or prevents it from starting up, then that input must be clearly identified by the system.

   Boilers and conveyers are two examples of operating equipment that have many permissives, any one of which could prevent the operation of the equipment if it were in an unsafe state. The person who has to do the troubleshooting and trouble correcting should be able to determine immediately where the trouble is. If too much time is spent trying to locate the permissive that is preventing operation, then the tendency is for the plant operators to order the system to be bypassed so they can get back into production.
5. If plant conditions are such that the operator has lost control of the situation, then all the final devices in all type 5 systems should go to emergency status on their own. Loss of control generally implies complete loss of power, both electrical and pneumatic.

There are some additional comments worth making with regard to the design rules just laid down. First of all, if the consequences of an unplanned shutdown, in production losses or hazard to equipment, are serious, then the advisability of including a preemergency alarm in the system should be evaluated. A *preemergency alarm* is a type 2 or type 3 alarm triggered by the same process variable or variables as the emergency shutdown but at a more conservative value. The idea behind the preemergency alarm is to alert the operator to the fact that there is a potentially dangerous trend developing and to give the operator some time to take action on his or her own to correct the problem before matters reach the point where the emergency shutdown is activated.

Second, an item that does not show up in the design rules concerns including the capability for the plant operator to decommission an automatic shutdown system. There are cases where this would be desirable and safe and other cases where it would not. This aspect certainly should be considered in the design stage, but with due regard to the potential problems. There are cases on record where an emergency system was bypassed for good reason, perhaps to do minor maintenance, but no one realized that the bypass had not been removed until a real emergency developed and the automatic protective action did not happen.

Finally, there is the question of the number of permissives. Since any one permissive can cause a shutdown, each permissive that is added increases the chances that an automatic shutdown will occur. After every conceivable set of unfavorable circumstances has been taken into account by designers, some plants end up so well protected that they are practically protected against operating.

The performance of plant operators is often related to production. If production is being held up by automatic gadgets, then the temptation will be to turn them off. Better to take the chance on running unprotected than not to run at all.

The point is, the system designer must take the responsibility for getting involved in this situation. Is the situation really serious or only awkward? Is automatic action really required, or should the plant operator be able to handle it on his or her own? Is it really necessary to have an automatic shutdown, or would a simple alarm be adequate?

9.3.2 Remote control, manual systems

There is a type of system that can be considered to qualify as an emergency system, although it does not function automatically. It functions instead at the plant operator's discretion. In this system, valves, dampers, gates, motors, or other devices are operated by remote control from an operations center. If there is a fire, for example, it may be necessary to close a key valve from a remote location. Getting to the valve to close it manually may take too long or may not be possible at all because of the fire.

This system can be implemented easily enough; the hardware is readily available. What often happens, however, is that after the system has been installed, it is decided that it would be a great convenience if the valve or other device also could be operated from a field station in the neighborhood of the valve itself. This modification seems reasonable. However, in addition to requiring field controls, it also requires a selector unit that will establish which set of controls, remote or local, will operate the device.

This is where the trouble begins. If the selector switch is to be located in the operations center, then another set of wires must be run all the way from the operations center to the field station. This turns out to be awkward, so the decision is made to locate the selector switch at the field station instead. The potential problem is predictable, and it has happened. The selector unit is left on Local, an emergency develops, and the operator in the operations center is unable to close the valve. The controls were locked out at the field selector station.

Obviously, to avoid this situation the control stations should be designed to operate in parallel, or the selector unit should be in the operations center under the control of the operator there. Either of these arrangements will require more wire and time to install, yet this cost will be small compared with the potential cost resulting from not being able to operate a critical valve in an emergency.

9.3.3 Energized and deenergized systems

A subject unto itself is whether or not automatic shutdown systems should be designed to operate normally energized or normally deenergized. Here the term *normally* means when process or plant conditions are normal. This interpretation does not always agree with the meaning of *normally* that is generally accepted by electrical people.

Opinion on this subject seems to go in one of three directions:

- Those who are totally convinced that emergency systems should function normally energized.
- Those who are totally convinced that emergency systems should function normally deenergized.
- Those who are not convinced that it is necessary or even desirable to standardize on either of the two practices above.

To attempt to prove that any one of these three positions is correct is essentially futile. Thus an explanation of the factors involved will have to suffice.

In a normally energized system, electric current will flow in the input circuit between the alarm switch and the logic and in the output circuit between the logic and the final device when process conditions are normal and safe. If an abnormal condition trips the alarm switch, it opens the input circuit to break the current flow. This alerts the logic to the abnormal condition. The logic then breaks the current flow to the final device, which causes it to go to emergency status. If the emergency system operates with pneumatic power rather than electric, current flowing is replaced by air pressure on.

A normally deenergized system functions in the reverse fashion. When process conditions are normal, there is no current flow, or zero air pressure, in both the input and output circuits. If an abnormal condition is detected, current or air pressure is turned on in the input circuit to alert the logic and then on in the output circuit to put the final device into emergency status.

The benefits of one philosophy are the weaknesses of the other. Just what these are matters a great deal to the ultimate production and safety of the plant.

Consider first the system that operates normally energized. The input and output signals (On for normal, Off for abnormal) are transmitted over wires or air tubes. Sometimes signal lines get broken. Digging in the wrong place can cause this. So can small rodents nibbling on the plastic insulation on wires. How does a normally energized system react to this?

On the input side, an open wire has the same effect as opening the alarm switch. Current flow is interrupted, and the system trips to emergency. On the output side, an open wire stops current flow and deenergizes the final device. Once again, the system goes into emergency. Thus a normally energized system responds not only to an unsafe process condition, as established by the alarm switch contacts, but it also responds immediately to a fault in the input or output signal lines. An energized system monitors not only the status of the alarm switch but also the status of its communication lines.

For a normally deenergized system, in which there is no current or no air pressure when conditions are normal, a signal line can be broken, the system would become inoperative, and no one would be aware of it, at least not until there was a real emergency and the protective system did not respond.

Furthermore, in an energized system, deenergizing the final device causes it to go to emergency status. If there is a major plant disaster that knocks out all electrical and air power, then all the final devices will go to emergency, as they should. In this respect, energized systems conform with design rule 5.

Normally deenergized systems do not conform with this rule because it is necessary to power up the final devices in order to trip them to emergency status. If a catastrophe knocks out the power sources, this cannot happen. The emergency system does nothing to help the situation. It may even make it worse.

This being the case, why are there so many deenergized shutdown systems? The answer lies, once again, in production. Since power on is necessary to sustain an energized system in its normal state, any interruption of power, even if very short, will cause a trip to emergency. As everyone knows, power outages can occur for reasons quite apart from operating conditions. Natural faults such as lightning, superimposed faults such as startup of a large power user in an adjacent plant, or human errors such as inadvertent opening of the wrong circuit breaker can all cause a normally energized system to trip to shutdown unnecessarily.

Shutdowns for any of these reasons are widely referred to among operations personnel as "nuisance shutdowns." They are generally hated for the operating problems and production losses that result. Some operations and design people feel strongly enough on this matter that they will veto any proposal to install emergency systems that operate normally energized.

The real problem with normally energized systems is not so much that they can cause nuisance shutdowns, rather it is that energized systems are not forgiving of faults that can occur for natural, superimposed, or people reasons. From the overall plant point of view, energized systems are undeniably the safest simply because they conform with design rule 5. Eliminating the problem associated with energized systems by writing them off is not reasonable. It would be better to retain normally energized systems and attack the associated problems instead.

Faults that are natural or superimposed can be virtually eliminated by selecting a 100 percent reliable source of power. This would likely be a completely dc system with battery backup. Problems due to the human factor can be minimized through better training, better labeling, and just being more careful.

Finally, while it may be a sad reflection on the state of things today, anyone who is faced with the responsibility for deciding whether emergency shutdown systems will operate normally energized or normally deenergized should think carefully before opting for the second method, inasmuch as a designer can now be held personally liable for damage or injury that is perceived to have resulted from a bad decision on his or her part.

9.3.4 Hardware

Finally, hardware components that are to be used in emergency shutdown systems must be selected carefully. If any component is located outdoors and is kept in an energized state for a year or more, one has to wonder whether it will trip if it were called on to do so. Will corrosion, dirt, residual magnetism, temperature variations, or some other factor cause it to fail to function? Failure for these reasons can be minimized by

- Using only top-quality alarm switches, solenoid pilot valves, and other components.
- Testing the system on a regular schedule. To test the system completely, it may be necessary to install a bypass around the final device. If this is actually done, then a type 1 alarm must be installed on the bypass to alert the operator to the fact that the bypass is in and the final device is inoperative.

Chapter

10

Process Control From Square One

Process control is not concealed inside black boxes called *automatic controllers.* Process control involves systems of components, of which the black box is just one, and not necessarily the most dominant one.

All the components in a process control system are active in the sense that when something is done to them, they do something in return. What is done to the component is called, in the language of process control technology, an *input.* What the component does in response to the input is called its *output.* Thus, when a control system component receives a new input, it usually reacts in some manner to produce a new output. An understanding of the reactions that cause the observed outputs in response to observed inputs is an important part of understanding how process control systems function the way they do.

One of the more intriguing reactions of a component is recorded in Chapter 4 of the Book of Exodus, where it is written

> Moses threw...[the staff] on the ground and it became a snake, and he ran from it. Then the Lord said to him, "Reach out your hand and take it by the tail." So Moses reached out and took hold of the snake and it turned back into a staff in his hand.

The reaction process that produced this output-input relationship would be difficult to describe mathematically. Fortunately, the reaction relationships encountered in process control are somewhat easier to comprehend. These relationships are usually described in terms of the *characteristics* of the components. There are two kinds of characteristics that are significant in process control. These are the *static* characteristics and the *dynamic* characteristics.

10.1 Static Characteristics

Static characteristics, also known as *steady-state characteristics,* are determined by sending a range of different inputs to the component and observing the resulting outputs. A plot of the outputs, from 0 to 100 percent, for the same range of inputs is the usual way to document the static characteristics of a component. Examples of static characteristics for two different components are shown in Fig. 10.1.

The first of the two examples is a straight-line graph, which is characteristic of a proportional, or linear, relationship between the output and the input. If x and y are the inputs and the outputs, respectively, a linear relationship can be described by

$$y = b + gx$$

In this equation, b is the value of the output y when the input x is zero. b is generally called the *bias.* g is the slope of the line, that is, the ratio of the change in y (Δy), to the change in x (Δx) that created the Δy. In the language of process control, g is called the *gain.*

The important aspect of a linear component, from the point of view of its performance in control systems, is the fact that the gain is constant over the whole operating range of the component. Components that have different gains at different operating points can cause problems in control, as we will later see.

A component whose output-input relationship plots as a curve is a nonlinear component. The second of the two characteristics in Fig. 10.1 is an example. This particular component has a static characteristic that happens to follow a square law. It is typical of the static characteristic of certain types of flowmeters for which the input is rate of flow and the output is differential pressure.

For a nonlinear component, the gain will be the ratio of $\Delta y/\Delta x$ as before, except that this ratio changes with the operating point of the component. To find the gain at any specified operating point, the procedure is to draw a tangent to the curve at the operating point (*OP* in Fig. 10.1) and then to determine the slope of the tangent.

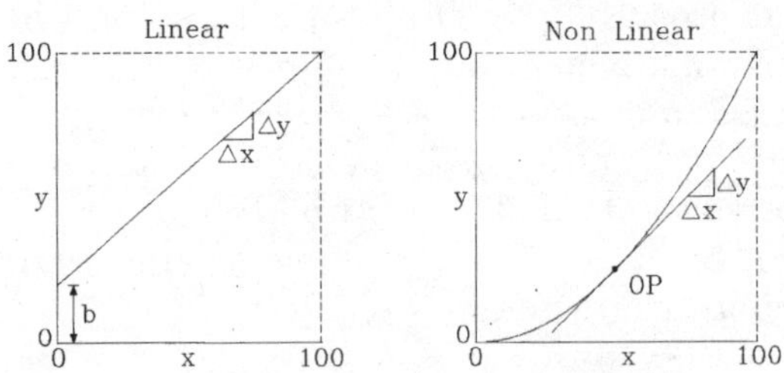

Figure 10.1 Examples of static characteristics.

A final point, and an important one, is the fact that when the static characteristics of a component are determined, *time is not a factor.* In other words, when the input is changed, the output is given all the time it requires to assume its new value. The input and output values that are recorded are strictly steady-state values. This is the single most significant difference between static characteristics and dynamic characteristics. The blessing of having all the time in the world to change its output to a new value is unlikely to be granted when the component is actually in service in a control system.

10.2 Dynamic Characteristics

In a dynamic test, the speed with which the output changes to its new value after the input is changed is highly relevant. In fact, the whole purpose of a dynamic test is to determine how quickly the output changes after the input is changed. This simply emphasizes that when the input to a control system component is changed, it is not only *how much* the output of the component changes but also *how quickly* it changes. How much is shown by the static test. How quickly is shown by a dynamic test.

To conduct a dynamic test, the component is given an input signal that follows a specified pattern with time. Thus the input in a dynamic test is really a test signal. Figure 10.2 shows three possible test signals that might be used in a dynamic test of a control system component. Other test signals for dynamic testing have been used, depending on circumstances, including the preference of the operator, and on the type of test information required.

In Figure 10.2, the first test signal is a sudden jump from one input level to another. The appearance of its graph gives this test signal its name—a *step change.* The step-change input test is widely used to determine the dynamic behavior of control system components for two reasons. It is a relatively simple signal to generate, and it yields just about all the dynamic information about the component that is of any practical significance in process control.

The second test signal is a *ramped input.* It requires a special signal generator that can produce a ramp output with time for input to

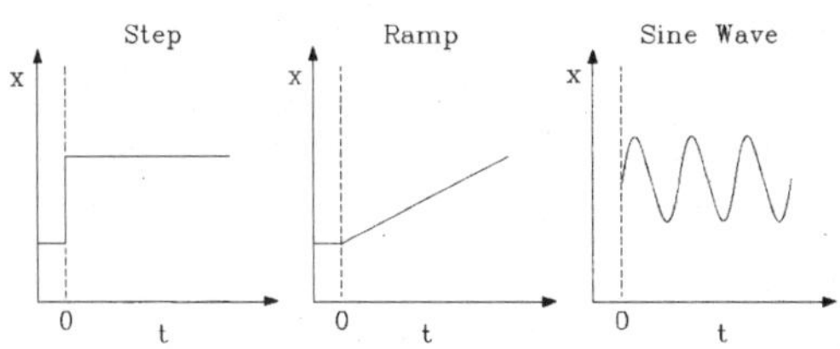

Figure 10.2 Test signals for dynamic tests.

the component under test. The resulting dynamic information obtained about the component is not significantly greater than that which can be obtained using a step change. Consequently, the ramp signal is not used often in field tests.

The third test signal varies in sine-wave fashion with time. The sinusoidal input test is the basis for the type of test that came to be called *frequency-response analysis.* Its virtue is that the sine-wave test signal is the only dynamic test signal for which the shape of the output signal is the same as the shape of the input signal.

At one time, many people thought that frequency-response analysis would become the primary method for determining the dynamic behavior of control system components. Such did not prove to be the case, however, for at least two reasons. First of all, the test requires a rather complicated signal generator that can generate a sine-wave signal over a range of frequencies, as well as over a range of sine-wave amplitudes. Second, nothing can really be predicted about the performance of the control system as a whole unless the dynamic behavior of all its components is known, and this includes the behavior of the operating equipment that is in the control system. Most operating people were not too keen on having a fired heater, a boiler, or a distillation column subjected to a succession of sine-wave disturbances at a variety of frequencies while they were trying to get production out of their equipment just so that dynamic data on the equipment could be collected.

For the purposes of this text, when dynamic characteristics are illustrated, it will be those characteristics which would show up in a dynamic test in which the test signal is a step change.

10.3 What Should We Learn From These Tests?

When a static (steady-state) test is run, the objective is usually to establish how linear or nonlinear the component really is. Control systems usually have fewer problems if the components are linear. It also may be desirable to determine the gain of the component at the anticipated operating point or how it changes over the whole operating range.

A special case of a static test is a calibration test of a meter. In this type of test, a further objective is to determine how closely the output values (usually the readings) the meter displays match the values it should be producing.

The knowledge being sought in a dynamic test is somewhat more complex. When a control system component is subjected to a dynamic test, it is usually found that the output signal lags behind the input signal in time. This shows that the component contains dynamic elements that tend to delay the reaction of the component. Just what these dynamic elements are and how they affect the performance of the component are the next matters that should be studied.

10.4 Dynamic Elements in Process Control

It is most important to understand the influence that dynamic elements have on the ultimate success of process control systems, but fortunately, there is good news. There are really only two types of dynamic elements that we need to study. They are *dead time* and *time constants.*

10.4.1 Dead time

Of all the forms of dynamic delay that can be encountered in process control, dead time is the absolute worst. Having made this assessment, two questions need to be answered.

First, in what respect is dead time the worst? Conscientious control system designers will try to design a control system so that when it is disturbed (and this does happen), the controlled process variable will come back on control as quickly as possible. The time required for a control system to get back on control after it has been upset and temporarily goes off control is called the *recovery time.*

Dead time will do more to increase the recovery time of a control system than will any other form of dynamic delay that could be present in the system. As a matter of fact, the recovery time of the control system will increase in direct proportion to the dead time. If the dead time is doubled, the control system will take twice as long to stabilize back on control.

Second, what is it about dead time that makes it so bad? The dynamic test data shown in Fig. 10.3 tells the story. The test signal is an input step change. To simplify matters, the steady-state characteristics for this element are such that the bias is 0 and the gain is 1. This means that the input and output start out at the same value,

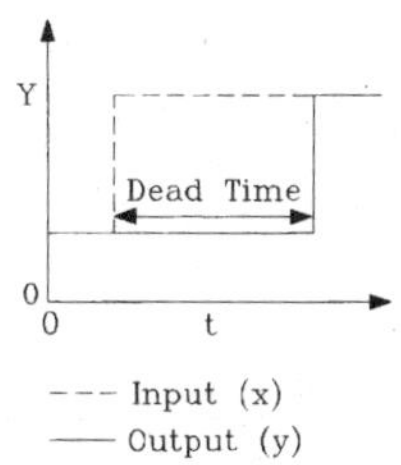

Figure 10.3 Response of a dead-time element.

and after the input is changed, the output will sooner or later come to the same value as the input. What Fig. 10.3 shows is that the output pattern is the same as the input pattern, except that it comes later by a gap in time equal to the dead time.

This is what makes dead time so harmful. There is *no response at all* until the dead-time interval is used up. While the output from a time-constant element may not change instantaneously to its final value at the moment the input is changed, at least there is some response. From a dead-time element, there is no response whatever.

In process control systems, dead time can be present as real dead time or virtual dead time. The effect on control system performance, overall, will be the effect of real dead time plus that of any virtual dead time.

Figure 10.4 shows a process application that has real dead time. The objective of this control system is to control the temperature of the liquid leaving the heater by regulating the amount of steam flowing to the heater through the control valve.

The logical place to locate the temperature sensor would be right where the liquid leaves the heater. However, due to concern about whether the liquid will be mixed to an even temperature at that point, for better accessibility, or for some other reason, it is decided to install the temperature sensor 30 m farther down the outlet pipe. If the liquid velocity in the pipe is 2 m/s, then any change in the temperature of the liquid created by the heater will not show up at the temperature sensor until 15 seconds later. This 15 seconds is real dead time.

Virtual dead time is created by time constants in series, as the forthcoming discussion on time constants will explain. The effect of virtual dead time on control system performance is the same as that of real dead time (equally bad).

10.4.2 Time constants

A *time constant* is created when there is a combination of resistance and capacitance. Capacitance has the ability to contain energy or

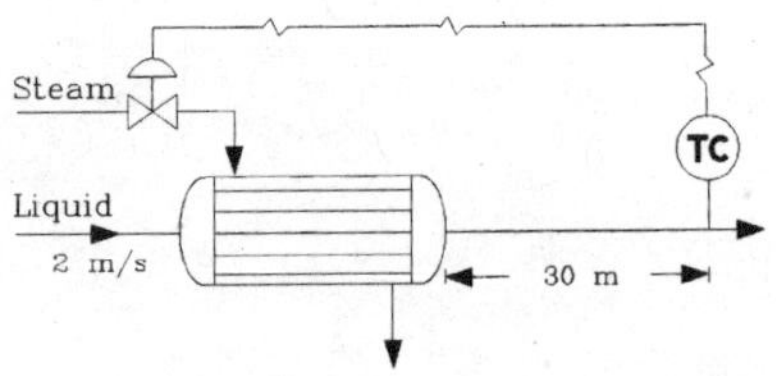

Figure 10.4 Real dead time in a temperature-controlled process.

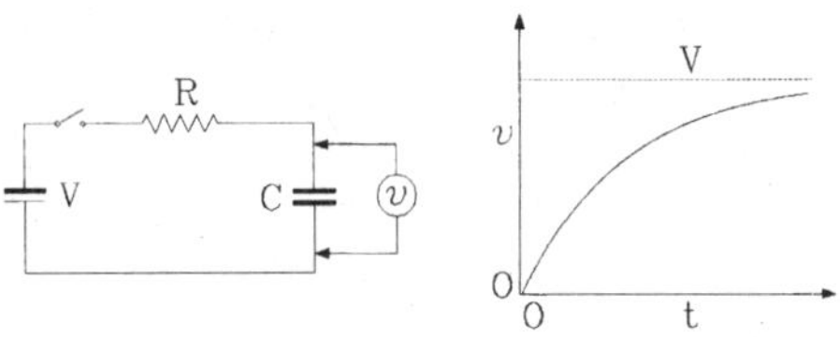

Figure 10.5 A time constant in a dc electric circuit.

material. Resistance will impede the flow of energy or material in a system. The purest form of a time constant shows up in a dc electric circuit like the one in Fig. 10.5. At time $t = 0$, the capacitor C is completely discharged, and the switch is open. Then the switch is closed.

Notice that when this happens, a step-change input from 0 to V volts is applied to the combination of the resistance and capacitance. The voltage V does not appear immediately across the capacitor but follows the path with time shown in the graph. This is due to the fact that before the voltage can build up in the capacitor, electric energy must flow into the capacitor. However, the flow of energy to the capacitor is impeded by the resistor. Thus it takes a finite time for the voltage across the capacitor to build up to its ultimate value V. In other words, there is a dynamic delay caused by the time constant.

The extent of the delay or, more specifically, the magnitude of the time constant depends on both the value of the resistance and the value of the capacitance. In fact, the time constant is equal to the product of R and C.

In electrical terms, R is equal to the voltage across the resistor divided by the current i, in accordance with Ohm's law. Similarly, C is equal to the charge q divided by the voltage across the capacitance. The units of the product $R \times C$ will be

$$(\text{volts/amps}) \times (\text{coulombs/volts})$$

Since an ampere is a coulomb per second, the units of RC reduce to units of time, which is, after all, what one would expect for a time constant.

In process control systems, it is usually simpler to evaluate time constants as the ratio of the capacitance C to the conductance Y, where conductance is the inverse of resistance. Capacitance is measured in terms of the change in material or energy stored in the component per unit of potential, while conductance is the change in material or energy rate of flow per unit of driving force across the barrier through which the material or energy is being transferred.

In a liquid storage vessel, the capacitance would be the number of cubic meters of liquid stored in 1 m of height in the vessel, whereas the conductance would be the change in outflow created by a 1-m change in height of the liquid in the vessel. In a heating system, the capacitance is the change in energy stored per unit change in the temperature of the component, and the conductance is the change in the rate of heat flow to the component per unit change in temperature across the barrier through which the heat flows.

There are many examples of time constants in process control. When a level sensor is mounted in a chamber external to the vessel, it is assumed that the level in the chamber is the same as the level in the vessel. However, for the level to change in the chamber, a certain amount of liquid must flow into or out of the chamber, owing to the fact that the chamber has capacitance. Furthermore, this liquid must flow into or out of the chamber through the connections on the chamber and the vessel and a short length of pipe. These impose resistance to flow. The combination of the resistance of the connections and the capacitance in the chamber creates a time constant and a dynamic delay. If the level in the vessel changes, the level in the chamber follows it, but not right away.

If a thermocouple is installed inside a thermowell, which is pretty standard practice, it is assumed that the thermocouple is measuring the process temperature in the pipe or vessel in which the thermowell is located. Unfortunately, this is not so. A thermocouple always measures the temperature of itself. If the process temperature increases, the temperature of the thermocouple will not follow it until a certain amount of heat (energy) has been transferred from the process to the thermocouple. This heat must pass through the walls of the thermowell, which has both the ability to retain heat (capacitance) and the ability to impede the flow of heat (resistance). There is additional resistance to heat flow in the film of fluid on the outside of the thermowell and the film of air on the inside. Thus there is a time constant in the thermowell, and this means a dynamic delay, sometimes of considerable magnitude, in the measurement of the process temperature.

When a valve positioner is installed on a control valve, the air output from the positioner goes to the top of the control valve through a short length of tubing. The valve top has capacitance, and the tubing has resistance. These combine to produce a time constant in the stroking of the valve.

Time constants show up in process control systems in so many places that their peculiarities must be studied. The voltage graph in Fig. 10.5 has been reproduced in Fig. 10.6, except that instead of voltage it has the general input and output variables x and y. We will assume that in this case the static characteristics are such that in the steady state y will have the same value as x. The input x in this dynamic test is a step change, shown as a dashed line. The output y is

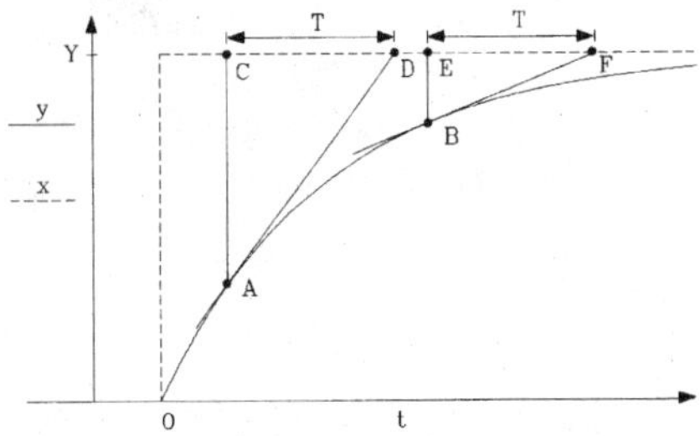

Figure 10.6 Properties of a time constant.

the solid line. Ultimately, the value of y should become the same as the value of x, but this does not occur immediately because of the delay created by the time constant.

The ultimate value that y should eventually attain, as a result of the step input x, is Y. The time constant is about 5 time units. If the time constant were shorter, the graph for y would have the same general form, but it would rise more steeply at $t = 0$, and it would lock in sooner to the ultimate value indicated by the dashed line. It is possible to write a mathematical relationship that describes the values of the output y with time. It turns out to involve an exponential.

$$y = Y(1-\varepsilon^{-t/T}) = Y - Y\varepsilon^{-t/T}$$

In this case, Y is the ultimate value of y, and T is the time constant, $R \times C$. In electrical textbooks, the point is usually made that when t becomes equal to the time constant T, the value of the output y will have reached 63.2 percent of its ultimate value Y. This can be proved easily be giving the ratio t/T a value of 1 in the preceding relation. In process control, however, this fact is of minor significance compared with other aspects of a time constant. These are revealed by taking the derivative with respect to time in the time-constant relation above.

$$\frac{dy}{dt} = -\left(-\frac{1}{T}Y\varepsilon^{-t/T}\right) = \frac{1}{T}Y\varepsilon^{-t/T}$$

However, from the original relation above, $Y\varepsilon^{-t/T} = Y - y$. Therefore,

$$\frac{dy}{dt} = \frac{1}{T}(Y-y) = \frac{Y-y}{T}$$

What this last line tells us is important. It says that the rate of change of the output from a time constant (dy/dt) will always be proportional to the distance that the output has yet to go $(Y-y)$ to reach its ultimate value Y. The proportional constant is the inverse of the time constant, that is, $1/T$. This physical rule, which describes the fundamental behavior of a time-constant element, is an important factor in the eventual behavior of the process control system.

Figure 10.6 shows the effect of this rule. *A* is any point on the time-constant curve. The distance *AC*, in terms of the output *y*, is $(Y-y)$. A tangent to the curve at point *A* has a slope that is equal to the rate of change of *y* at *A*, or dy/dt. The slope of the tangent is also equal to the ratio of the rise to the run, or AC/CD. Consequently,

$$\frac{dy}{dt} = \frac{(Y-y)}{T} = \frac{AC}{DC}$$

Since *AC* is equal to $(Y-y)$, then *CD* must be equal to *T*, the time constant. Select any other point *B* on the time-constant curve, and the same situation results. The tangent will intersect the ultimate value of *y* at a time *T* later. Thus a prime characteristic of the behavior of a time-constant element is that, *at any instant, the value of the output is changing at the rate that would cause the output to reach its ultimate value one time constant later.*

A useful result of this physical rule is that by applying the principle, we can plot a graph of the output from a time-constant element without having to calculate points, plot them, and then draw a curve through the points. The way this is done is shown in Fig. 10.7.

Along the horizontal axis of this graph there are 10 time intervals, all 1 time unit in size. The time intervals have been made fairly wide so that the effect of drawing the output graph is more apparent. The time constant in this case is 3 time units. The steady-state characteristics are such that the output will have the same value as the input when the output has attained its final value. The input to the time constant is a step change that will cause the output to change from zero to an eventual value *Y*.

The rule governing the behavior of the time constant is that at any moment the output will be changing at a rate that would cause the output to reach its ultimate value one time constant later. Therefore, during the first time interval, the output will be heading toward point

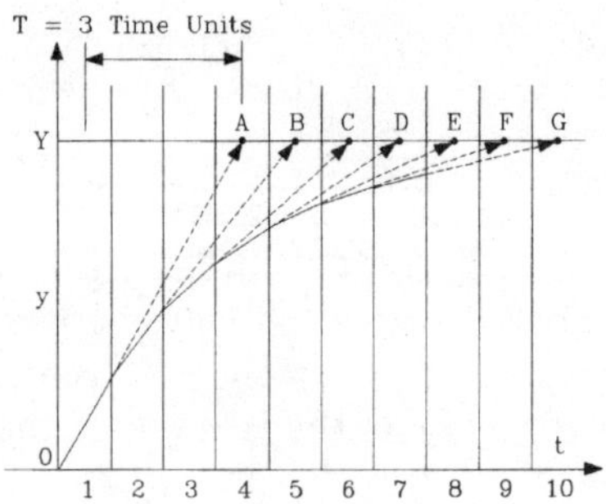

Figure 10.7 Graphing the output from a time constant.

A, which is its ultimate value during the first interval, delayed by 3 time units. The segment of the line from 0 to point *A* that falls in the first time interval thus becomes the first segment of the output graph.

Over the second time interval, the same rule applies. The output will be changing at a rate that would cause its to reach its ultimate value during the second interval, only 3 time units later. This is represented by point *B*. The next segment of the output graph is therefore that segment of the line which starts where the first segment left off and ends at point *B*.

Figure 10.7 shows the first seven segments of the output graph, each one drawn by following the rule that in any time interval the output will be changing at a rate that will cause it to reach its ultimate value of that time interval, only one time constant (3 time units) later. The smaller the size of the time intervals, the closer the output graph will approach a smooth curve.

It is worth one's time to obtain a sheet of graph paper and a straightedge and experiment with this method of generating the graph of the output from a time-constant element. Incidentally, I do not claim the credit for originating this technique. It was acquired from an acknowledged expert in process control, J. G. Ziegler.

By applying the same method, it is possible to see what the output graph would be like if there were two time-constant elements in series. The input to the first time constant is a step change, as before. The output from the first time constant will have the conventional time-constant form shown in Figs. 10.6 and 10.7. However, the input to the second time constant, and its ultimate value at any point in time, will *not* be the step change. It will be the output from the first time constant.

Nevertheless, the output from the second time constant can be graphed by applying the rule that at any moment the output will be changing at a rate that would cause it to attain its ultimate value one time constant later. The result is shown in Fig. 10.8.

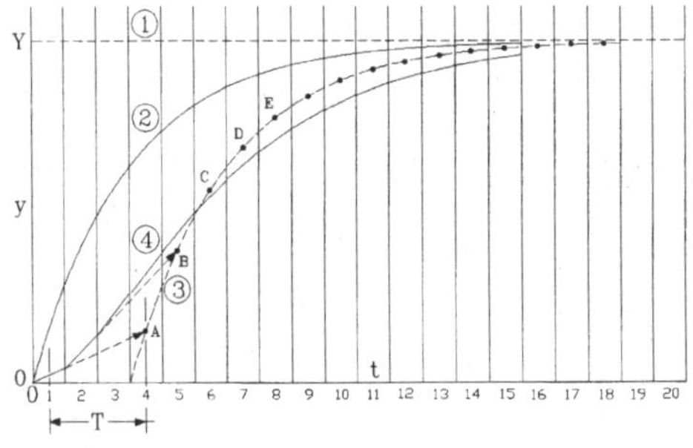

Figure 10.8 Output from two noninteracting time constants in series.

In Fig. 10.8, the dashed line (1) is the step-change input to the first of the two time-constant elements. The first time constant has a value of 3 time units, so its output is shown by the curve marked (2). Curve (2) becomes the input to the second time-constant element, which has a time constant equal to 3 time units also. Curve (3) is actually curve (2) moved to the right 3 time units or 1 time constant on the time scale. Curve (4) is the graph of the output from the second time-constant element.

During the first time interval, on average, the output of the second time constant will be increasing so as to reach the average value of curve (2) over this interval, projected one time constant (3 time units) later. This will be point *A*. During the second time interval, the output of the second time constant, on average, will be aiming at point *B*, which is the average value of the input to the second time constant and its ultimate value during the second time interval, but projected 3 time units or one time constant later.

During the third time interval, the output will be aiming at point *C*, and during the fourth and fifth time intervals, it will be aiming at points *D* and *E*, respectively. This process will continue from interval to interval; each time the output will be aiming at the point on curve (3) represented by the dot that is located one time constant later.

Certain features of curve (4), the output from the second time-constant element, are important. For example, whereas the output from a single time constant, such as curve (2), starts out at its maximum rate of change at $t = 0$, the output from the second time constant, in series with it, does not reach its maximum rate of change until some time later, possibly during the third time interval in the case of Fig. 10.8. Thus the output from the second time constant has a pattern that is somewhat S-shaped. This S-shaped output graph is an identifying characteristic of a system in which there is more than one time-constant element in series. Just how long it takes for the output from the second time constant to reach its maximum rate of change and how pronounced the S shape will be will be determined by the relative values of the time constants for the two elements.

The explanation of the S shape lies in the fact that at $t = 0$ the input to the second time constant is zero; therefore, the ultimate value that the output from the second time constant is attempting to attain is also zero. Thus the output from the second time-constant element, and in from fact all the others that come after, must start out tangent to the horizontal baseline.

The same graphic procedure can be used to produce the output from the third of three time-constant elements in series. The S-shaped output from the second time constant, which was graphed in Fig. 10.8, will be the input to the third time constant. It also will be

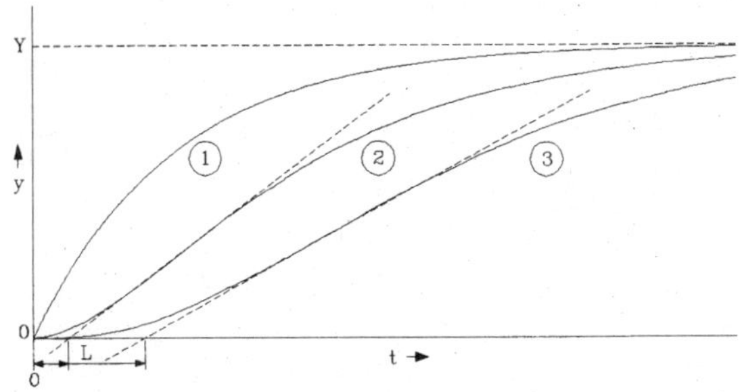

Figure 10.9 Output graphs for one, two, and three time-constant elements in series.

the ultimate value, at any instant, which the output of the third time constant will be trying to attain. The effect is shown in Fig. 10.9.

This figure shows the results of a dynamic test of three time-constant elements in a chain, that is, connected in series, so that the output of one becomes the input to the next. In this test, each of the time constants has a value of about 5 time units. It is not necessary for the test that all the time constants be the same, and in real life, they rarely are. In addition, for the purposes of this test, it is assumed that the static characteristics of each of the time-constant elements are such that they have a gain of 1 and there is no bias. This simply means that whenever the input is changed from one steady state to another, the output and input start out and end up at the same value.

The input to the first of the three time constants is the conventional step-change input, at $t = 0$, from $y = 0$ to $y = Y$. The output from the first time constant, and the input to the second, will be the standard time-constant curve (curve 1). Curve (1) is also the graph of the input to the second time-constant element.

With curve (1) as its input, the second time constant produces output graph (2). It has the S shape, which is the signal that more than one time constant is present. The output from the second time constant, curve (2), becomes the input to the third time constant, which causes it to produce output graph (3). Notice that each time another time constant is added to the chain, the S shape becomes more pronounced.

The question that needs to be dealt with now, having drawn all of these intriguing dynamic test curves for time constants, is, What does this have to do with process control, if anything? While it is relatively easy to describe the output from a single time constant (e.g., curve 1) with a mathematical equation, any mathematical analysis of the S-shaped curves that result when there are more than one time constant is a very complex matter, well beyond the capabilities of the majority of people who deal with practical process control. The approach usually taken, therefore, is to approximate the dynamic effect typified by the S-shaped curve with a combination of two dynamic elements whose characteristics can be readily described. The combination is dead time and a single time constant.

If suitable values for the dead time and the time constant can be established, then all the time-constant elements could be removed from the process control system and replaced with the dead-time and time-constant combination without any significant change in control system performance.

10.4.3 Values for the dead time and the time constant

The appropriate dead-time value is found by locating the point on the output graph where the slope is steepest, in other words, the place where the rate of change is maximum. At this point, a tangent is drawn to the graph and is extended downward to the baseline that represents the original $t = 0$ value of the output. The time interval between $t = 0$ (the start time for the dynamic test) and the time point where the tangent intersects the baseline is the *dead-time value.* This is represented by L in Fig. 10.9.

The dead time L so measured is *virtual dead time.* It is virtual because it is dead time that results from applying the approximation of dead time plus a single time constant. It is not dead time that is actually present in the control system components themselves. Nevertheless, it will have the same effect as real dead time on the performance of the control system.

Notice that for a single time constant, the output changes at its maximum rate right at $t = 0$. Thus the tangent to the curve would intersect at $t = 0$, and the virtual dead time would be zero. Because of the S shape of the following output graphs, there will be some virtual dead time, and its magnitude L will become greater with each time-constant element added to the chain or with an increase in size of any one of the time constants.

A single time constant in a control system is highly desirable because it contributes no virtual dead time. In fact, the capacitance of the first time constant adds stability to the system. It is really the second, third, and other trailing time constants that do the damage. It follows, then, that a practical way to improve the performance of any real process control system is to try to eliminate or at least reduce all the time constants except the largest one.

In the real world, there are no pure single-time-constant control systems, although there are some control systems in which the largest time constant is much bigger than the others. A vessel that had a large capacity for liquid, compared with the amount of liquid that was being run through it, would be a very easy level-control application.

Similarly, a pressure vessel that held a large volume of a gas, compared with the amount of gas being used and replaced, would be an easy pressure-control application. Since these control systems contain one time constant that is considerably larger than all of the others, they can function at a high gain, will be quite stable, and will recover quickly if disturbed.

If a dynamic test of either of these two processes were run, the output graph would show very little dead time, which is the main reason for their superior performance. In the final analysis, the recovery time of a control system, after it is disturbed, will be in direct proportion to the dead time, and by dead time is meant *the total of any real dead time plus any virtual dead time.* Increasing dead time or adding time constants over and above the first one will both extend the recovery time of any process control system.

Real-life processes can be tested by sending a step-change input to the process. This is done by switching the automatic controller to manual control, allowing the process variable to stabilize, and then abruptly moving the control valve from one steady-state setting to another. Figure 10.10 shows the results of a dynamic test using a step-change input at the control valve for a temperature-control system.

The output signal that comes back via the measuring system for the process variable, as a result of the step-change input disturbance, will contain all the dynamics contributed by dead time and time constants that are present in the control valve, the process, and the measuring system. In the Fig. 10.10 case, the step-change input, after working its way through all the dynamic elements that were present, emerged as output graph (1). Notice that this graph has the familiar S shape. The overall change was from a starting point of about 265°C

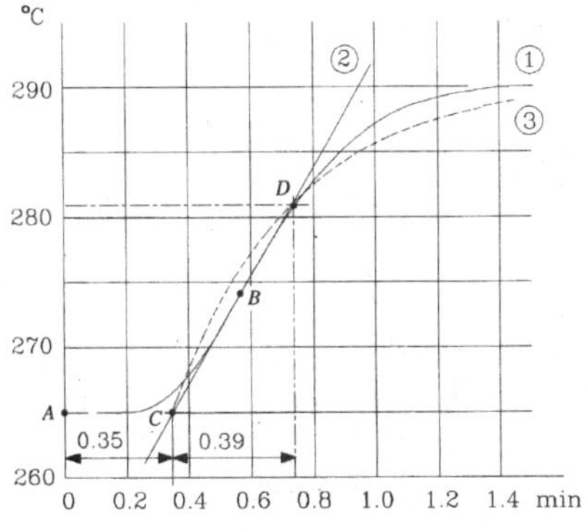

Figure 10.10 A dynamic test on a temperature-controlled process.

at $t = 0$ to an ultimate value of about 290°C, which is a net change of 25°C. The output reached its maximum rate of change at point *B*; consequently, the tangent (2) is drawn in at this point.

The tangent intersects the 265°C baseline at point *C*. The effective dead time is therefore the time interval represented by *AC*, or about 0.35 minutes. This dead-time value will include any real dead time in the system plus any virtual dead time contributed if there are two or more time constants in the system.

The time value for the single-time-constant element is established by locating the point on the output graph at which the output has reached 63.2 percent of its ultimate change. This implies that enough of the output curve must be tracked that the value at which it will finally level off can be predicted. The time-constant value that should be used is the time interval from $t = 0$ to the time point when the output has attained 63.2 percent of its ultimate change, minus the observed dead time.

The overall temperature change is 25°C, and 63.2 percent of this is 15.8°C. The point where the output attains 63.2 percent of its overall change will be at 265 + 15.8, or 280.8°C. This would be point *D* on the output curve (1). Thus the time-constant value that is appropriate for the single-time-constant approximation is the time interval between points *C* and *D*, or about 0.39 minutes.

The broken curve (3) is an output curve for a single time constant, starting at point *C*, and with a time-constant value of 0.39 minutes. This is drawn in to illustrate how closely the dead time plus a single-time-constant approximation matches the actual reaction of the process.

The concept that the behavior of a process control system would not be significantly altered if all the dynamic elements were removed and replaced by dead time and a single time constant, with the correct time values, was proposed by J. G. Ziegler and N. B. Nichols in 1941. Inasmuch as the effect of dead time plus a single time constant on control system performance is not difficult to predict, their method has been most helpful to the average person striving to understand why process control systems behave as they do.

10.5 Self-Regulation and Integrating Processes

Process control people need to be aware that apart from their peculiar static and dynamic characteristics, processes come in two types. Processes of the first type will have a determinable static characteristic. For each position of the control valve, in its stroke, the controlled process variable will settle out at some definite value. In technical terms, this says that each process input produces a particular process output. Many temperature-controlled processes behave in this way. Such processes are said to have *self-regulation.*

A different situation prevails in the case of a liquid storage vessel from which the liquid is drawn off at a constant rate. The level in the vessel will remain constant only if the inflow is exactly equal to the outflow. If the inflow is less than the outflow, the level will not stabilize at some new value. It will keep falling until the vessel is empty. Similarly, if the inflow is greater than the outflow, the level will keep on rising until the vessel overflows.

What is happening here is that the level in the vessel is actually *integrating,* over time, any difference between the inflow and the outflow. Another common example of an integrating process is a pressure vessel that has either a constant inflow or a constant outflow of gas. The pressure in the vessel at any moment will be the time integral of any difference between the inflow and outflow of the gas. The regenerator of a fluid catalytic cracker is a petroleum industry case in point.

Integrating processes do not have self-regulation. If one tried to determine the static characteristics for such a process, it could not be done, since any difference between the material or energy input to the process and the requirement of the process would drive the process variable off scale.

Whether or not a process has or lacks self-regulation is a characteristic that should be established at the outset, since this could be a contributing factor in a control problem. One problem of this type is limit cycling, which is described in Chap. 15.

10.6 A Process Reaction Curve Horror Story

The process reaction curve shown in Fig. 10.10 is typical of the reaction of a great many industrial processes to a step-change input. The main differences from process to process will be the magnitudes of the dead times and the time constants.

There are, however, a very few processes that are really mean and miserable, since they have an annoying habit of reacting to a step change in the wrong direction before they begin to react in the right direction. The graph of the process variable following a step change at the control valve, shown in Fig. 10.11, is typical. These processes have a *reverse-reaction curve,* and it is virtually impossible to achieve stable control using a conventional feedback loop. If the automatic controller makes a sensible movement of the control valve but the

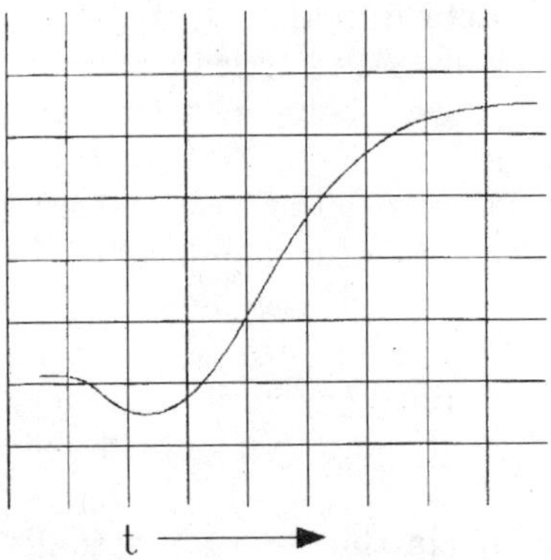

Figure 10.11 Output from a process with a reverse reaction.

process output goes the wrong way before it goes the right way, the controller becomes confused.

Three real-life examples will be described here. The first, and probably the best known example, is the case of a level-control system on a steam boiler, as sketched in Fig. 10.12. The level controller measures and controls the level in the boiler drum, with the level-control valve on the feedwater line to the boiler.

A step-change movement of the control valve in the opening direction should cause the level in the boiler drum to rise. However, the drum does not contain boiling water alone. What is actually in the drum is a froth of water and steam bubbles. When the control valve is opened and more hot water is admitted to the drum, its initial effect is to collapse some of the mass of bubbles in the drum, which causes the level of the mixture in the drum to fall. Eventually, the normal effect takes over, and the addition of more water causes the level in the drum to rise.

Feedback control alone will not work for controlling the level in a boiler drum. To maintain the correct inventory of water in the drum, it is necessary to material balance the boiler. For every kilogram of steam that comes out of the boiler, a kilogram of water must go in. Accordingly, the steam flow out of the boiler is measured, and the measurement signal sets up the flow of boiler feedwater to the boiler.

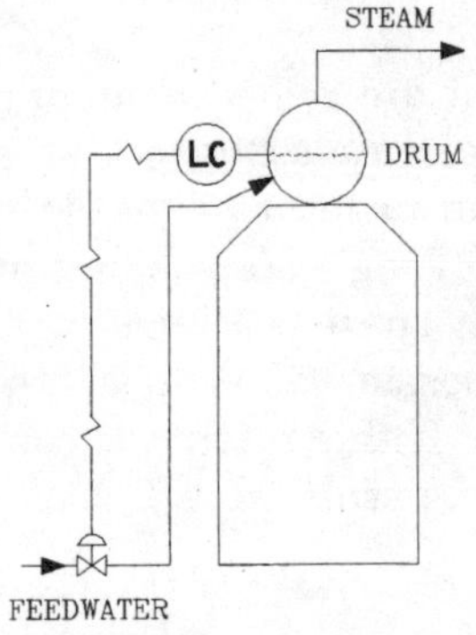

Figure 10.12 A boiler drum level-control system has a reverse reaction.

A level controller is used as well, but its function is only to trim the boiler feedwater inflow to compensate for boiler blowdown, possible errors in the measurement of the steam or feedwater flow, and other factors that are not taken into account in the material balance. This method of controlling the drum level in boilers has been around for a long time and has come to be called a *3-element boiler feedwater control system.* The three elements involved are the steam flow, feedwater flow, and drum level.

The second example concerns pressure control for a lights ends fractionator and is sketched in Fig. 10.13. The problem here is created by a combination of two factors: The overhead vapor from the fractionator is totally condensed (there are no noncondensables), and the condenser is mounted at grade. The pressure controller is on the fractionator, with the control valve on a line that bypasses vapor around the condenser and into the distillate drum. This control arrangement may be called a *hot gas bypass control system.*

Once again, a process reaction test run on this process will show the process reacting in the wrong direction before it reacts in the right direction. If a step change in the opening direction is sent to the control valve, the immediate reaction of the fractionator pressure is to drop, because the flow of vapor out of the fractionator is increased. What really determines the pressure in the fractionator, however, is the temperature of the liquid in the distillate drum. Since the liquid is a pure component, the pressure in the drum will be the vapor pressure of the liquid, which is determined by the temperature of the liquid.

When the control valve opens, vapor flows into the drum and condenses on the liquid. This releases the latent heat of the vapor and eventually raises the temperature of the liquid in the drum, which increases the vapor pressure in the drum. The increase in drum pressure then reflects back through the condenser to increase the pressure in the fractionator.

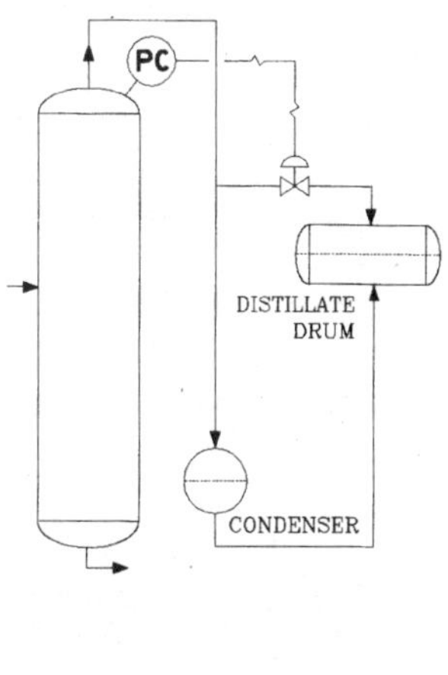

Figure 10.13 Reverse reaction in a distillation column pressure-control system.

While feedback control has virtually no hope of working under these circumstances, the correction for this problem is so simple it is almost ludicrous. Move the pressure controller to the distillate drum. Then both the short-term and long-term reactions will be in the same direction.

The final example is the case of a mysteriously unstable orifice flow controller. In order to get the required straight pipe for the meter run, and to improve accessibility for the control valve, the mechanical department decided to break with tradition and install the control valve upstream of the orifice meter run. Instrument maintenance personnel knew that this can sometimes cause problems, one of which is a noisy differential pressure signal coming off the pressure taps at the orifice flange. To counter this, they installed a dampener in the high-pressure impulse line between the upstream pressure tap and the HP side of the DP meter.

When the flow controller was switched to automatic it oscillated wildly. No matter how much the gain and reset rate of the controller was lowered, the oscillations continued. It simply would not run on automatic without cycling. During the course of analyzing the problem, the suggestion was put forward that the control valve was too large and should be replaced.

What really happened was that when the high-pressure impulse line was dampened, a reverse reaction process was inadvertently created. When the control valve opened the flow increased, as did the differential pressure at the orifice. At the same time, however, the system pressure at the orifice increased owing to the control valve upstream taking less pressure drop. As a result, the pressure at both the orifice taps increased, although the increase on the downstream side was less than on the upstream side, due to the increased flow.

With the dampener in the high-side impulse line, the pressure at the HP side of the DP meter increased more slowly than the pressure on the LP side. Consequently, when the flow increased, the differential pressure sensed by the DP meter decreased first, then increased, as the pressure on the HP side gradually built up to its new value. Result: a reverse reaction.

The moral is that if unequal damping is used in the impulse lines for an orifice flow controller or any head type flow controller for that matter, then the heavier damping should be on the impulse line opposite from the side where the control valve is located.

It is worth noting that the potential for a reverse reaction exists on any application in which control is based on the difference between two measurements. It would take only a difference in the response times of the two measurements, in the unfavorable direction, to cause the problem.

The reader now has the benefit of the examples of reverse process reactions known to me, but there are, no doubt, other reverse-reacting processes out there. This simply serves to point out the value of running a process reaction curve on any unfamiliar process to see just what kind of dynamic behavior one is up against. A control specialist who is unaware that a reverse reaction is present can spend a great deal of time, and experience a great deal of frustration, trying to attain stability in a control system which has a reverse-reacting process.

10.7 The Fundamental Laws of Process Control

All the basic sciences have their fundamental laws. In physics there are the laws of motion. Electricity has Ohm's law and Kirchoff's laws. In thermodynamics there are the first and second laws. Process control, while one of the newer sciences, nevertheless has its basic laws too. Yet how many university or technology school graduates who have presumably studied process control can confidently answer the question, What are the fundamental laws of process control?

It is interesting to reflect that while it is generally assumed that the starting point in studying most sciences is to become familiar with the fundamental laws, in process control many people seem to believe that these laws are so self-evident that they need not be concerned about them. This is a mistake, one that will sooner or later have its consequences.

Process control has three fundamental laws.

> *Law 1:* All control starts with measurement, and the quality of control can never be better than the quality of the measurement that goes into it.

Of course, no one is going to argue this point. However, if this law has universal acceptance, how do we explain these situations? In the process instrumentation field, textbooks on control outnumber textbooks on measurement by two or three to one. The same applies to technical articles on instrumentation in the journals devoted to this subject.

When any large exhibit of process instrumentation is held, why is it that the booths containing digital systems that display process data in grandiose fashion are crowded with interested visitors, while in the booths exhibiting measurement sensors the attendants have a hard time keeping awake? And in many large petroleum and petrochemical plants, the measurement sensors being used for level, pressure, flow, and temperature are the same ones that were in use 40 years ago. Organization of the instrument maintenance effort is likewise virtually unchanged.

In data processing there is the well-known and oft-quoted axiom, "garbage in, garbage out." There are, unfortunately, too many data-processing experts who would agree wholeheartedly with this saying yet who would be unlikely to recognize garbage process measurement input if they saw it.

At this point in time, our ability to process data has far surpassed our ability to get good data to process. There is a need to recognize this situation, as well as a need for volunteers who are willing to try to correct it.

> *Law 2:* When the material or energy needs of the process are exactly balanced by the material or energy provided to the process through the control valve, then the controlled variable will become stable.

This is what makes a process control system work. It is also a prime requirement for success. Liquid-level, pressure, and fluid-flow control systems function on a basis of material balance. Temperature-control systems function on energy balance. What this law is saying is that in any level-control system, the level will become constant when the inflow to the vessel is equal to the outflow. In a pressure system, the same applies, except that the fluid will be a gas or vapor. And in a temperature-control system, a constant temperature is an indication that the energy flowing to the process through the control valve is in balance with the energy that the process is consuming.

The priority job of the system designer, therefore, is to ensure that this balance can happen. When a control system fails to keep the controlled variable on control, a frequent cause is that the control valve cannot deliver the right amount of material or energy to balance the needs of the process. The symptom of this predicament will be that with the controller on automatic, the controller output drifts up to maximum or down to zero, indicating that the control valve has gone to one of its two extremes while the controlled variable drifts away from the desired value.

When a control system is on control, the necessary balance is in place. If the control system is disturbed, the balance is upset, and the controlled variable deviates from the setpoint. The automatic controller must then find a new output and a new valve position to rebalance the system and bring it back on control.

Note that there are two separate factors here. It must rebalance the system *and* bring it back on control. The controlled variable can be stabilized by simply restoring the balance between material or energy provided and that required by the process. However, it may not be stabilized at the desired value. To get back on control, the automatic controller will have to temporarily unbalance the system in the opposite direction to get the controlled variable back to the setpoint and then stabilize it there by restoring the material or energy balance.

> *Law 3:* The number of variables that can be controlled is always limited to the number of variables that can be manipulated.

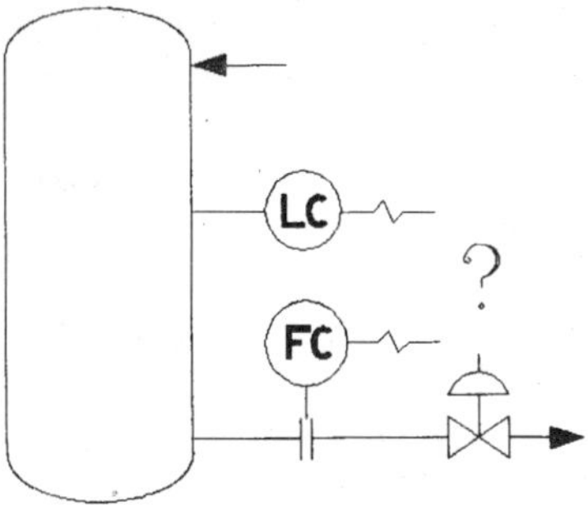

Figure 10.14 Which controller, level or flow, should be connected to the control valve?

Some process control theorists like to call the available manipulated variables *degrees of freedom.* Like the first two laws, this one seems obvious enough, yet designers who should know better will sometimes try to design around it.

Consider the feed surge drum in Fig. 10.14. There are actually two process variables that are of concern here: the level in the vessel and the outflow to the following process unit, which requires a constant feed rate. Both these variables should be on control. Unfortunately, there is only one control valve, which means only one manipulated variable. A decision has to be made, consequently, as to which variable has priority, the level or the flow rate.

Sooner or later, someone will try to circumvent the problem by specifying a cascade control system, with the level controller positioning the setpoint of the flow controller. The presence of the two controllers seems to give the impression that both level and flow are being controlled. In a sense they are, but the flow rate is not being independently controlled. The setpoint of the flow controller cannot be selected arbitrarily. It must be readjusted constantly so as to keep the level-control system on control. This scheme turns out to be, consequently, only an extravagant method for controlling the level in the feed surge drum.

These are the three fundamental laws of process control. Although they should be self-evident, they are often overlooked, discounted, or ignored by people who consider themselves to be knowledgeable in process control design. However, just like other basic laws of science, they are not very forgiving if they are broken or even bent. Skeptics are invited to try it and see what happens.

Chapter

11

Nature of Process Control Systems

Any study of the nature of process control systems should really be an examination of process control fundamentals. In too many cases, problems with control systems that are apparently not performing as they are expected to can be attributed to failure to understand, and accept, fundamental principles. The objective of this chapter, therefore, is to clarify just what we have a right to expect from automatic process control systems and, equally important, what we should not expect.

By far the most common type of control system in use in industry is the feedback control system. Feedback control has two virtues that tend to make it attractive. First, implementation of a feedback control system involves a minimal monetary investment in hardware. An automatic controller and a control valve, along with their hookup, are all that is required.

Second, the design of a feedback control system does not carry a heavy requirement for engineering investigation. If it is known that a change in rate of the material flowing through the control valve will result in a consistently reproduced change in the variable that is to be controlled, then it can be assumed, with reasonable assurance, that the control system will work in an acceptable manner. Since this fact is usually not difficult to confirm, the design of process control systems is often thought to be simple enough that it can be dealt with by persons with limited background.

In every feedback control system there are three principal components. These are

- The process
- The automatic controller
- The control valve

The features of each of these components should be studied separately, before the matter of how they combine in a control system is considered.

11.1 The Process

Process control specialists put a different interpretation on the word *process* than do process designers and operators. To the second group, a *process* is a collection of operating equipment that produces a product. However, when process control people talk about a *process,* they mean only that part of the operating equipment which is incorporated into the control system. This distinction should be recognized. In the process control sense, the process could be anything from an elaborate reactor or fractionating column to a plain piece of pipe. The process in a flow control system is, in fact, a section of pipe. A storage vessel for a liquid or a gas and a heat exchanger are other examples of control system processes.

Along with the process there will be a process stream that flows into or through the process. Associated with the process stream, there will be some variable that should, for the good of overall operations, be held at a constant value. The most frequently controlled process variables are the four basic operating conditions: liquid level in vessels, pressure, flow rate in conduits, and temperature. In a relatively few but nevertheless increasing number of cases, the objective will be to control the properties of product streams, or computed variables such as the circulation rate of catalyst, or even the cost per hour of operating a process unit.

11.2 The Automatic Controller

The automatic controller is an instrumentation component. It has two main sections: a measuring section and a control section. The function of the measuring section is to provide the controller with up-to-date information on the actual value of the process variable that is being controlled. The term *actual* is interpreted to mean "as accurately as the controlled variable is being measured." It is a basic maxim of control that the quality of control will never be any better than the quality of the measurement information provided to the automatic controller. Unfortunately, the importance of this fact is often overlooked in the rush to develop new and interesting data-processing systems or control algorithms.

The measurement side of the controller will display the measurement of the process variable as it receives it. The display may be in the form of an analog indication (moving pointer, moving scale, bar graph, or ink record) or a digital readout.

The control section of the controller will display other information and will have some manual functions. One manual function is the setpoint input, through which the desired value of the variable under control is established in the controller. The setpoint must be displayed in some form so that it can be set correctly. There also will be an indication of the controller output that goes to the control valve in a conventional feedback control system. The output signal is usually graduated in percentages.

Finally, the controller will have a means of switching from *automatic* control to *manual* control, in the event that the plant operator elects to take control of the situation himself or herself. If control is set on manual, there must be some way to manually adjust the output to the control valve.

The function of the automatic controller is to produce the right output signal to the control valve, the right value being that value which positions the control valve so that the controlled process variable is held as closely as possible to the setpoint value. This simply confirms that the automatic controller has to know, with reasonable accuracy, what the real value of the process variable is.

There is a group of simple controllers that control plant variables without displaying the value of the measurement, even though they still have to have a measurement section in order to control. This feature has led to their being called *blind* controllers. If the value of the controlled variable is required as well, then a separate measuring instrument must be installed.

Should there be any difference between the actual value of the controlled variable and the setpoint, then this difference constitutes a *control error.* Control error may be measured in percentage of the measurement range span or in actual units of the process variable. It also may be positive or negative depending on whether the value of the controlled variable is higher or lower than the setpoint. Using the concept of control error, therefore, the function of the automatic controller can be said to be to reduce to zero, as quickly as possible, any control error that may appear.

11.3 The Control Valve

The control valve is the system component by means of which the controller affects the value of the controlled variable. Variations in the flow of material or energy through the control valve to the process will have an influence on the process and cause the process variable under control to change, hopefully in a predictable and productive manner.

There are two principal differences between a control valve and a conventional manually operated shutoff valve. First, a control valve will need a special actuator so that it can be positioned from an air-pressure or electrical signal rather than by a manually operated hand wheel. Second, a control valve is designed to have good throttling characteristics. This simply means that as the valve is stroked from closed to open, the flow through the valve increases smoothly from zero to the maximum flow the control valve can pass.

Usually, the ability to vary the flow smoothly is achieved at the expense of being able to close tightly. If tight shutoff and smooth throttling are both required, it may be necessary to install two valves, a control valve for throttling, and a manual valve for shutoff.

11.4 The Development of the Control System

The process variable to be controlled will be linked in some way with the process stream. The measurement of the process variable will consequently be made on the process stream, either as the process stream leaves the process or within the process itself. For liquid-level and pressure control systems, the measurement is usually made inside of the process. For flow and temperature systems, the measurement is usually made on the process stream as it leaves the process. Both possibilities are shown in Fig. 11.1.

The next step in the development of the control system is to identify all the variables that are known to have the ability to affect the value of the process variable being controlled. This is to say, if any one of these other variables were to change in value, it would cause a change in the value of the controlled variable. In some systems there are several such process variables.

Consider, for example, the fired heater in Fig. 11.2. In this case, the controlled variable is the temperature of the oil as it leaves the heater. The list of variables that can affect the value of the outlet oil temperature includes fuel gas flow rate, gas pressure, gas composi-

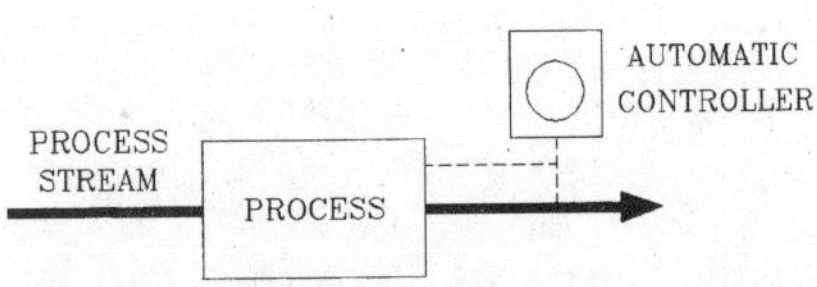

Figure 11.1 Feedback control: beginning of the development of the control system.

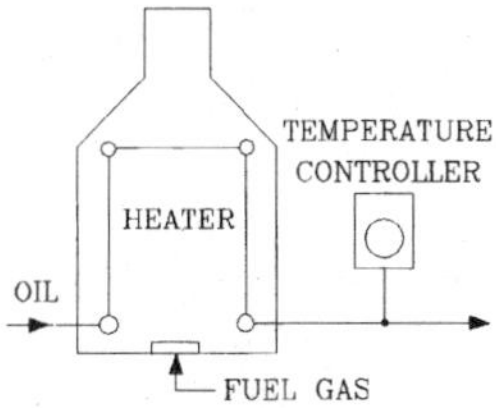

Figure 11.2 What factors can affect the temperature at the outlet of the fired heater?

tion, oil flow rate, oil inlet temperature, oil composition, heater draft, condition of the heater tubes, and ambient air temperature. A change in any of these variables would sooner or later cause a change in the oil outlet temperature.

These influential variables must be identified, because at this point a decision that is crucial to the success of the control system is made. This decision concerns the selection of one of the identified variables for manipulation by the automatic controller so as to provide it with a means of holding the controlled variable at the desired value, or setpoint. As a starting point, it is necessary to eliminate from the list of the influential variables those which cannot be manipulated by means of a mechanical device such as a control valve. In the case of the fired heater, this would preclude selecting the condition of the heater tubes, the weather conditions, and very likely the composition of the oil or the fuel gas. The choice of the manipulated variable is then made from those variables which remain.

It is a fact of life that one of the common reasons for failure of feedback control systems to perform as they were expected to is failure to select the best available manipulated variable. What makes one variable a better choice than the others as the manipulated variable? From the control point of view, the variable that should be selected to be manipulated by the automatic controller is the one that can make *the largest and the fastest changes in the variable that will be controlled.*

The reasoning behind these criteria is simply this: Once one of the previously identified variables has been selected for manipulation by the automatic controller, the others do not go away. They will still be present, and if any one of them changes, it will cause the value of the controlled variable to change. The remaining variables consequently become *disturbance variables* in the control system.

Herein lies the potential problem. If any disturbance variable can make larger or faster changes to the controlled variable than the automatic controller can by varying the designated manipulated variable, then the automatic controller will be hard put to hold the controlled variable at the setpoint with any consistency.

The point has been made that the variable selected to be the manipulated variable should be the one that can cause the largest and fastest changes in the controlled variable. If one variable makes the largest changes but another makes the fastest, which one should be chosen? In this case, the question of how much greater or how much faster the effect created really is should be taken into consideration. Other considerations being equal, however, the priority should be on the speed of the reaction rather than on the degree.

11.5 The Form of the Control System

As a general rule, disturbances in feedback control systems will come from four sources:

- Variables that affect the properties or the rate of flow of the process stream
- Environmental variables around the process
- Physical or chemical changes occurring inside the process
- Variables that affect the properties or the availability of the material or energy that flows through the control valve to the process

Changes from these sources, along with the possibility that the setpoint could be altered, constitute the ways in which a process control system can be disturbed.

In Fig. 11.3, the manipulated variable has been selected, and the control valve is shown in place. The black arrows represent the disturbance variables. Notice that if any one of the disturbance variables were to change in value, then the automatic controller would have to alter the manipulated variable in order to compensate for it.

The fundamental manner in which all feedback control systems function can now be set forth. The automatic controller is continuously striving to balance the effect that the manipulated variable has on the process against the combined effects of the various disturbance variables so that the controlled variable is held at the desired value.

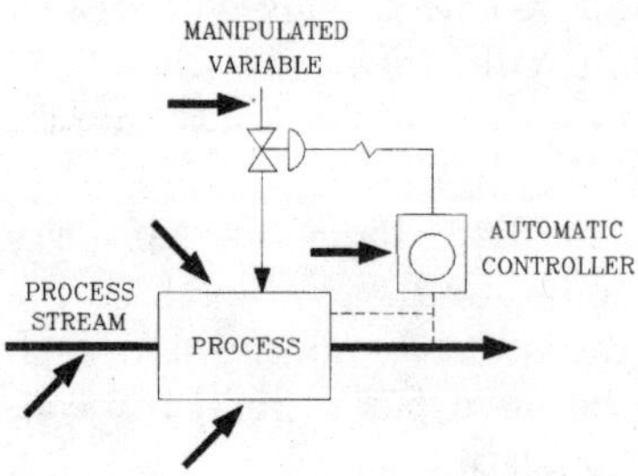

Figure 11.3 Disturbance variables in the feedback control system.

If the level of any of the disturbance variables changes, the balance will be upset, and the controlled variable will drift away from its setpoint value. It will then be up to the automatic controller to find a new value for the manipulated variable that will rebalance the system with the controlled variable at the setpoint.

What also comes to light is that feedback control has a fundamental weakness: Specifically, *corrective action always comes too late.* The automatic controller does not act until after the disturbance has occurred, the process has been affected, and the controlled variable has started to deviate from the setpoint. This state of affairs has two consequences that must be accepted because they are important feedback control facts of life.

- Whenever a feedback control system is disturbed, it has to go through an interval of upset and recovery before it can get back on control.
- The automatic controller cannot hold the controlled variable at the setpoint. It can only bring the controlled variable back to the setpoint after a disturbance has occurred.

If the controlled variable is at the setpoint, where it should be, the controller does nothing. Corrective action by the controller is taken only when the balance has been upset due to a change in one of the disturbance variables.

A control system that had no disturbances to cope with would have no need for an automatic control system. It would require only a meter and a hand valve, i.e., the devices required for manual control. The hand valve would have to be set once to bring the controlled variable to the right reading on the meter. After that, no further adjustment would be necessary, since there would be nothing to upset the balance in the system and cause the controlled variable to deviate from its setpoint.

Sometimes this point is not taken into consideration when a comparison is made between the performances of two separate control systems. The fact that one controls well and the other badly is not hard proof that one is superior in its design or has better hardware than the other by any means. It probably means that the better-performing control system has few, if any, disturbances to deal with, while the other has to cope with frequent or severe disturbances that require a great deal of corrective action.

Figure 11.4 shows the completed control system block diagram, but without the disturbance variables. Each component of the control system has an input and an output. The output of the process is the value of the controlled variable.

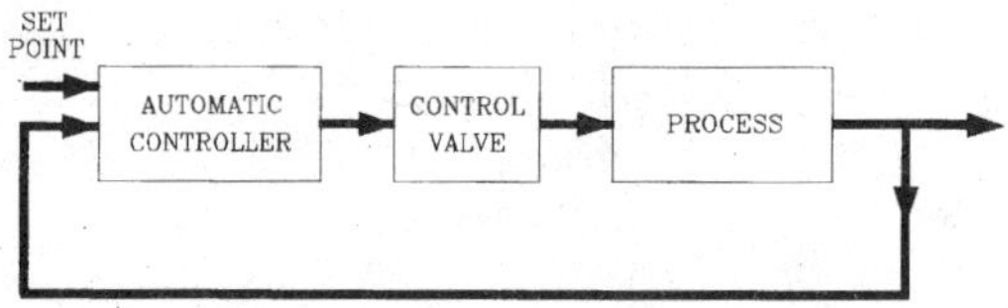

Figure 11.4 The completed feedback control system block diagram.

Furthermore, in this system, the output of one component becomes the input to the next. The output of the process is the input to the automatic controller through its measuring system. The controller output is the input to the control valve. The control valve output is the value of the manipulated variable, which is the input to the process. All three components are consequently tied together in a closed loop—hence the name *closed-loop control.*

The controller output sets the value of the manipulated variable, which, in turn, affects the process and the value of the controlled variable. The value of the controlled variable is fed back to the controller through the measuring system so that it can be compared with the setpoint to determine whether there is a control error and therefore a need for corrective action. This is where the name *feedback control* originates.

11.6 Stability

Whether feedback is used in process control or in some other type of system, it is characteristic that part of the output of the system is fed back into its input. This creates the conditions under which oscillations can occur. Impulses that travel around closed-loop control systems do in fact take the form of oscillations or cycles having a sine-wave form. The ultimate performance of the control system will be closely linked to the frequency of the oscillations, as well as to whether they continue on or eventually die out

Because the output of the control system is fed back into its input via the measurement path, the behavior of the control system, when it is disturbed, invariably follows an oscillatory pattern. Every component in the loop will oscillate at the same frequency, although probably not in the same phase. If the measurement of the controlled variable is recorded, the cycles will show up on the chart.

A train of oscillations, such as those shown in Fig. 11.8, is typical. The oscillations swing back and forth across a baseline that usually coincides with the setpoint of the controlled variable. For any one oscillation, the distance from the baseline to the top of the peak or to the bottom of the valley is the amplitude of the oscillation.

The term *stability* describes whether or not the oscillations in the system eventually die out. A *stable system* is one in which the oscillations die out in due time. In an *unstable system,* oscillations continue on indefinitely. There are, in addition, varying degrees of stability. If the oscillations die out quickly, the system is highly stable. If the oscillations continue for an extended period but eventually die out, the system is only slightly stable.

In process control system analysis, it is desirable to be able to put a number on the degree of stability of the control system. This is done by determining the ratio of the amplitude of any one oscillation to the amplitude of the oscillation that preceded it. For the oscillations shown in Fig. 11.8, the amplitude ratio is about 1/4. This number would be obtained by taking the ratio of the amplitude of oscillation 2 to the amplitude of oscillation 1, or the ratio of the amplitude of oscillation 3 to the amplitude of oscillation 2, and so on. In any chain of oscillations, the amplitude ratio will be constant all down the chain.

Figure 11.5 shows other degrees of stability, along with the amplitude ratio *AR*. Anything in between these examples is also possible. If the control system cycles continuously, with the oscillations becoming neither larger or smaller in amplitude, the amplitude ratio will be 1. An amplitude ratio of less than 1 is an indication that the oscillations will die out and the system is stable. It is also possible to have a situation in which the amplitude ratio is greater than 1, in which case the system would not only be unstable, but the oscillations would be getting larger.

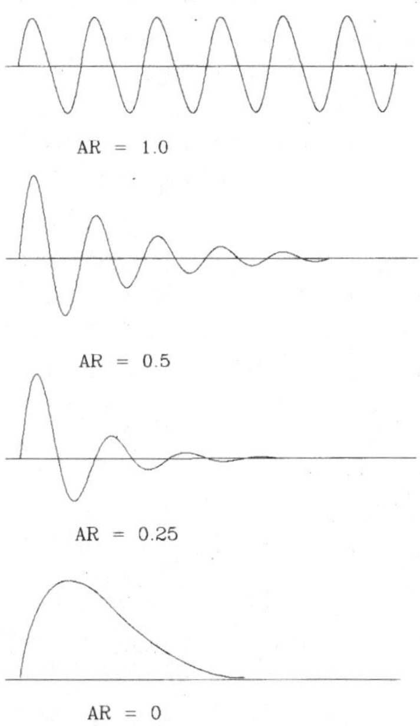

Figure 11.5 Varying degrees of stability in a feedback control system.

In the majority of process control systems, stability will be a requirement. The overall stability of a control system will be determined by the combined effect of the gains of all the individual components of the system. However, in one particular component, the automatic controller, the gain can be adjusted. This means that varying the gain of the controller will vary the gain of the whole control loop. Therefore, *the stability of the system can usually be set to any degree desired.*

The optimal value is generally considered to be attained when the amplitude ratio is 1/4, as shown in Fig. 11.8. If the amplitude ratio is greater than 1/4, the control system takes too many oscillations and too much time to get back on control. On the other hand, it is possible to be stable to the point of sluggishness, as in the case of the last example in Fig. 11.5.

Before leaving the subject of oscillations and stability, a word of warning. Sine-wave pattern cycling, caused by too much gain, is not the only kind of cycling that can occur in a feedback control system. Another type of cycling, called *limit cycling,* is also possible. Limit cycling is not caused by high gain; therefore, lowering the gain does not make it go away.

If a control loop is cycling, it is crucial that the type of cycling be identified correctly. Otherwise, the wrong action will be taken, it will prove ineffective, and frustration will set in. Limit cycling is described in detail in Chap. 15.

11.7 Control System Load

It is easy to state that when the control system is disturbed, the controller should find a new value of the manipulated variable that will rebalance the system with the controlled variable at the setpoint. In reality, the only thing that the controller can do is change its output and move the control valve, hopefully in a productive manner. Accordingly, a prerequisite for keeping the controlled variable on control must be that the required balance can be achieved at some controller output between 0 and 100 percent, since no other output levels are possible. This is equivalent to saying that the balance in the system must be achievable with the control valve at some intermediate position between wide open and closed. Control people would say "with the control valve in a throttling position."

This introduces the concept of *control system load.* The load on the control system is quantified by how far the control valve is open when the controlled variable is on control. *On control* implies a steady condition with the controlled variable at the setpoint. If the control valve is 60 percent open when the controlled variable is on control, the control system load is 60 percent.

The extent to which the control valve is open is measured in terms of how far the control valve is stroked toward its open position, not in terms of the percentage of its potential flow capacity flowing through the valve. This interpretation is used because percent valve stroke is equivalent to percent controller output. Thus, to determine the load on a control system, it is only necessary to check the controller output level, on automatic control. It is not necessary to inspect the control valve itself.

Another way to describe the requirement for achieving a balanced condition in the control system is to say that the load on the system must be between 0 and 100 percent. In fact, one of the more common reasons for control system failure is the inability of the controller to generate the correct output for balancing the load. Once the controller has moved the control valve to either its open or closed position, it cannot do anything more. There is no such thing as 110 percent valve opening.

An important fact, often overlooked, is that in most cases the automatic controller has a direct influence on the value of the controlled variable in one direction only, either upward or downward. In order to be able to change the value of the controlled variable in the other direction, the controller depends on the control system being loaded, i.e., there being a process requirement for material or energy.

In the fired heater control system in Fig. 11.6, the controller can drive the outlet temperature up by opening the control valve and firing more gas. However, it will not be able to bring the temperature down by closing the valve unless the heater (the process) is consuming heat energy.

Similarly, in the level-control application in Fig. 11.7, the controller can raise the level by opening the control valve and admitting more liquid makeup, but it could not bring the level down unless liquid is being drawn off, even if it closed the valve completely. Thus zero load on a control system, the condition that exists if the process has a zero requirement for material or energy, is also an unworkable situation.

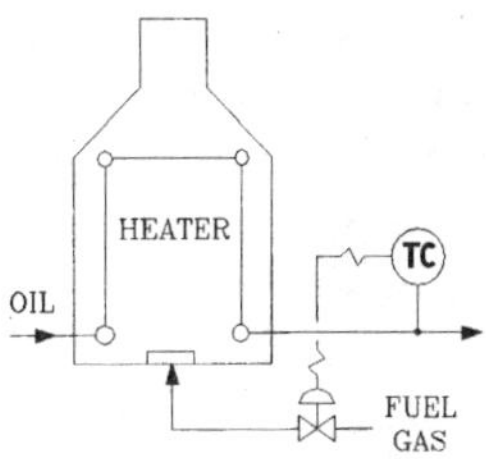

Figure 11.6 Closing the control valve will not lower the outlet temperature if the heater is not consuming heat energy.

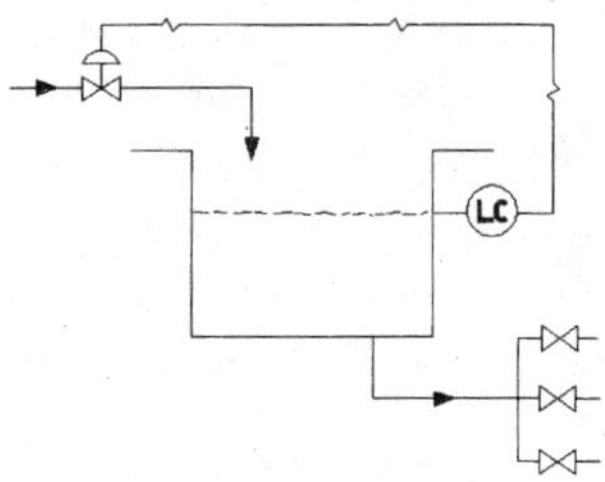

Figure 11.7 Closing the control valve will not cause the level to fall if there is no liquid flowing out of the holding tank.

11.8 Period of Oscillation

While the stability of the control system, determined by whether oscillations will die out in a reasonable time, is a key factor in the performance of the system, there is another even more important consideration, namely, the frequency with which the oscillations occur. This frequency is a unique property of the system, since there is only one frequency which the oscillations can have once the control system has been put together.

In process control systems, the frequency of oscillation is very low, sometimes of the order of 1/3000 Hz. It is so low, in fact, that process control people find it more convenient to talk instead in terms of the period of oscillation, which is the time required to make one oscillation. The *period of oscillation* is the inverse of the frequency. The upper diagram in Fig. 11.8 shows a train of oscillations in which each oscillation takes about 4 minutes. This is the period of oscillation for this control system.

Every feedback control system will have its own inherent period of oscillation, and the oscillations in the control system will always occur with the same period. The period of oscillation is consequently an important property of any feedback control system. Unlike system stability, the *period of oscillation cannot be altered to any significant extent* by changing the settings in the automatic controller.

The importance of the period of oscillation is twofold. First, if the amplitude ratio of the system is adjusted to be 1/4, then it will take two or three oscillations, on average, for the control system to get back on control after it has been disturbed. The time required to make one oscillation therefore has a direct bearing on the recovery time of the system.

Second, the automatic controller has to be tuned so that the corrective action it takes is not any greater or any faster than the process can absorb. Overcorrecting in feedback control systems invariably causes oscillations. Since the period of oscillation of the control system is determined by the combined reaction times of all the components in the system, the period of oscillation can be used to establish the correct tuning settings for the controller.

While ultimately the combined reaction times of all the components in the system establish the period of oscillation, the dynamic reaction of one of the components, the process, always seems to dominate. The manner in which the process responds, on a time basis, usually will contribute more to the eventual period of oscillation, and hence to the recovery time, than the combined effects of the automatic controller and the control valve.

It is for this reason that arguments about whether electronic control is better than pneumatic or digital is preferable to analog become somewhat academic. No matter what form of control is selected, the control system is still a feedback system, and its period of oscillation and recovery time will be determined mainly by the dynamic characteristics of the process, which is not an instrumentation component and which usually cannot be modified the slightest amount to improve automatic control.

11.9 Recovery

Figure 11.8 shows a plot of a good recovery for a feedback control system following a disturbance. The upper graph is the graph of the controlled process variable. However, when the control system is oscillating, every component in it oscillates. The lower graph shows the pattern that might be followed by the control valve. Notice that prior to the disturbance, the load on the system, as evidenced by the position of the control valve, was 65 percent. After the system had gone through its characteristic oscillatory pattern to get back on control, the load on the control system was 55 percent.

The change from a 65 percent load to a 55 percent load constitutes a 10 percent load change on the system. The size of the load change is sometimes used as an indication of the size of the disturbance for which the automatic controller had to correct.

The point has been made that feedback control systems go through a period of upset and recovery after they are disturbed. It follows, then, that the recovery should be the best attainable. Accordingly, a question to be answered is, "What constitutes a good recovery?" Two factors are used as criteria of the excellence, or lack of it, of a control system recovery. The first is the amplitude of the first oscillation, i.e., the height or depth of the first swing away from the setpoint. It is on this excursion that the controlled variable deviates farthest from the setpoint. The oscillations that follow will be smaller.

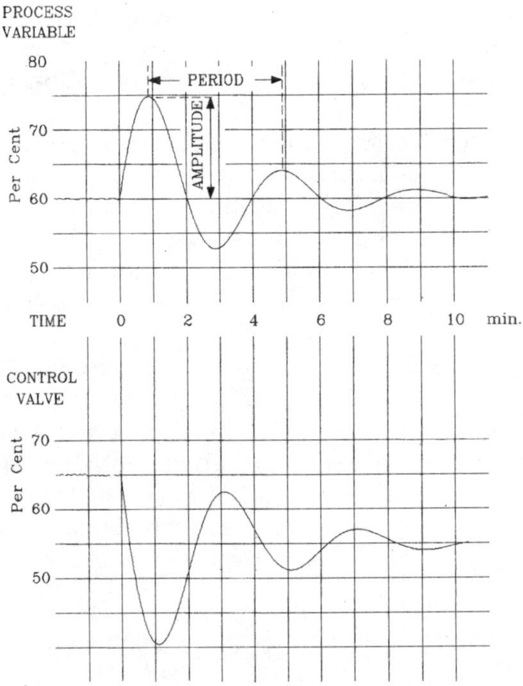

Figure 11.8 Oscillatory recovery in a feedback control system.

The second is the overall recovery time. This is the time required for the system to settle down once again with the controlled variable at the setpoint or within the tolerance allowed around the setpoint. Considering these two criteria together helps to explain why the optimal control system stability is an amplitude ratio of 1/4. If the controller gain were increased so that the amplitude ratio was greater than 1/4, then the amplitude of the first oscillation would be smaller (an improvement), but the number of oscillations the system would go through before they died out would increase, and so would the recovery time (a degradation).

If the controller gain were set lower so that the amplitude ratio was less than 1/4, the number of oscillations required to settle down would be less and the recovery time would be shorter, but the amplitude of the first cycle would be greater. Theory, combined with experiment on real control systems, has shown that the best compromise for optimizing both the amplitude of the initial oscillation and the recovery time is an amplitude ratio of 1/4.

A valid test of whether an automatic controller is really controlling effectively is not the fact that the controlled variable does not change when the controller is on automatic. This is more likely to be an indication that there are no significant disturbances. The real test is to see if the control system will recover properly after an upset. Since one form of disturbance is a setpoint change, a practical test would be to move the setpoint about 5 percent and verify that the controlled variable recovered to the new setpoint with an amplitude ratio of 1/4.

On one occasion, I had been discussing some of the basics of feedback control with a plant operator. After hearing me out, the operator commented that he had been an operator for some years, and in all that time he had never seen, on any recorder chart, the sine-wave recovery pattern, with the cycles getting smaller on an amplitude ratio of 1/4. This was just one more of the many situations that seem to be a characteristic of process instrumentation, wherein the theory in the textbook says that one thing should happen, but out in the plant, it does not work out that way.

The usual reaction is to conclude that the theory is wrong or misapplied. Yet, if time is taken to really study the problem, the right answer generally will emerge. In this case, the missing fact is that the recovery pattern in Fig. 11.8 is the response of a feedback control system to a disturbance that has a step pattern with time. In many plants, disturbances hardly ever occur in this way, so the Fig. 11.8 recovery curve does not show up. The plants in which this operator worked were probably in this category.

If the disturbances spread themselves out over a period of time, in ramp form, or if they wander around in random fashion with time, then they do not cause oscillations as they are shown in Fig. 11.8. The only time that any of the control systems are hit with a step-type disturbance may be when an operator inadvertently makes a quick change to the setpoint of one of the automatic controllers.

11.10 Predicting Control System Performance

The stability and recovery time of a feedback control system, which are its two major criteria of performance, usually can be predicted if the characteristics of the individual components are known. Two different sets of characteristics are required: the *static* (steady-state) *characteristics* and the *dynamic characteristics.*

The product of the gains of the three major components of the control system, at the operating point, along with the gains of their interconnecting signal lines, if any, will establish the gain of the control system as a whole. The control system gain is the factor that establishes whether the system will oscillate and how much. Thus *the static characteristics* of the components of the control system *will determine the stability* of the system.

If the static characteristics graph of any component is nonlinear, then that component will have different gains depending on the operating point. Therefore, it is quite possible for a control system to have the correct stability at one operating point and be either unstable or too stable at another.

The dynamic characteristics of a component are shown by a plot of both the output and the input of a component on the same time base. The dynamic characteristics describe how the component responds on a time basis to changes in its input. The characteristics of the individual components and their connecting links can be combined to establish the dynamic characteristics of the complete control system. *The dynamic characteristics* of the control system *will determine the period of oscillation and,* consequently, *the recovery time.*

The impression one gains from this is that predetermining the performance of the control system should be no great problem. There is, however, a complication. The static and dynamic characteristics of the instrumentation components, the automatic controller and the control valve, are easily obtained in laboratory tests. The data that are hardly ever available, however, are the characteristics of the most important component, the process.

It invariably happens that the first static or dynamic information that process control people obtain about the process comes at the startup time of the plant. At this late date, attempting to predict the control system performance is not very profitable.

11.11 Summary

Most process control systems in use in industry are feedback control systems. The weakness of feedback control is that the corrective action always comes too late, so the system must go through a period of upset and recovery following each disturbance.

Feedback control generally will produce the desired results if either of two situations prevail:

- The period of oscillation of the control system is short, and the system recovers quickly. In this case, the disturbances do not matter much.
- The disturbances are small or infrequent, or they spread themselves out over a period of time so that even if the period of oscillation is long, the control system will still hold the controlled variable reasonably close to the setpoint.

If, however, the period of oscillation of the system is long and the disturbances are severe, then some other form of control more elaborate than feedback control will be required.

Chapter

12

Controller Responses

Another title for this chapter might be, "How to Explain Proportional, Reset, and Derivative, Without Using Math." Query a plant instrument technician about why automatic reset is used in an automatic controller, and he or she will undoubtedly tell you that without it the controller would not be able to hold the setpoint. If he or she cares to elaborate, the technician may explain that with proportional response alone, an offset will occur when conditions in the control loop are disturbed. The term *offset* applies to a difference between the setpoint value and the actual value of the controlled variable which the automatic controller seems unable to correct.

The question that logically follows is, "Why do we get this offset in automatic control systems?" If your instrument technician has trouble explaining this to you, then the chances are that his or her instructor did not do a very good job of explaining it. In the end, the technician may simply settle for the assertion that automatic reset is necessary to get rid of offset and that anyone who knows anything about automatic control knows this.

For those who want a somewhat better explanation, the description of automatic controller responses, including proportional action, automatic reset, and some others, tends to get bound up in the math. This need not be so. Equations with integral and derivative signs are not inevitable when the purposes underlying the use of proportional action, automatic reset, and derivative response are explained.

The way in which an automatic controller changes its output signal, or "responds," is related to the control error in the loop. The *control error* is the difference, at any point in time, between the measured value of the controlled variable and the desired value, or setpoint. The control error may be positive or negative depending on whether the value of the controlled variable is greater than or less than the setpoint value. The function of the automatic controller is to reduce to zero, as quickly as possible, any control error that may arise.

If a control error is present, the automatic controller recognizes certain aspects of it and changes its output to the control valve accordingly. This poses a basic question. In order to be able to function as an automatic controller at all, what aspects of the control error, at a minimum, must the automatic controller be able to recognize? The answer is that to be a functioning automatic controller, it must be able to recognize at least two properties of a control error. These are

- Its existence
- Its direction

If the automatic controller could not detect the existence of the control error, then it would not take any corrective action when in fact action was required. If it could not detect the direction of the control error, the controller's response, moving the control valve, might be in the wrong direction, which would make the control error worse instead of tending to reduce it to zero.

12.1 On-Off Control

If the ability to recognize the existence and the direction of a control error is the minimum requirement for an automatic controller, then it follows that the simplest controller of all is one that responds only to these two aspects of the control error. This simplest of all controllers will react by moving the control valve as soon as a control error is detected, and it will move the control valve in the right direction. However, this is as far as it can go.

The extent of its response will not be related in any way to the size of the control error. Whether the control error is large or small makes no difference. The simplest controller always moves the control valve the same amount, and this means all the way to its wide-open or its tight-shut position, whichever is appropriate. Since the response is of this elementary nature, the simplest automatic controller has come to be called an *on-off controller.*

On line, an on-off controller will keep the process variable under control, but continuous cycling will occur (see Fig. 12.4 on page 197). If the size of the cycles were on the order of ±5 psi around 100 psig in a compressed air receiver, the cycles would probably be acceptable, as they would be on certain other simple process applications. However, in most process control applications there is a requirement for stability.

Industrial processes that operate continuously have a built-in requirement for either material or energy. The fluid that flows through the control valve provides this material or energy. To stabilize the controlled variable, a necessary condition is that the material or energy flowing through the control valve must exactly match the material or energy needs of the process. Since the on-off controller can produce only two discrete control valve positions, wide open or closed, the controlled variable will cycle back and forth across the setpoint as the controller alternately hits the process with zero or maximum material or energy. The required balance between supply and demand is never attained.

12.2 Proportional Control

When stability of the controlled variable is a requirement, then clearly the extent of the output change from the automatic controller must be related to the size of the control error. This is another way of saying that in addition to being able to recognize the existence and the direction of the control error, the automatic controller must be able to recognize its magnitude. With this capability, the controller will respond when there is a control error and in the right direction. In addition, it will make small changes in output, and in control valve position, when the control error is small and large changes when the control error is large. The change in controller output signal will actually be in proportion to the size of the control error—hence the name *proportional controller.*

A proportional controller can create any output signal level, and therefore any control valve position, between zero and maximum. Thus the proportional controller can position the control valve at that point where the material or energy passed on to the process through the control valve exactly balances the needs of the process. The controlled variable will then become steady. The problem of stability accordingly has been resolved.

On line, the proportional controller does the job expected of it, at least in the initial stages. The control error is reduced to zero, and the controlled variable becomes stable. Then, after a time, the process is disturbed, and the control loop is upset. The loop recovers in cycling fashion back to a condition of stability, but not with the controlled variable at the setpoint (see Fig. 12.4). The much-publicized offset appears, and we have an anomaly.

The function of the automatic controller is to reduce to zero, as quickly as possible, any control error that appears. However, here we have a case in which the controlled variable is under control and stable but not at the setpoint. There is an obvious control error, and the proportional control is not doing anything about it.

Apparently there is a limitation to the manner in which a proportional controller responds that we did not foresee. Looking deeper into the matter, this is what we find: When a proportional controller is placed on automatic control, four controller parameters have been set and are fixed. These are

- The setpoint
- The controller action (direct or reverse)
- The controller gain (or proportional band)
- The output signal at which the controller was aligned

The *alignment* is the signal level at which the controller output has been set when the control error is zero, in other words, when the measurement signal matches the setpoint. It is really a calibration adjustment in the controller. In the absence of any better information, the manufacturer of the controller usually aligns the controller at 50 percent output, i.e., 9 psig or 12 mA.

Once these control parameters have been set, there will be a fixed relationship between the controller output (or control valve position) and the measurement signal the controller receives. Figure 12.1 shows one possibility for this relationship. In this figure, the controller output decreases as the measurement signal increases, which is characteristic of a reverse-acting controller. For a direct-acting controller, the output increases with an increasing measurement, as would be shown by a rising line.

The words *fixed relationship* are the key to understanding the proportional controller's inherent limitation. Since the proportional controller must abide by this relationship, *it cannot change its output signal,* and therefore the position of the control valve, *unless it gets a new measurement signal.*

The level-control system shown in Figure 12.2 illustrates this point. A water tank supplies three users, and each draws about the same amount of water when the valve in the supply line is opened. The water level in the tank is controlled by a proportional level controller that positions a control valve in the water makeup line. Increasing

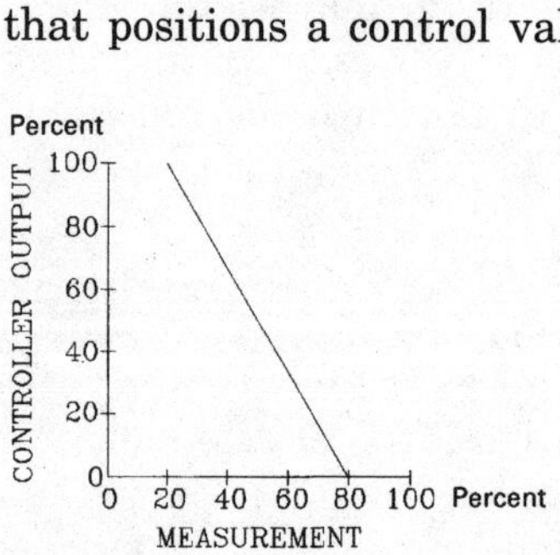

Figure 12.1 Output graph for a reverse-acting proportional controller.

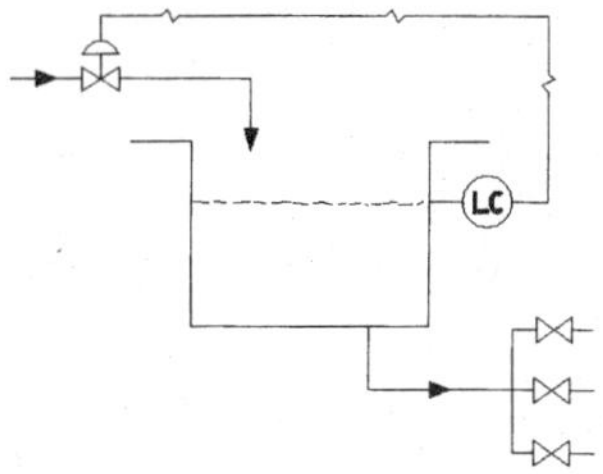

Figure 12.2 Proportional control for a water reservoir tank.

controller output opens the control valve, so the controller operates in reverse action.

Assume that there are two users taking water and that the level-control system is stable on automatic with zero control error. This means that the material requirement of the process, in this case water, is being exactly balanced by the quantity of makeup water flowing to the tank through the control valve.

Then, sometime later, a third valve is opened. Now there is more water leaving the tank than is coming in through the control valve. The balance of material supplied versus material required is upset, with the result that the level in the tank begins to fall. Obviously, if the level continues to drop, the tank will empty. What the level controller must do is get the control valve open wider so that there is as much water flowing into the tank as is flowing out. This will once again stabilize the water level in the tank.

Remember, however, that the proportional level controller cannot change its output signal, and consequently the control valve opening, unless its gets a new measurement signal. Therefore, the level controller has to let the level fall below the original setpoint, *and stay there,* in order to get the valve open farther and regain the balance between inflow and outflow. The difference between the new point of control and the original setpoint, which has not been changed, is the offset.

There are some additional points worth noting. If one of the three users closed his or her valve, the point of control would go back to the original setpoint, since the level controller would have to let the level rise in order to partly close the control valve and achieve a new balance of inflow against outflow. For the same reason, if two of the outflow valves were closed, then the level controller would have to allow the level to rise above the setpoint and stay there in order to get the control valve closed far enough to restore the material balance and stabilize the level.

On many process applications, it is not necessary to hold the setpoint exactly. There may be an acceptable band around the setpoint so that offsets of 10 or even 20 percent would be considered satisfactory control, provided that the controlled variable is stable. Many level-control loops are in this category. In these cases, proportional control by itself will do the job.

The size of the offset that occurs depends on the controller gain (or proportional band) setting. This is so because the gain setting establishes how much of a measurement signal change is required to produce a control valve movement of the required amount. The higher the gain (or narrower the proportional band) that can be set without causing the control loop to cycle, the smaller the offset will be after a process load change has occurred.

12.3 Automatic Reset

In order to correct the problem of offset, assuming that for a particular control loop it needs to be corrected, the automatic controller must be freed of the limitation that constrains the proportional controller, namely, that it cannot change its output signal unless it gets a new measurement signal. What is needed is the ability to alter its output even if the measurement signal does not change. Such a controller would be able to produce any output signal from zero to maximum, and therefore any required control valve position, with the controlled variable stable and at the setpoint value.

The proportional controller based its response on the existence, direction, and magnitude of the control error. If it is to eliminate offset, the automatic controller must be able to respond to one further property of the control error—its duration.

The implications of this are that the longer the control error is present, the more the controller output signal will change, and the more the control valve will be moved in the appropriate direction. If the control error persists despite the efforts of the automatic controller to correct it, then the control valve will eventually drift (be reset) right to its wide-open or its tight-closed position. Assuming that the control valve does remain within operating limits, the drift, or resetting action, will continue, in the correct direction, until the control error is eliminated.

The controller response that functions in this manner is called *automatic reset. Integral action* is another name sometimes used for the same response. With automatic reset added to a proportional controller and the controller gain and reset rate set at their optimal values, the recovery of the control loop after a disturbance should be of the form illustrated in Fig. 12.4, as well as Fig. 11.8. Note the pattern of control valve movement on the same time base, particularly the original and the final steady-state values, in Fig. 11.8.

In pneumatic and electronic analog controllers, automatic reset is never used alone. It is always used in conjunction with proportional response—hence the term *proportional plus reset response* (abbreviated P + R). A control response that approximates automatic reset alone is sometimes used in hydraulic controllers, where it is called *floating control.* Automatic reset by itself is possible in digital systems through the development of an appropriate control algorithm.

In general, automatic reset in the automatic controller is required if

- The controlled variable must be held at the setpoint under all foreseeable conditions of load (any offset is unacceptable).
- The automatic controller must be able to assume different outputs and control valve positions in order to hold the controlled variable at the same setpoint, owing to a process requirement for material or energy that varies with time and circumstances.

It would appear that having provided the automatic controller with the ability to respond to the duration of the control error, as well as to its existence, direction, and magnitude, the controller can then cope with all situations. This would be subject only to the condition that the quantity of material or energy required by the process should fall within the capability of the control valve to provide it.

However, it turns out that automatic reset, while solving one problem, may create another. The drifting effect of the controller output that is characteristic of automatic reset tends to destabilize the control loop. In order to obtain satisfactory stability during recovery, a P + R controller will usually have to be set at a lower gain setting than if the controller had proportional response alone. This results in somewhat poorer recoveries following equivalent disturbances. Specifically, the first transient away from the setpoint is larger, and the overall recovery time is longer.

12.4 Reset Windup

The behavior pattern of automatic reset is such that it causes a drift in the controller output, in the right direction, that is proportional to the length of time a control error is present. This can create a problem that is peculiar to automatic-reset response.

The output range for automatic process controllers is now standardized at 3 to 15 psig for pneumatic controllers and 4 to 20 mA for electronic controllers. These output ranges correspond 1:1 with the standard 3- to 15-psig air-pressure range that moves the control valve through its full stroke.

The design of many controllers, however, makes it possible for the controller to generate an output that is less than 3 psig or 4 mA or greater than 15 psig or 20 mA. Putting this another way, if the automatic reset in the controller were in the process of decreasing the controller output to compensate for a control error, and the error did not, as a result, decline to zero, then the controller output could eventually be driven all the way down to 0 psig or 0 mA by reset action. The same thing can happen at the high end. Automatic reset could drive the output above 15 psig or 20 mA, depending on the air pressure or the dc voltage with which the controller happened to be powered.

Proportional response will not do this, nor will derivative response, which is discussed next. This idiosyncrasy is peculiar to automatic reset. If, due to the action of automatic reset, the controller output is driven downward past 3 psig or 4 mA or upward past 15 psig or 20 mA, the controller is said to have "wound up." Since only automatic reset can cause this to happen, the problem has come to be called *reset windup.*

Automatic process controllers do not possess magical qualities, even though at times they may appear to do so. All that any controller does is react to a control error, in some logical manner, by altering its output to the control valve. For various reasons, doing this may have no effect on the process. One reason is that the controller has been disconnected from the control valve so that the control valve can be positioned by a second controller that is controlling a different process variable. This situation is discussed in Chap. 14 under "Selective Control." Another reason is that the fluid flow which the control valve is supposed to be manipulating has been shut off at some point in the process other than at the control valve. In any event, if the controller detects a control error, and reset action is ineffective in removing it, then sooner or later the controller output will end up at one extreme or the other, and reset windup will have occurred.

Figure 12.3 is an example of the kind of operating problem that can be created by reset windup. In this application, a pneumatic level controller LC is controlling the level in a storage vessel by throttling the flow of liquid makeup into the vessel. The level controller has proportional response and automatic reset. At the particular moment, the draw-off valves are all open, but the block valve ahead of the control

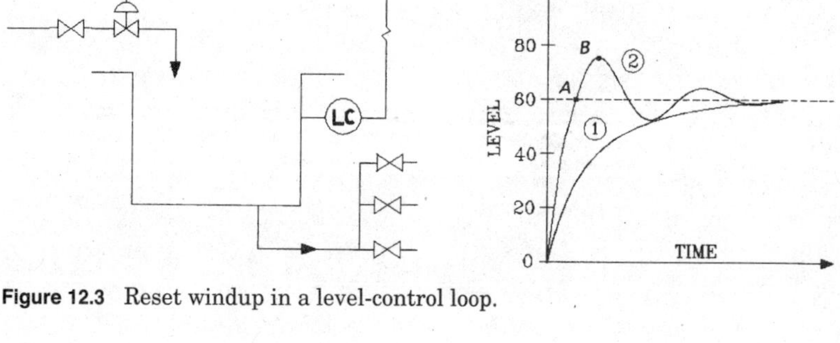

Figure 12.3 Reset windup in a level-control loop.

valve has been closed. Thus the liquid runs out of the vessel, and the vessel empties. The level controller detects that the level is below the setpoint and changes its output so as to open the control valve wider. This has no effect on the level in the vessel, and the automatic reset continues to increase the controller output past 15 psig, at which point the control valve is wide open. Eventually, the output reaches 20 psig, which is the air supply pressure for the controller.

Meanwhile, back at the control center where the level controller is located, the output indicator for the level controller reads 100 percent. While the controller output indicator generally will have one or other of the two standard ranges (3 to 15 psig or 4 to 20 mA), it is customary for it to be graduated 0 to 100 percent output. The point is that even though the output of the controller may go higher than 15 psig or 20 mA, the output indicator will not read higher than 100 percent. In the same way, the output indicator will not read less than zero, even though the controller output is less than 3 psig or 4 mA.

Returning to the example, the output gauge reads 100 percent, but the controller output is actually 20 psig due to reset windup. Then the block valve ahead of the control valve is opened, allowing liquid to flow into the vessel. The graph at the right shows two plots of the vessel level with time. The curve that is tagged 1 is what many plant operators would expect to see. Unfortunately, curve 2 is what they will see, and it does not exactly impress them.

Bearing in mind that until the level in the vessel actually reaches point *A* on curve 2, it is still below the setpoint, 60 percent, as the level controller sees the situation. Consequently, the automatic reset has no reason to act, and the output stays at 20 psig. As soon as the level passes point *A*, the control error becomes positive instead of negative, and the automatic reset reacts to this and begins to drift the output downward to close the control valve. However, the control valve does not begin to close until the controller output has dropped from 20 to 15 psig, and this takes a finite time. As a result, the level in the vessel continues to rise past point *A*.

As if this were not bad enough, the change in controller output from 20 psig down to 15 psig is not readable on the controller output gauge. Thus, to the operator, it appears that even though the level has passed 60 percent and is still rising, the level controller is not doing anything about it. This will naturally cause some operators to question whether the controller is working.

If the controller is left on automatic, it will gradually reduce its output and close the control valve to the point where the inflow is the same as the outflow. This is point *B* on the curve, where the level finally stops rising. From there on, the level will oscillate back to the setpoint in the characteristic feedback control fashion.

Summarizing this, reset windup is a phenomenon peculiar to automatic reset, not to the other responses. It can cause the controller output to wind up or down past the operating limit of the control valve. Aside from any anxious moments this can cause a plant operator, the real concern is that for some processes or operating equipment the oscillations that occur as the automatic controller tries to get back on control may exceed the danger limits for the equipment and for the people who are on the job. Oscillations of the type illustrated may be of little consequence if they represent liquid level, but if they are oscillations in temperature, pressure, flow rate, or machine speed, the results could be disastrous.

The obvious way to circumvent problems with reset windup is to remove the automatic reset from the controller. This may sound heretical to some, but there definitely are control applications in which the offset that accompanies proportional control alone is not really important. It also may be possible to eliminate reset windup through the use of special controller circuitry. (See Chap. 14).

12.5 Derivative Response

The addition of automatic reset is necessary to cancel out the offset that usually occurs when proportional response alone is used. Unfortunately, automatic reset in the controller has an unsettling effect that may result in poorer recoveries, relative to proportional control, from a disturbance of the same size and form. Can anything be done to retrieve the loss of stability and the ability to recover?

The answer is that usually, but not always, a significant improvement can be made by providing the automatic controller with the capability to react to one further property of the control error—its rate of change. This most complex controller of all will consequently respond to the existence, direction, magnitude, duration, and rate of change of the control error. The ability to respond to the rate of change of the control error is called *derivative response,* or *rate action.*

The derivative effect is additive to the controller output signal change and control valve movement created by the proportional and reset responses. The faster the control error is changing, the greater is the increment of output signal change contributed by derivative response. If the control error is not changing, the derivative effect is zero. Thus derivative response strives to discourage the control error from changing. As a consequence, it tends to stabilize the controlled variable. The potential for improvement in recovery varies with the control system dynamics but could be like that shown in Fig. 12.4.

Notice that the derivative response is not concerned with whether or not the controlled variable is at the setpoint. Any value of the con-

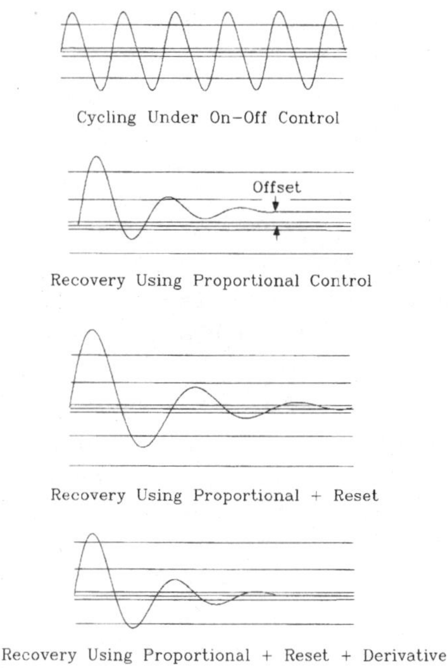

Figure 12.4 The effects of the various controller responses on recovery.

trolled variable is acceptable, so long as the value is not changing. This is the contribution that derivative response makes to overall control. It is the job of the automatic reset response to take whatever action is necessary to reduce the control error to zero and get the loop back on control.

There are a number of points relative to the use of derivative response that are worth consideration. How much good does derivative response really do? Considering that in some plants the derivative in most of the automatic controllers is turned off and left that way, one could conclude, not much.

In theory, the addition of derivative response can improve recovery in *any* process control loop, provided that the derivative action in the controller can be set at the correct value. Current design electronic controllers can have the derivative response set as low as 0.002 min. This would be the optimal value for a control loop that had a period of oscillation of 1.2 s. No industrial feedback control system oscillates this fast. While this suggests that derivative response would be useful on any process control loop, even flow-control loops, there are some practical realities.

- The tuning problem. Derivative action improves control only if it is tuned to the correct value. Adding the third response, derivative, to the controller complicates controller tuning and increases the chances that the settings will not be optimal.

- The degree of improvement possible through the use of derivative response depends on the kind of dynamic elements present in the loop. In process control loops, the dynamic elements are usually a combination of time constants and dead time. The more dead time that is present relative to the time constants, the less improvement one will get from using derivative response, and vice versa. This is why demonstrations of controller responses in which the controller is coupled to a simulator often show phenomenal improvement in recovery when derivative response is added to the controller. The dynamic elements in most simulators are almost entirely time constants.
- Unfortunately, some process measurements come overlaid with sudden transients which instrument people call "noise." Noisy measurements often turn up in liquid flow systems. If noise on the measurement signal can't be removed using some suitable filtering technique, then the use of derivative in the controller may be precluded, since derivative response is based on rate of change, and it will tend to amplify the noise effect and pass it on to the control valve.
- Many process control loops that could benefit from derivative response and which actually have a controller that has derivative response, are operated with the derivative turned off. The reason is that some plant operators and maintenance people do not like the sudden control valve movements that derivative response causes. They would prefer to accept the somewhat poorer recovery they will get with proportional plus reset without derivative for the sake of having the control valve move in a more conservative manner.

Inasmuch as the ultimate choice of which is worse, the problem or the cure, should be up to the "patient," plant personnel have every right to make this decision, much as it may seem pusillanimous to the proponents of automatic control.

The response of derivative is in proportion to the rate of change of the control error. One way to create a control error, and create it quickly, is to move the setpoint suddenly. The increment of controller output due to derivative action in this case would be large, possibly large enough to momentarily drive the control valve to the end of its travel. When a feedback control system is so badly upset that the control valve is hitting either or both of it limits, then it is unlikely that the control loop will be able to stabilize itself. It will probably have to be switched to manual control, temporarily, to induce it to settle down.

Some of the more modern automatic controllers have the control algorithm designed so that derivative action occurs on a change in measurement but not on a change in setpoint. This is a good idea.

Finally, although derivative response is almost always added on to proportional plus reset, an interesting possibility, often overlooked, is the use of proportional response and derivative response, without automatic reset. Automatic reset is used to eliminate offset. Derivative response can reduce offset, not necessarily to zero, but perhaps to an acceptable level, because of its ability to improve the stability of the control loop. This permits increasing the controller gain, which, in turn, reduces the size of offset. If the offset that remains is acceptable, then from the point of view of the process, the end result can be achieved without using automatic reset at all.

12.6 Derivative on the Measurement

One aspect of derivative which has received less attention than it deserves is the possible use of derivative on the measurement signal. This could cancel out the equivalent of one time constant in the measurement, and would be particularly beneficial on temperature measurements where considerable measurement lag is contributed by the thermowell which is invariably used. The end result: better measurement means better control.

This idea was proposed by J. G. Ziegler (1950). The Taylor Instrument Companies actually produced a gas thermometer temperature transmitter which had adjustable derivative on its output. I tried it as a second measurement for the temperature in a cat-cracker reactor, and with the derivative set at the optimum value, it showed interesting variations in temperature which did not show up on the record produced by the primary temperature sensor, which was a traditional thermocouple.

The idea died, unfortunately, for two reasons. First, the temperature transmitter was a pneumatic unit; it was never made in the electronic form, although the electronic form could have been simulated by using the pneumatic transmitter and a converter.

Second, some instrument maintenance people never really grasped the concept of derivative in a transmitter. They had been brought up to believe that derivative belongs in the controller and there only. Thus, they were reluctant to use it, and, as is generally the case when a new development is not supported by maintenance, it did not survive. It is possible, however, that the advent of computer-based control systems will revive the opportunity to develop the derivative on the measurement technique.

12.7 Units for Controller Responses

Most present-day automatic controllers have the proportional response calibrated in terms of gain, which is the ratio of the change in controller output to the change in input that produced the output change. Thus the units of gain will be milliamperes output per milliamperes input, or pounds per square inch output per pounds per square inch input, or percentage output change per percentage input change. Since the units for both terms of the ratio are the same, gain becomes a dimensionless quantity and, as such, will have no units.

Percent proportional band is an older method, still in use, to quantify proportional response. Proportional action in the controller will cause the control valve to be in a throttling position over a definite part of the controller's measuring range. For example, proportional action might cause the control valve to be open when the measurement input is at 30 percent of the measuring range, closed when the measurement is at 80 percent, and in a throttling position in between these points. The region between 30 and 80 percent through which the control valve is in a throttling position is called the *throttling band.*

The *proportional band* is that fraction of the measurement range occupied by the throttling band, expressed as a percentage. In the example, the control valve is in a throttling position over half the measurement range, so the proportional band is 50 percent. A proportional band greater than 100 percent is possible. It simply means that the measurement would have to change by an amount greater than the range span of the controller to cause the control valve to move through its full travel. The relationship between gain and proportional band is

$$\text{Percent proportional band} = \frac{100}{\text{gain}}$$

The automatic reset or integral response produces an output drift whose rate with time is proportional to the control error. It turns out, however, that in many of the earlier analog controllers, the automatic reset section did not sense the control error directly. It detected it by sensing the output change due to proportional response. The units for reset response were therefore milliamperes per minute per milliamperes of proportional response change. At some point, the term *milliampere* per *milliampere* came to be called a *repeat,* and the automatic reset rate was measured in repeats per minute.

Some controllers have automatic reset expressed as reset time rather than reset rate. Reset time is measured in minutes. The value will be 1/(reset rate).

Derivative response, or rate action, creates an additional output change proportional to the rate of change of the control error and has a leading effect on the controller output. The result is that with derivative response active, the controller output will attain any particular value sooner than it would have due to proportional action alone by an amount equal to the derivative time.

Once again, in the earlier analog controllers, the derivative section did not sense the rate of change of the control error directly. It sensed it from the rate of change of the output from the proportional section. Thus the units of derivative response became milliampere per milliampere per minute, which simplifies to minutes. Minutes are the only units used for derivative response in present-day process controllers.

12.8 The Future

Admittedly, the discussion of controller responses in this chapter has been devoted to proportional response, automatic reset, and derivative response only. All analog controllers will be limited to these three responses. Since pneumatic controllers come in analog only, this comment can be extended to say all pneumatic controllers will be limited to P, I, and D responses.

The advent of digital controllers has opened up new control possibilities that are just now being exploited. Newly designed controllers will have the ability to generate appropriate dynamics, both lagging and leading effects, so that their behavior can be matched dynamically to the behavior of the process they are controlling. They will have the ability to filter out extraneous effects that inhibit the accuracy of control. There will be controllers that can figure out their own "optimum" settings. And there will be controllers that try out their control action on a built-in model of the process and compare the results with the reaction of the process. If there is any difference, the controller will modify its output or even correct the model of the process with which it is working.

It is worth remembering, however, that the proponents of digital techniques have been predicting the demise of PID control for years. It is not going to happen. The new techniques will augment the methods that we now have, but they will not supersede them.

One also has to wonder if the radically new methods of control will turn out to be so complex that the average plant person will not be able to understand them well enough to use them properly. A well-tuned PID controller will perform better than a badly tuned state-of-the-art model-based digital controller. All this remains to be seen. Fascinating developments and experiences are in store.

Chapter

13

Optimum Controller Settings

Most writers dispose of the subject of "optimum" controller settings in two or three pages. That being the case, one would be entitled to ask, "Why is this chapter so long?" The methods for determining controller settings are indeed described in this chapter in a few pages. However, space also has been devoted to a question that hitherto has not been treated all that well. Why is it that so many people experiment with some formal method of finding the best controller settings, conclude that the method does not work in the real world, and give up and go back to finding settings by trial and error?

The following incident has probably occurred at some time or other in a number of plants. A keen younger person in the engineering department happens to read an article on how to determine optimum controller settings by making a dynamic test on the process. She decides that the method has promise and should be tried out. With the reluctant assistance of one of the instrument technicians, she runs the test on a key control loop, obtains the necessary data, and calculates the settings. They find that the gain, reset rate, and derivative settings which are currently in use are more conservative than the calculated optimum values.

The instrument technician is hesitant to change the settings, but at the insistence of the engineer, he does so. They spend and hour or so watching the record of the controlled variable, and sure enough, recovery from disturbances appears to be faster and control is tighter than it was before. The engineer then departs, believing that she has made a significant contribution to improved process operations.

Next morning the engineer goes down to the control center again for another look at the results of the "improved" controller settings. She finds to her dismay that the settings are once again back to what they were originally. Rather upset, she locates the instrument technician and asks him why all their apparent good work had been so quickly discarded. In answer to her question, the technician shows her the record of the controlled variable through the previous night. At one point there was a major upset, after which the controlled variable cycled wildly until someone finally cut back on the proportional (P), integral (I), and derivative (D) settings. This allowed the control loop to stabilize.

This experience completely disillusions the engineer. Back in her office, she throws the article on optimum settings in the waste basket and vows to waste no more time on the subject. On the other hand, the technician feels somewhat gratified by the turn of events, since it vindicates the trial-and-error method that is standard procedure for determining how controllers should be tuned in this plant.

What really went wrong in this case? Was the conclusion—that calculating the controller P, I, and D settings from data taken from a dynamic test does not work—the correct one? Is controller tuning by trial and error the only realistic way to do it?

13.1 Ziegler and Nichols

The cornerstone paper on determining optimum controller settings was authored by J. G. Ziegler and N. B. Nichols, then with the Taylor Instrument Companies, and was presented by Ziegler on December 3, 1941 at the annual conference of the American Society of Mechanical Engineers (ASME). At that time, the Instrument Society of America (ISA) had not yet been born, and technical papers that dealt with process instrumentation and control had to be presented at conferences sponsored by other engineering disciplines such as the ASME, the American Society of Chemical Engineers (AIChE), or the Institute of Electrical and Electronics Engineers (IEEE).

When a record has sold a million copies, a gold record is often made and presented to the artists. The same does not apply to technical papers, but if it did, then "Optimum Settings for Automatic Controllers," by Ziegler and Nichols, would have been become a gold paper. The Taylor Instrument Companies probably lost track of the number of times they reprinted it. It has been distributed all over the world.

Since 1941, many other persons have generated papers, talks, and articles on the same subject. I have personally contributed my share, with minimal success when it came to converting operating and maintenance people. Some authorities have proposed procedures that are more elaborate than those of Ziegler and Nichols. In any event, in all the verbiage which this subject has produced, an important detail has been missing. It is the fact that explains why the controller settings that were determined from process test data apparently did not work.

The theory underlying the calculation of controller settings using data obtained in a dynamic test of the process is based on an assumption, namely, that *as the control loop recovers from a disturbance, all the components in the loop will remain within their operating limitations.* That is to say, the oscillations in the loop will not push any of the components to saturation, or up against a limit.

The control valve is the component that is most likely to have this trouble, and this is what probably happened in the case described. While the engineer and the technician were watching the controlled variable, the disturbances were relatively minor. The loop recovered with the oscillations small enough that the control valve and the other components in the loop stayed within their limits. However, during the night, there was a severe upset, which created oscillations large enough that the control valve began to run into either its wide-open or tight-closed stop.

When this happens, the control theory that is the basis for calculating the controller settings breaks down. The loop continued to cycle until there was manual intervention. The appropriate corrective action would have been to put the control loop on manual control until the controlled variable had settled down to a reasonable degree. However, instead, it was assumed that the new controller settings were the cause of the problem, so they were discarded in favor of the old values. *Anyone who tunes an automatic controller on a real-life process control application, by whatever method, should be aware of this limitation.*

The original methods proposed by Ziegler and Nichols have been criticized for years by process control experts for various reasons, such as the process model they used as a basis for the development of the theory was not correct or their constants for the proportional, reset (integral), and derivative equations are not the right ones. Ziegler and Nichols never claimed that their method was rigorously correct, although this is the basis on which they are most often criticized. Their objective was simply to give the average plant person who was stuck with tuning PID controllers something that would give him or her a better chance of getting the control system to work properly than simply turning the knobs and watching what happened. If the control improved, then he or she should turn the same knob some more. If it got worse, then he or she should turn the knob back to where it was and turn some other knob.

The trial-and-error method is still popular, and it may eventually produce a usable result. The point that Ziegler and Nichols were trying to make was that the same or better results can be obtained a lot sooner by making a relatively simple test of the control loop.

The reaction-curve method proposed by Ziegler and Nichols did presume that the process had no self-regulation. This was justified by the fact that it simplified the test procedure: Only two bits of test data had to be determined, rather than three, which would be the case if the process had self-regulation. Furthermore, if the process did in fact have self-regulation, then the settings found by Ziegler and Nichol's method would produce the optimum reset and derivative, which are the hard ones to determine, and a gain or proportional band setting that was conservative but still satisfactory.

As far as the constants for their equations were concerned, we should remember that at the time Ziegler and Nichols produced their famous paper, the only automatic controllers around were pneumatic. The Taylor Fulscope controller that Ziegler and Nichols were obliged to promote had the reset and derivative needle valves in parallel. This meant that it would be next to disastrous if someone inadvertently set the two needle valves at the identical restriction. Then air would flow in and out of the reset and derivative bellows at the same rate, the pressures in these two bellows would always be the same, the reset and derivative actions would cancel each other out, and the controller would go into on-off control.

For the Taylor controller, the actual settings that would produce the same restriction in both needle valves were the combination that made the product of the reset rate and the derivative time equal to 1.0. Consequently, Ziegler and Nichols had to come up with relationships for the derivative and reset settings that avoided this possibility. Using the formulas in their original paper, the product is actually 0.25.

In the contemporary Foxboro M40 controller, the reset and derivative needle valves were in series. Thus, when tuning a Foxboro controller, one did not have to worry about a combination of reset and derivative settings for which the product was 1.0. As a matter of fact, in the Foxboro model, this was the best combination. The controller was usually shipped with the reset and derivative settings locked together so that their product was 1.0.

The Taylor and Foxboro controllers of the 1940s are still around, but there are many other models now, including those which are electronic, both analog and digital. If the circuitry or software for generating the controller PID algorithm is changed, the form of the optimum settings equations does not change, but the constant in each of the equations may have to be revised.

The optimum reset and derivative settings still depend on the period of oscillation, or the apparent dead time. The proportional or gain setting depends on the ultimate gain, or on the dead time and the reaction rate in combination. If the terms *apparent dead time, ultimate gain,* and *reaction rate* are not familiar, they will be explained farther on in this chapter.

13.2 Trial-and-Error Tuning

Considering all the publicity that optimum settings methods have received, why is it that the majority of instrument technicians still set controllers by trial and error? There are at least three reasons.

Reason 1: The method was tried, but it resulted in the classic experience in which the control loop would not stabilize after a severe upset that caused the control valve to go to its open or closed limit.

As a result, it was concluded that the method was no good, and it was never used again. (This situation has now been clarified.)

Reason 2: The method is too cumbersome.

Determining the period of oscillation requires a stop watch, which most technicians do not carry. It would take a trip back to the shop to get one. Running a process reaction curve generally requires two people, and instrument technicians generally work alone.

Reason 3: Trial and error works; we have proved it many times.

This can be deceptive, particularly in oil refineries, where almost half the control loops are flow control, and the controller has proportional action and reset but no derivative. A P + R controller on a flow-control loop will work at just about any gain and reset rate as long as neither setting is too high. If the settings are too conservative, the control loop may take 12 seconds to recover from a disturbance instead of 6, but this will not concern the average process operator.

The problem arises when the conclusion, that trial and error under these circumstances works OK, is extended to *all* control loops, including those which have derivative response. There are quite a number of stories around about plants that have been made to run noticeably better by doing nothing more than tuning all the controllers properly.

Reason 4: The calibration of the controller settings is so far off that it is not worth one's time to run the tests that yield the data for calculating the optimum values.

For many analog controllers, this complaint is valid. In the era of pneumatic controllers, the numbers on the dials for proportional, reset, and derivative were unlikely to be better than within 25 percent of the true values. When one considers that the various control actions were produced by means of variable airflow restrictors that were not individually calibrated, it is not difficult to believe that this was the case.

When electronic analog controllers became available, however, one of the benefits many users expected to get was accurately calibrated controller settings, inasmuch as the necessary dynamic delays were now created by resistors and capacitors that could be selected accurately, rather than variable air restrictors. Unfortunately, it was not to be. As things turned out, the accuracy of the controller settings in the electronic analog controllers was no better than it was for the earlier pneumatic models.

This being the case, to do a complete job, the user really should check, and adjust if necessary, the calibration of the P, I, and D settings on the controller to ensure that the proportional, reset, and derivative actions are actually what he or she believes he or she is dialing in. Instructions for checking the calibration of the controller settings have been included at the end of this chapter.

It is also worth noting that in recognizing the indifferent accuracy of P, I, and D settings in analog controllers, we have inadvertently introduced an argument for digital control, in which the settings are entered as software instructions.

13.3 The Methods

The methods that will be described here are essentially those proposed by Ziegler and Nichols in 1941. The only variations are the constants in some of the equations. At the outset, we have to accept the fact that *it is necessary to have two methods.* Anyone who believes, or proposes, that one method can handle all situations is mistaken.

We must never lose sight of the fact that the tests from which the necessary data are obtained are done in an operating plant. The plant was not put there so that process control experts could run interesting experiments. It is there to make a product that hopefully will result in a profit. Accordingly, whatever test is used, it must create a minimum of disturbance to the plant. The idea that one should cause a plant to cycle for a few hours in order to find its period of oscillation is simply not realistic.

The two methods described involve either finding the period of oscillation of the control loop or making a process reaction curve. The first choice is usually to find the period of oscillation, provided that this can be done quickly and without serious process upset. For nearly all flow-control loops, pressure loops on liquid systems, or gas-pressure and liquid-level systems that do not have a great deal of storage capacity, the period of oscillation will likely be a matter of seconds. If the required data from a period-of-oscillation test can be obtained in a few minutes, the appropriate procedure is to find the period of oscillation.

On the other hand, if the period of oscillation of the control system is on the order of 30 to 60 minutes, or even longer, which it can be on some temperature-control loops and can certainly be on plant stream analyzer loops, then finding the period of oscillation has to be abandoned in favor of running a process reaction curve.

If tuning automatic controllers really is difficult, what makes it so? A television set has a number of tuning controls, each of which must be set correctly for the best picture. The difference, however, is that an incorrect setting of any of the TV set controls produces its own peculiar problem pattern on the screen so that it is reasonably easy to establish which control should be adjusted in order to correct it.

An automatic controller can have up to three different settings. If any one of them is set at too high a value, the effect on the control loop is the same—it goes into oscillations. Instability always occurs in a feedback control system if corrective action is too great or if it is introduced faster than the process can absorb it. Thus, when a control loop cycles, it is next to impossible to tell from looking at the pattern which of the P, I, or D settings is excessive.

This does tell us, however, that if we knew that two of the three settings were correct, then the third one must be the one that is causing instability or poor recovery. A practical approach, therefore, is to use one of the formal methods to tie down the optimum values for the reset (integral) and the derivative and then to find the gain setting that produces the desired stability by trying different values and watching the recovery pattern.

Finally, when the time and effort have been spent to run a process test, to analyze the data obtained, and to determine the optimum controller settings, then the remaining step is to record the results someplace where they can be retrieved at a future time. This will circumvent the potential problem of having to do the whole test over again because the controller settings were mysteriously altered.

13.3.1 Optimum settings from the period of oscillation

This test is done with the automatic controller on automatic. Note that this is different from the test that requires making a process reaction curve. The basic procedure involves inducing the loop to oscillate by turning up the controller gain, and then timing the period of the oscillations. However, if the control loop is already in operation on automatic (it usually is) and is observed to be cycling, *do not get trapped into assuming that the oscillations must be caused by high gain in that loop.* It is only prudent to make the following preliminary check.

If the controller has automatic reset, the loop could be limit cycling (see Chap. 15) or the oscillations may be a disturbance imposed on the loop by some other control loop in the plant that is oscillating. First, if automatic reset is active, turn it off. If this quiets the oscillations, then limit cycling not high gain is the likely culprit. Next, put the loop on manual. If the cycling ceases, it can be assumed that the cycling is being caused within the loop under consideration. If cycling is still observed, then the oscillations are coming from somewhere else in the plant. Obviously, this period or any period caused by a limit cycle will not be correct for the loop whose controller is to be tuned.

After satisfying oneself that there are no gremlins present, one should use the following procedure:

1. Set the reset and derivative action at zero, or minimum, so that only the proportional action is working.
2. Disturb the control loop to cause it to oscillate. One convenient way to do this is to move the setpoint about 5 percent, and then when the controlled variable begins to follow the setpoint, move the setpoint back to its original value.

The objective is to see the control loop oscillate continuously, with an amplitude ratio of 1.0. If the oscillations die out or increase in amplitude, make the appropriate adjustment to the controller gain setting. Those who are new at the game tend to make too small a change in the gain setting when a change is required. On most process control loops, the oscillations will not change much in amplitude unless the gain is doubled or halved.

3. When the desired state is achieved (continuous oscillations, amplitude ratio close to 1.0), determine the period of oscillation (P seconds) by timing it. The accuracy can be improved by timing several oscillations and dividing the measured interval by the number of oscillations timed. Record the value of the period.

An important requirement of this test is that while the control loop is oscillating, none of the components of the loop, especially the control valve, is running up against a limit. If the control valve is hitting its wide-open or closed stops, then the oscillations are too large. This may result in an incorrect value for either the period of oscillation or the ultimate gain, which are the required measurements.

4. Record the gain setting that produced continuous cycling. This is the ultimate controller gain G_u for the loop.

For a control loop that has proportional control only, the optimum controller gain setting will be

$$\text{Optimum gain} = 0.5 \times G_u$$

However, this relation is hardly ever used, since a proportional-only control loop will have a controller that has only one knob to set. Tuning this control loop is merely a matter of adjusting the gain knob until the right degree of stability is obtained.

For a controller that has proportional and reset responses, the optimum values will be

$$\text{Optimum reset setting} = \frac{72}{P \text{ repeats per minute}}$$

where P is the period, measured in seconds and

$$\text{Optimum gain} = 0.45 \times G_u$$

These values will be close to the correct proportional and reset settings for just about all control loops and all types of controllers: pneumatic, electronic, analog, or digital. If the stability is not quite that desired, leave the reset at the calculated value and trim the controller gain. Do not be surprised if the reset rate turns out to be 20 repeats per minute for flow-control loops. These loops practically run on automatic reset, and 20 is actually the optimum value.

If the control loop has proportional, reset, and derivative action, then the optimum values can be obtained from the equations in the table in the Appendix. Some words of caution, however. If the automatic controller really needs derivative response to control effectively, then the implication is that the period of oscillation of the control system is relatively long, so long in fact that the period-of-oscillation test should not be used to determine the optimum controller settings. The period of oscillation of some temperature-control loops can be 15 minutes or even longer. The idea of cycling a control loop with a period of oscillation of this duration on an operating process unit to determine the period and the ultimate gain is completely unrealistic.

Note also that when P, I, and D are all used, the constants in the equations for the different settings will change depending on the way the control algorithm is generated, which depends, in turn, on the design of the controller circuitry. This means that in addition to running a control system test, something may have to be known about the design of the automatic controller.

13.3.2 Making a process reaction curve

The process reaction curve was given its name in Ziegler and Nichols' original 1941 paper. Hence the impression that may be given is that process reaction curves are only useful when determining controller settings. This is not so by any means. A process reaction curve is a really dynamic test in which the input test signal is a step change, as explained in Chap. 10. A dynamic test using this particular input pattern is a very useful one, in that it is easy to generate, and it yields all the dynamic data that one could ever require to make a practical analysis of a process control system.

A process reaction curve is made with the automatic controller on manual. Note that this is different from the period-of-oscillation test. The procedure is

1. Set the controller on manual, and allow the control system and the controlled variable to stabilize. Record the manual output to the control valve.
2. Step the manual output to the control valve quickly to a new value. The size of the step change must be great enough that the controlled variable, when plotted either manually or automatically, exhibits the familiar S-shaped curve, as shown in Fig. 13.1. If no better information is available, then a 5 percent change in output to the valve is a reasonable number to start with. Record the size of the step change Δm in percentage output signal to the control valve.

Once again, one has to keep in mind that this test is being done on an operating plant. Consequently, it can only be done with the concurrence of the plant operator. One usually finds that if the operator can be convinced that running the test will lead to better control, he or she usually will agree to the test. On most process control systems, the equivalent data will be obtained whether the step change drives the controlled variable up or down. Ask the operator which he or she would prefer.

3. The test is begun when the step change in output to the control valve is made. Starting from this moment, record corresponding readings of time and the controlled variable. Continue taking readings until it is obvious that the rate of change of the process vari-

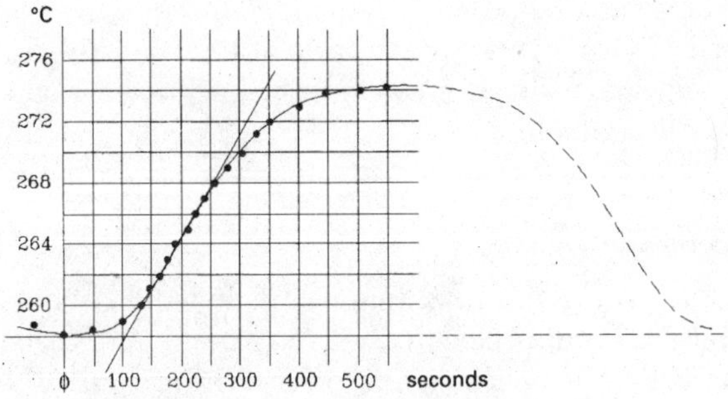

Figure 13.1 Graph from a process reaction curve test.

able is slowing down. The readings should be spaced closely enough that they can be plotted later to produce a reasonably representative process reaction curve.

4. When step 3 is completed, set the output signal to the control valve, still on manual control, back to the original value recorded in step 1. If this procedure is followed, the only disturbance which the process will see is a hump, upward or downward, in the controlled variable, as shown in Fig. 13.1.

Before describing what should be done with the data obtained, some points on technique for taking the data are worth mentioning. The documentation is made easier if the data sheet is set up beforehand. This is where another problem can be inadvertently created. What would seem to be a logical approach would be to concentrate on the independent variable, i.e., the time values. When the data sheet is laid out, the times, usually at about 15-s intervals, are written in. Then, when the test begins, it is up to the team member who is tracking the process variable to call out the values as his or her partner calls out the times.

We now have the potential for a real confrontation. If the process variable is displayed on an analog indicator or recorder and the value is in between two of the divisions, it may be difficult to decide quickly enough what the value of the process variable really is. In fact, it may happen that before the team member who has this job has figured out the value at one time point, his or her partner is calling for the next value. At this stage, there may be some pointed criticism about the capabilities of the harassed team member, with the result that he or she quits and goes back to the shop.

This problem should not occur if it is recognized that the time device, stop watch or otherwise, is the easier instrument to read. Some timers even have a digital readout. Consequently, a better procedure is to set down even-division and middivision values of the controlled variable and, as the test proceeds, record the values of the times at which the controlled variable attains these values.

The ultimate arrangement, of course, would be to have a high-speed recorder to record the values of the controlled variable on a time base over the course of the reaction curve run. Unfortunately, most instrument departments will not have this equipment. The alternative method usually requires a two-person team. One person watches and reports the designated values of the process variable, while the other reads and records the corresponding time values. One person with a cassette recorder also might be able to do the job adequately.

Since the measuring system of an automatic controller contributes its share to the overall dynamics of the control loop, the readings of the process variable that are taken must come from the measuring system of the automatic controller, not from some other meter that may be measuring the same process variable and which may be easier to read.

13.3.3 Processing the reaction curve data

When all the data are in, it should be processed in the following manner:

1. On graph paper, plot the data points for the process variable over the time base that was used. Draw the best possible smooth curve through the points. It should exhibit the S-shaped appearance of the first half of the graph in Fig. 13.1.
2. On the curve, locate the point where the slope is steepest, i.e., the point where the rate of change of the process variable was greatest. At this point draw in a tangent to the curve, as in Fig. 13.1.
3. Using any two convenient points on the tangent, determine the rate of change of the process variable in the applicable units of measurement (°C, psig, kPa, etc.) per minute. Divide this value by the range *span* of the controller's measuring system, and multiply by 100. This will give the process reaction rate R, in percent per minute.
4. Determine the time interval between the moment when the test was begun and the time point where the tangent intersects the horizontal line representing the initial value of the process variable. This is the apparent dead time L, in minutes. Optimum settings for a PID controller will then be

$$\text{Derivative} = 0.4 \times L \text{ minutes}$$

$$\text{Automatic reset rate} = \frac{0.6}{L \text{ repeats per minute}}$$

$$\text{Gain} = \frac{1.6\Delta m}{RL}$$

In these equations, Δm is the size of the step change made at the control valve (percent), L is the apparent dead time (minutes), and R is the process reaction rate (percent per minute), taken from the slope of the tangent.

Depending on the circuitry used in the controller to generate the PID algorithm, the gain determined from this relationship may not result in an amplitude ratio of exactly 1/4. Some trimming of the gain may be necessary. However, the values calculated for the derivative and the reset should be about right.

In the example (Fig. 13.1), the following data were obtained:

The test points, as indicated by the black dots.

A step change of 5 percent to the control valve created the process reaction curve (= Δm).

The measurement range of the controller was 200 to 400°C. The range span was consequently 200°C.

After plotting the test points and drawing in the best smooth curve, the tangent to the steepest point of the curve was found to have a slope of 4.08°C/min. Dividing this by the range span, 200°C, and multiplying by 100 gives a reaction rate of 2.04 percent per minute (= R).

The initial value of the process temperature was 258°C. The tangent intersects the horizontal line that represents this temperature at $t = 105$ seconds, or 1.75 minutes. This is the value of the apparent dead time L.

The optimum controller settings for this application will be

$$\text{Derivative time} = 0.4 \times 1.75 = 0.70 \text{ minute}$$

$$\text{Reset rate} = \frac{0.6}{1.75} = 0.34 \text{ repeat per minute}$$

$$\text{Gain} = \frac{1.6\times5}{2.04 \times 1.75} = 2.2$$

13.3.4 Further revelations from the process reaction curve

As a means of visualizing how a feedback control system is going to behave and the controller settings that will be required for optimum performance, process reaction curves are underrated. Figure 13.2 is a generalized version of the process reaction curve shown in Fig. 13.1. Comparing this curve with Fig. 13.1, the interval *AB* in minutes is

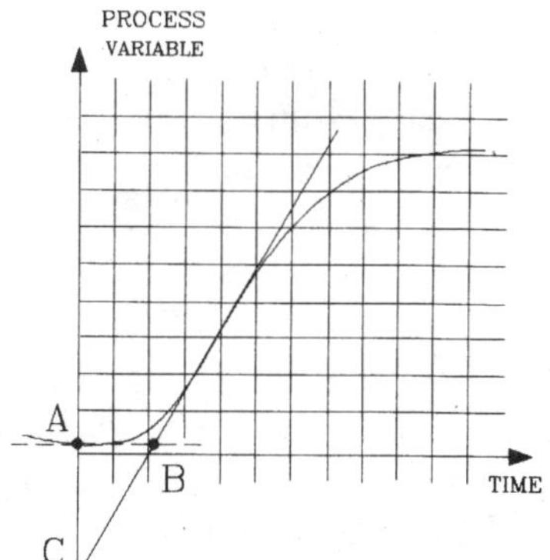

Figure 13.2 The required controller gain is inversely proportional to the intercept *AC*.

the dead time L. The slope of the tangent, in percent per minute, will be the process reaction rate R. The optimum gain for the controller, from page 214, will be

$$\frac{1.6\Delta m}{RL}$$

This relation shows that the optimum controller gain, i.e., the gain that will produce recoveries with an amplitude ratio of 1/4, varies inversely as the product RL. As RL increases, the controller gain should be set at a lower value.

From Fig. 13.2, the slope of the tangent R is equal to the tangent of the angle ABC, or AC/AB. Since $AB = L$ (the dead time), then $AC/L = R$, and $AC = RL$. Apparently, then, the interval AC, measured down the vertical axis, is related to the required controller gain. The greater the interval AC, the *lower* is the required controller gain.

This helps to explain, for example, why flow-control loops tend to run at very low controller gains. If a process reaction curve were made on a flow loop, it would show a short, but measurable, dead time, followed by a very fast rise to a new flow rate. This would result in a very steep tangent so that the point C in Fig. 13.2 would fall well below point A, an indication that a low controller gain is required.

Furthermore, the short dead-time interval means a short period of oscillation. The period of oscillation will be, in fact, about 3 to 5 times the dead time, depending on the relative values of the dead time and time constant. The optimum settings formulas show that the required reset rate goes up as the period of oscillation goes down. Thus a short dead time requires a high reset rate. Anyone who has tuned a proportional plus reset controller on a flow-control loop knows that these loops invariably run on lots of reset but not much proportional action. The process reaction curve confirms that this is so.

Process reaction curves run on temperature-control loops generally show more dead time but a much slower rate of rise. The tangent to the curve at its steepest point cuts the vertical axis such that the distance AC is shorter. From the viewpoint of the controller settings, what are the conclusions? The greater dead time AB means that less reset and more derivative are required, while the smaller intercept AB requires a higher gain.

13.4 Calibration of Controller Settings

Determining controller settings by testing the dynamic behavior of the control loop and then making a calculation based on the data obtained is a practical way to deal with the problem. The case for using this method is unfortunately weakened by the roughness with which most analog controllers have their settings calibrated. Tolerances of ± 25 percent are not uncommon. Therefore, if the recommended optimum settings procedures are to be effective, it will be necessary to check, and possibly adjust, the calibration of the controller gain, automatic reset, and derivative settings to ensure that the numbers on the dials really mean what they say.

To add to the problem, there are few, if any, automatic process controllers available today that have any amount of adjustability in the calibration of the controller settings. In a pneumatic controller, the spacing of the numbers on the dial is supposed to conform with the characteristics of the needle valve or other restriction that is the basic means of making the setting. In an electronic controller, the numbers should be compatible with the characteristics of the component resistances and capacitances that create the responses. Usually, the best that can be hoped for is that it will be possible to offset the pointer with respect to the scale so that the calibration of the gain, reset, or derivative setting will be accurate at one point and reasonably good at the others.

Generally, it will be best to calibrate the controller settings for maximum accuracy at the following points:

- For the proportional response setting, at a gain of 1.0 or a proportional band setting of 100 percent, whichever applies.
- For the automatic reset, at 1.0 repeat per minute or 1.0 minute, whichever applies.
- For the derivative setting, at 1.0 minute.

If the construction of the controller is such that it is not even possible to offset the pointer with respect to the scale so that the calibration is accurate at one selected point, then the best that can be done is to make a scribe mark on the scale at the real position of the chosen calibration point.

Before checking the calibration of the settings, the output of a pneumatic controller should be blocked off. An electronic controller should have its output connected into a dummy load.

13.4.1 Test equipment required

- A signal generator for synthesizing the controller measurement signal and displaying the value
- A signal generator for synthesizing the controller setpoint and displaying the value
- An accurate meter for reading the controller output
- A timing device such as a stop watch

13.4.2 Calibration of proportional action

1. Using a signal generator, input a 50 percent signal (12 mA for an electronic controller, 9 psig for a pneumatic controller) to the measurement port of the controller.
2. Set the derivative, if present, to its lowest value, right off if possible.
3. Set the automatic reset to maximum: maximum reset rate or minimum reset time.
4. Set the proportional action at a gain of 1.0 or 100 percent proportional band, whichever applies.
5. By manipulating the setpoint, bring the controller output to 50 percent. Ensure that the controller output remains steady at 50 percent for at least half a minute, even though the reset section is active.
6. Turn the automatic reset back to the minimum reset rate or maximum reset time (right off if possible). This locks the reset section at 50 percent of the controller output range. (The steps taken so far will render the reset and derivative sections of the controller inoperative, leaving only the proportional action section alive. It may be possible to achieve the same result by breaking connections inside the controller itself. The alternate procedure is just as good.)
7. The calibration of the proportional action setting can now be checked. With the controller gain setting at 1.0 or a proportional band of 100 percent, make a setpoint change from 25 to 75 percent (8 to 16 mA or 6 to 12 psig). This should produce a 50 percent change in the controller output signal. If it makes a significantly larger or smaller change, then an adjustment to the calibration of the proportional setting should be made.

13.4.3 Calibration of automatic reset

The timing device will be required to check the calibration of the automatic reset response. The procedure is as follows:

1. Fix the measurement input at 50 percent. Cut out the derivative action as completely as possible.
2. Set the proportional gain at 1.0 or the proportional band at 100 percent. Set the reset rate at maximum or the reset time at minimum.
3. By juggling the setpoint, bring the controller output to 40 percent and stabilize it there (no discernible drift for half a minute). Then set the reset rate at minimum (or reset time at maximum), right off if possible.

4. Juggle the setpoint to bring the controller output to 50 percent. There is now a 10 percent driving force between the reset section of the controller and the output. If the reset setting is changed, the reset section should add this 10 percent difference to the 50 percent value of the output a number of times per minute equal to the new reset rate setting.
5. Quickly change the reset rate setting from minimum (or off) to 1.0 repeats per minute (or the reset time from maximum to 1.0 minutes). At the same moment, start the timing device. If the automatic reset setting is calibrated accurately, the reset action should add 10 percent to the 50 percent starting point after 1.0 minutes. In other words, the controller output should pass through 60 percent after 1 minute has elapsed. If the controller output increases faster or slower than this, the calibration of the reset action should be adjusted.

13.4.4 Calibration of derivative response: Preferred method

Calibrating the derivative is a more complicated task than calibrating the other two responses. If it is to been done accurately, a signal generator that can generate a ramp signal is required. A generator of this type usually can be made up by the plant instrument shop.

The requirement is for a signal that increases with time at the rate of 20 percent per minute. This can be achieved by a rotating cam in a pneumatic signal generator or with the right combination of resistances and capacitances in an electronic generator.

The output of the signal generator will be the input to the measurement section of the controller. Reasonably accurate meters are required so that the controller measurement and output signals can be compared on the same time base.

The calibration procedure is as follows:

1. With the signal generator connected into the controller measurement, run the output of the signal generator up to 50 percent, and hold it there.
2. Set the controller in direct action. Set the controller gain at 1.0 (proportional band at 100 percent). Set the derivative at minimum (off if possible). Set the reset at maximum repeats per minute or minimum reset time.
3. By juggling the setpoint, bring the controller output to 50 percent and stabilize it. Then set the reset to minimum (off if possible) to lock the reset section at 50 percent.

4. Start the ramp signal generator. The controller output should follow the output of the signal generator, both increasing at the rate of 20 percent per minute.
5. Change the derivative setting to 1.0 minute. This should cause the controller output, at any moment, to be 20 percent greater than the measurement signal coming from the signal generator, since the controller output is leading the measurement by 1.0 minute due to the action of the derivative.
6. If the controller output is not 20 percent greater than the measurement, change the derivative setting until it is. This will be the real 1.0-minute setting of the derivative.

13.4.5 Calibration of derivative response: Approximate method

If a ramp signal generator is not available, there is an alternate method that can be used, although the results cannot be expected to be as reliable as they would be using the first method.

The basis of this second method is the fact that if a sudden movement of the setpoint is made, the output from a proportional plus derivative controller will jump to zero or maximum and remain there for a definite interval of time before returning. The duration of this interval, at the particular signal levels used, is an indication of the derivative time. The procedure is as follows:

1. Set the controller in direct action. Set the controller gain at 1.0 (or proportional band at 100 percent). Set the derivative at minimum (off if possible). Set the reset at maximum repeats per minute or minimum reset time.
2. By juggling the setpoint, bring the controller output to 50 percent and stabilize it. Then set the reset to minimum (off if possible) to lock the reset section at 50 percent.
3. Fix the measurement signal to the controller at 50 percent.
4. Set the derivative time at the minimum value.
5. Set the controller gain at 2 or the proportional band at 50 percent.
6. By experimenting with the setpoint, determine the setpoint values that produce controller output levels of exactly 55 and 45 percent, respectively. Set the derivative at 1.0 minute.
7. Start with the setpoint that produces 45 percent output. Then move the setpoint suddenly to the value that produces the 55 percent output. At the same time, start the timer.

8. The controller output will drop quickly to zero, remain there for a short period, and then rise quickly to 55 percent. Stop the timer when the output signal comes back on scale. The derivative value will be

$$\text{Derivative time (minutes)} = t \times \frac{M + 25}{\Delta m}$$

where t = timed interval (minutes)
M = average output signal level (percent)
Δm = change in output level (percent)

Using the suggested values, 45 and 55 percent, the derivative time will be equal to $7.5t$ minutes. A derivative setting of 1.0 minute should give a timed interval of 1/7.5 minute, or 8 seconds. The actual interval should be within 1 second of this value; otherwise, an adjustment to the derivative calibration should be made.

13.5 Self-Tuning Controllers

An interesting development in process control has been the introduction, in 1988 or thereabouts, of automatic controllers that are claimed to be able to figure out their own optimum settings. There have been several different versions of these self-tuning controllers developed by manufacturers of process instruments. And there are at least two significant ways in which they may differ.

- Does the controller create and base its results on a model of the process? If it does, then a key question is, "What assumptions are made regarding the dynamic form of the process, i.e., dead time plus a single time constant, two time constants, or what?"
- Does the controller create and input a disturbance signal to the process to obtain the information it needs to describe the process, or does it wait for the naturally occurring transients to come along and evaluate the reaction of the process to them?

A relevant fact, not to be overlooked, is that all the modeling and calculations, including the eventual PID output of the controller to the process, are done in the digital realm. These controllers are digital controllers, and as such, they operate from sampled data. A pertinent question to be asked in the evaluation of all self-tuning controllers is, "How is it possible for the controller be fooled?"

In 1989, the University of Alberta Chemical Engineering Department made a study of the then-available self-tuning controllers for the purpose of discovering the principles that were behind their ability to tune themselves. The following is a brief summary of what was found out.

Controller no. 1 created a model that appeared to have the form of two time constants. The model was intended to predict the process output. The model parameters that matched the actual process performance were calculated by the controller, based on a comparison of the output of the model with the output of the actual process. The model parameters were then used to generate PID controller settings such that the model had the desired recovery characteristics. Adjustments to the P, I, and D settings were based on how the model behaved, not on how the process behaved.

The actual control loop had the conventional feedback form. The model parameters and the controller settings were continuously calculated. If the model were perfect, the controller would always be optimally tuned.

Controller no. 2 based its estimation of the optimum PID settings on the output from the actual process. It did not use a model. While this is more true to life, the calculations could only be run after a transient (process disturbance) of some type had gone through the process. Thus the controller would always be tuned according to what was "learned" from the last transient.

The controller was started up with PID settings that were known to be conservative. Ten or more trials by the controller might have been required before it locked onto the PID values that give acceptable recovery.

The controller might not tune itself properly if the process refused to overshoot or oscillate. The controller output, over time, was not used in determining the controller settings; only the process output was used. Thus it was possible for the controller to "gain windup" to a very high value.

The designers recognized this potential problem and included as an optional feature the capability of the controller to decrease its gain if oscillations in the controller output were sustained. However, if this option were used, it might fight the primary self-tuning routine.

Controller no. 3 used a model that it created by sending an oscillatory perturbation signal through the process. It then measured the magnitude ratio and the phase shift of the process output. The frequency chosen was that which occurred at the breakpoint of the magnitude ratio curve of the process frequency response characteristics.

By testing the controller settings on the model, it anticipated the correct controller settings to be used on the process.

Controller no. 4 also created a model of the process. It did this by switching the setpoint of the controller alternately plus and minus a prescribed amount. It had to be started up with some reasonable PID settings already in place. Two inputs that had to be provided by the operator were

- The process response time to reach 90 percent of the ultimate process change in response to a step input.
- The size of the setpoint change that was permissible in perturbing the process.

The controller disturbed the process (on control) using the allowable setpoint changes (plus and minus). On the basis of the observed reaction of the process, it created a model of the process. When the model had been completed, a signal was given. The operator could then discontinue model development and have the controller settings calculated and set on the basis of the prototype. Or he or she could allow the controller to go on updating the model and recalculating the PID settings.

If the response time was set faster than the process could actually follow, then overshoot would result.

This summary of the characteristics of these examples of self-tuning controllers should only be considered valid as of 1989. Obviously, any instrument manufacturer who concludes that a weakness, real or perceived, in its controller is putting the company at a disadvantage with the competition will take the appropriate measures to correct the situation. The value to be derived from this summary, therefore, is not in the comparison of the different brands per se but in the comparison of the different approaches taken to the design of a self-tuning automatic controller.

The key questions seem to be, “Is it better to have the dynamic behavior of a process tested by people or by instruments?” and, “Is it better to have this done once, and the resulting controller settings set and left, or to have the testing and setting go on repeatedly?”

Presumably, if the cost of the self-tuning controller is not much more than the cost of a regular controller, then the additional self-tuning feature might as well be included in the controller package. It would by unwise, however, to adopt this philosophy without an in-depth understanding of what the self-tuning controller will be doing to the process and what might happen as a result.

In any event, self-tuning controllers come in the digital form only, which means that self-tuning will not be a consideration in pneumatic or analog electronic controllers.

Chapter

14

Variations on Feedback Control

So far our study of process control has been aimed at understanding feedback control systems, the way that they function, and their limitations. As we now know, the fundamental weakness of any feedback control system is that corrective action always comes too late, sometimes well after the system has been disturbed. Thus the control system has to go through a period of upset and recovery in order to get back on control.

The four sections of this chapter deal with ratio control, cascade control, selective control, and multivariable control. Before discussing these individually, however, two features that are common to all of them should be pointed out. First, these four process control schemes are all variations of feedback control. The basic characteristics of feedback control will apply to ratio, cascade, selective, and multivariable control as well.

Second, in any simple feedback control loop, the setpoint of the controller is usually set manually, by turning a knob or by pressing keys. Three of the four types of control to be covered next require a controller designed to have the setpoint set by an incoming signal generated by some other control system component. The type and range of the setpoint signal will be the same as they are for measurement and control signals, i.e., 3 to 15 psig or 4 to 20 mA.

Most process controllers have a transfer switch that permits the plant operator to switch from automatic to manual control if he or she determines that this is necessary. The special controller that can accept an external setpoint signal will have a second transfer switch on the incoming signal. The two positions of this switch are usually marked "Local" and "Remote." In the local position, the operator sets the setpoint manually in the usual manner. In the remote position, the setpoint is set by the incoming setpoint signal.

14.1 Ratio Control

In some operations it is necessary to control one process variable at a value that depends on the value of a second process variable. Some examples are:

- In potable water systems, chlorine is injected into the water supply at a controlled rate to kill bacteria. The rate at which chlorine should be added will depend on the rate at which the water is being drawn through the system.
- In a building that has a hot water heating system, the value at which the temperature of the circulated hot water is controlled must be raised as the outdoor temperature goes down. This example serves to illustrate that ratio control systems do not always involve flow control.
- Fuel oil for ships is blended to the proper viscosity by cutting back heavy fuel oil with the right amount of light oil diluent. The rate of addition of the diluent oil depends on the rate of flow of the heavy base oil.

Control systems that function in this manner are called *ratio control systems.* In each case, the pattern is the same: one variable is being controlled, but the setpoint of its controller is being automatically adjusted in relation to the value of a second variable. The second variable may be uncontrolled and often is. Control systems of this type consequently require a controller whose setpoint can be set from an external signal. Furthermore, in ratio control systems, the external signal is always some function of a *measurement* signal. This feature, the controller setpoint dependent on the measured value of some other process variable, is the main distinguishing characteristic of a ratio control system.

The process variable that is under control is usually called the *ratio-controlled variable* or sometimes the *dependent variable.* The second process variable whose measurement will eventually create the setpoint of the ratio controller is usually called the *independent variable.*

The measurement signal for the independent variable comes from a standard process transmitter with an output signal range of 3 to 15 psig or 4 to 20 mA. In a relatively few situations, this signal will go directly to the setpoint of the ratio controller. However, in most cases, the signal passes first through a ratio function module that modifies the signal in some way before sending it on to the ratio controller's setpoint. The function module may be a separate component, or it may be built right into the ratio controller itself. In a computer-based system, the ratio function is usually generated in software.

The modifications which the ratio function module may make to the original measurement signal are not very elaborate. In fact, there are only four possibilities. The function module may

- Change the gain of the signal
- Bias the signal (output different from zero when the input is zero)
- Put a limit on the signal
- Invert the signal

The ratio function will be of the form

$$S_c = X_0 + RX$$

where S_c = setpoint of the ratio-controlled variable
X = measured value of the independent second process variable
R = ratio factor (actually, the gain)
X_0 = value of the setpoint S_c when the independent variable X is zero. X_0 is sometimes called the *bias*.

Figure 14.1 illustrates these possibilities. In each case, the input X is the original measurement signal, and the output S_c is the ratio controller's setpoint signal. The dashed line shows what the output would have been if the ratio module had not been used and there were no modifying effect.

The ratio module may make any of these modifications, or any combination of them, to the setpoint signal that proceeds to the ratio controller depending on the relationship required between the ratio-controlled variable and the independent variable.

The design difficulties associated with a ratio control system depend on how elaborate the ratio module's output function has to be. A second consideration is whether or not it is necessary, from an operations point of view, to be able to set the ratio at different values. The simplest design, and the easiest to devise, is the one in which the ratio controller always operates at one ratio setting only, and there is

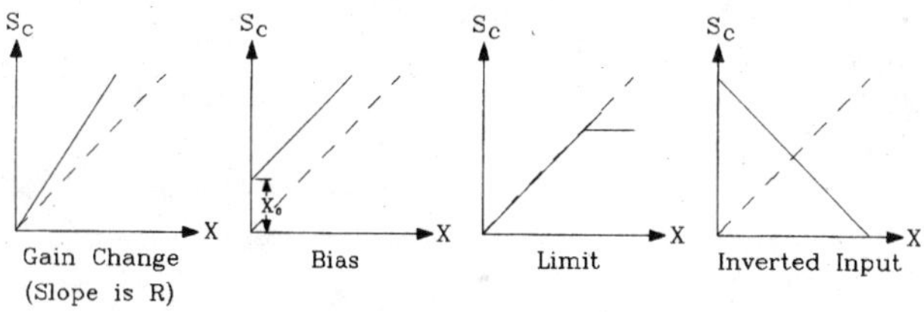

Figure 14.1 Any of these functions, or a combination, can be generated by a ratio module.

no requirement for bias, a limit, or inversion. In this case, a ratio control system can be constructed without using a ratio module at all. The output of the measurement transmitter for the independent variable goes directly to the setpoint of the ratio controller, and the required ratio is achieved by a proper selection of the measurement ranges for the independent variable and the ratio controller. The ratio R that will prevail will be the range span of the ratio controller divided by the range span of the independent variable.

A common example of a ratio control system is one in which the setpoint of a flow controller on one stream is varied by the rate of flow in another stream. This example occurs wherever the requirement is to proportion an additive into a principal stream. It is also possible to have more than one stream proportioned into the main stream to make a running blend, as happens when gasoline or lube oils are blended in line. In these cases, the flow of the additive or blended component is nearly always zero when the flow of the main stream is zero so that $X_0 = 0$ and the ratio relationship is $S_c = RX$.

In the design of ratio flow control systems, two questions frequently come up:

1. What happens if the flowmeters used are not linear measurement devices, such as, for example, if they are orifice meters? Will the ratio setting R change as the flow rates change?
2. What value for R should be set on the ratio module in order to achieve the required ratio of the two flow rates? The ratio module does not understand flow rates. It only understands signal levels.

If the ratio-controlled stream and the independent stream are measured with nonlinear meters, the ratio R will be unchanged across the entire flow range, provided that *the nonlinearity is of the same type for both flowmeters,* and provided that *the ranges of both meters start at zero,* which is generally the case for flowmeters. However, when the meters are orifice meters, it is actually differential pressure that is being ratioed rather than flow rate. The flow rates are proportional to the squares of the differential pressures. Thus, if the ratio module has a linear scale, the ratio R should be set at the square of the desired ratio of the flow rates. Some ratio flow modules have the square root extracted on the dial of the ratio adjustment so that the dial reads flow ratios directly instead of differential pressure ratios.

On all flow ratio control systems, the actual (operating) ratio of flow rates will depend on the value of R that is set on the ratio module *and* on the ratio of the flow ranges selected for the two flow measurements. Since the flow rates of the independent stream and the ratioed stream are often quite different, it is not unusual for the two flowmeters to have different ranges. If R_r is the ratio of the range of the ratio controller's flowmeter to the range of the independent stream's flow measurement, then the operating flow ratio R_o that actually will be obtained will be

$$R_o = R_r \times R$$

This means that to obtain the desired operating flow ratio R_o, the R value on the ratio module should be set at R_o/R_r. For example, suppose that the independent stream is measured by a flowmeter with a range of 0 to 100 l/s, while the flowmeter for the ratio controller has a range of 0 to 10 l/s. If the R value on the ratio module were set at 1, the measurement of the independent flow rate would go directly to the setpoint of the ratio controller, and the operating flow ratio would already be 1 to 10, based on the ratio of the ranges of the two flowmeters. If it were desired to have an operating flow ratio of 1 part ratio-controlled stream to 5 parts of the independent stream, then the R value on the ratio module should be set at

$$\frac{R_o}{R_r} = \frac{1/5}{1/10} = 2$$

14.1.1 Ratio flow blending systems

When two or more liquid components are to be combined into a blended product, either of two methods can be used. The first consists of measuring out the correct amounts of each of the components and dumping them into a holding tank. This is batch blending. It is an accurate method of blending, and no controls are required (meters only). However, the quantity that can be blended at any one time is limited by the size of the tank, the job is time-consuming, the personnel requirement is relatively high, and to ensure that the blend is consistent, the tank must be stirred. Batch blending is usually limited to small-volume blends.

For high-volume blends, in-line blending is the generally accepted method. In in-line blending applications, all the components are blended into the main transfer line at the same time, and the finished stream flows continuously to its destination. Automatic ratio flow controls are needed to ensure that the proportions are correct.

The in-line operation was first performed using flow-rate controls. Current technology, however, favors flow control on a cumulative basis. The distinction between the two methods has come to be equivalent to the distinction between an analog blending system and a digital blending system. Persons who believe that anything digital is better than anything analog will naturally favor cumulative blending on

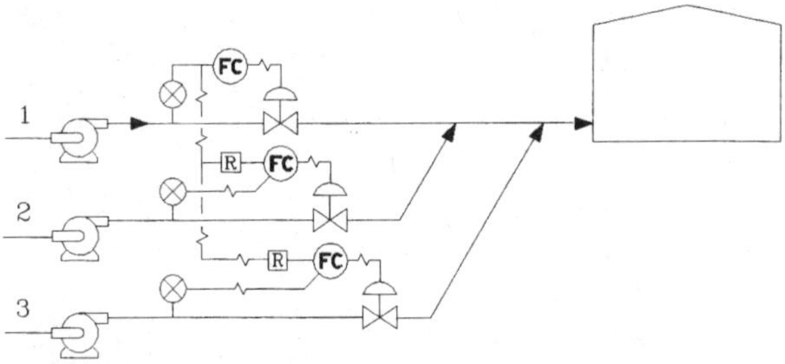

Figure 14.2 An in-line rate-of-flow blending system.

this basis alone. Nevertheless, there is merit on both sides, and it is worthwhile to understand how the two methods compare.

Figure 14.2 shows an in-line blending system for three components using rate-of-flow controls. Component 1 is the principal component. Its flow rate essentially governs the flow rate of the whole blend. The setpoint of the flow-rate controller for component 1 is set in the conventional way, by turning a knob in the controller.

The setpoints of the flow controllers for components 2 and 3 are set from the measurement signal for component 1. In both cases, the setpoint signal that each controller receives is the component 1 measurement signal modified by a ratio module *R*. All three components end up in a receiving tank. When the tank is full, the blended product is usually tested for accuracy of the blend and then passed for sale.

This in-line system has two fundamental weaknesses. First, accurate blending depends on accurate flow measurement, and the types of meters available for rate-of-flow measurements are, for the most part, not as accurate as flowmeters that measure on a cumulative basis.

The second problem is that the rate-of-flow system has no memory. Suppose that while the blend is proceeding, the pump for component 3 stops, and it requires 10 minutes to restart it or to bring on a spare pump. While this is going on, components 1 and 2 are still going into the tank. When component 3 is finally restarted, its rate of flow will be the same as it was prior to the outage. The control system does not recognize that when the tank is full, the blend will be deficient in component 3 by the amount that should have gone into the blend during the 10-minute outage.

More accurate meters, along with the capability to incorporate memory into the system, are what favor the digital (cumulative) blending system. What detracts from it is the higher cost. Bearing in mind that a rate-of-flow system lacks memory, the difference between the two systems can be illustrated by a somewhat simplistic example.

Suppose you intend to drive from Edmonton to Fort McMurray, a distance of 500 km. You want to get there by 6 P.M., but you cannot leave before 1 P.M. This gives you 5 hours. You have cruise control on your car, so you set the control at 100 km/h when you start off. Everything is on schedule until you get to Wandering River, where you decide to stop for coffee. This takes half an hour. Following the coffee break, you get back in your car and continue on with the cruise control set at 100 km/h as before. The result, predictably, is that you arrive in Fort McMurray at 6:30 instead of 6 P.M.

This is an exact parallel to the case of the rate-of-flow blending system. The cruise control only understands the speed (the rate) at which the car is moving. It does not take into account, and does not allow for, any breaks or shutdowns in the operation of the car.

Now consider another possible set of controls, as illustrated by Fig. 14.3. The first scale, denoted the "Pace" scale, shows the plan, or schedule, on which the car is required to operate. The pointer on this scale is clock-driven. The clock is started when the trip begins, and the pointer is driven at a constant rate so as to move from 0 to 500 km in exactly 5 hours.

The second scale, denoted the "Road" scale, shows the number of kilometers actually traveled. The pointer for this scale is connected to the odometer of the car. The setpoint of the cruise control is adjusted by the control system so that the road pointer stays in line with the pace pointer. During the coffee break, the pace pointer will continue to move, whereas the road pointer is stopped. At the end of the break, the pace pointer will be ½ hour, or 50 km, ahead of the road pointer.

When the car is once again back on the road, the control system will increase the setting of the cruise control to a value higher than 100 km/h so as to eventually bring the two pointers back in line. How quickly it will do this depends on how high a speed is practical for the

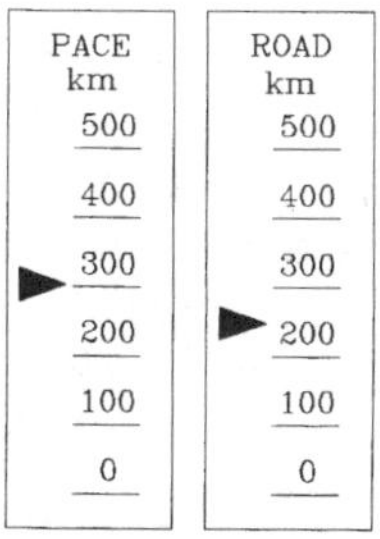

Figure 14.3 In a control system which has memory, the automatic controller sets the car speed so as to keep the road kilometers in line with the pace kilometers.

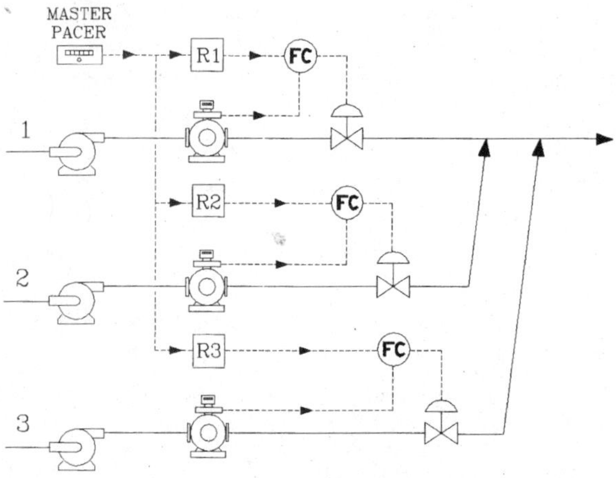

Figure 14.4 A digital blending-control system for three components.

car. When the road pointer has caught up with the pace pointer, the cruise control will be throttled back to 100 km/h. The end result is that despite the coffee break, you still arrive at Fort McMurray, on schedule, at 6 P.M.

Digital blending systems work on the same principle. Figure 14.4 is a sketch of a digital blending system with three components. Additional components can be accommodated by adding more flow-control stations to the system. A pacer that is clock-driven generates a digital signal. The pulses from the pacer represent volumes of liquid that should be going into the blend. The signal from the pacer goes to the control station for each of the blend components. At each control station, the pace signal is modified, actually scaled, by a ratio module whose output becomes a digital signal representing the number of volumes of that component that should have gone into the blend at any particular moment.

The flowmeter for the control station is a cumulative-flow type, i.e., one that tallies the number of volumes of the component that have gone into the blend rather than the rate of flow of the component. The automatic controller for the station compares the cumulative-flow signal it receives from the meter against the pace count it receives from the ratio module. If there is any difference, the controller moves the control valve in the appropriate direction to bring the amount of the component that has gone into the blend in line with what should have gone in, as required by the pace signal.

If there is a temporary outage of any component, the pace count continues on while the cumulative-flow signal from the meter stops. When the flow is restarted, the controller for that component realizes that it is a certain number of volumes behind, and it opens the control valve wide to catch up. When the measurement count is once again in line with the pace signal, the controller throttles back on the control valve.

The digital blending method is the most accurate of all currently known methods for two reasons. The system has memory, which allows it to compensate for outages, while at the same time the cumulative flowmeters that are used are the most accurate meters available for flow measurement. The drawback to the digital system is its higher cost. It can be argued, however, that blending is sufficiently accurate that the receiving tank is no longer necessary. What can be saved on the tank will easily make up the cost difference for the digital blending system.

14.1.2 Ratio control with adjusted ratio

Figure 14.5 shows a ratio flow-control system for blending fuel oil. The desired end result is that the viscosity of the blended fuel conforms to a specified value. In the ratio module *R,* the ratio is set manually. Presumably, if the proportions of the diluent and the base oil are maintained by ratio control, then the viscosity of the blended fuel oil also will be correct.

However, ratio flow control will only take care of the flow rates. It will not take care of variations in the quality of the base oil or the diluent, both of which can upset the viscosity of the product fuel oil. If there are quality variations, then the ratio will need to be changed to compensate for them; otherwise, the final viscosity of the blended fuel will deviate from the specification value.

On-line viscosity measurement usually requires a sampling system, so the measurement is dynamically slow. Attempting to control the viscosity of the blended fuel oil by having a viscosity controller on the blended product, with its output going to the control valve on the diluent, would likely be unsuccessful because of the delay associated with measuring the viscosity.

Viscosity control of the blended product can be attained if the output of the viscosity controller goes to the ratio module so that the

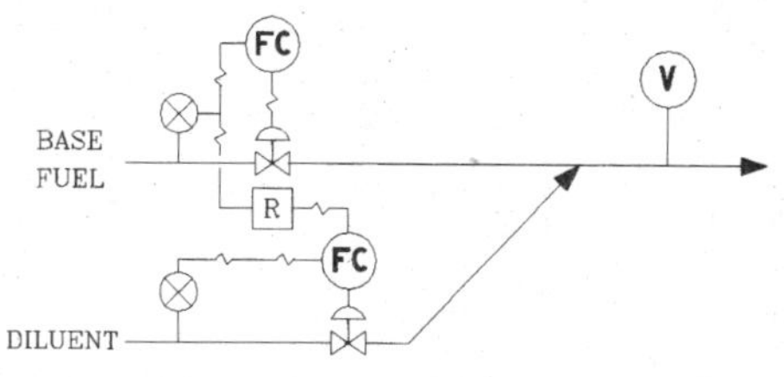

Figure 14.5 A ratio flow control system for blending fuel oil.

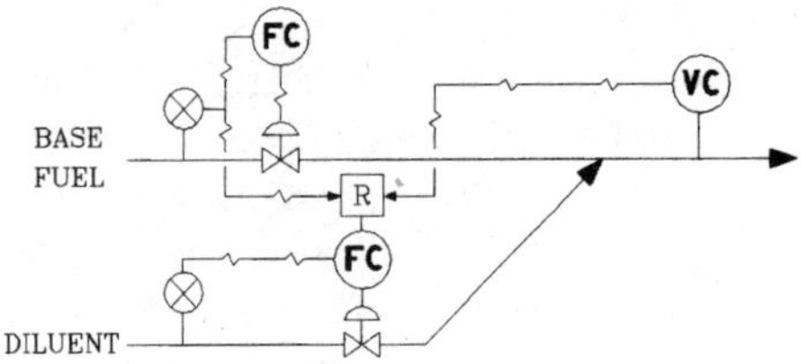

Figure 14.6 Ratio control with adjusted ratio for blending oils.

ratio factor is automatically adjusted to the right value for maintaining the viscosity of the blended fuel. This scheme is shown in Fig. 14.6. Ratio flow control proportions the flow of diluent to the flow of base fuel oil, while the function of the viscosity controller is simply to trim the ratio over the long term, should this be necessary. As such, the viscosity controller will operate at a fairly low gain and reset rate.

When the output of the viscosity controller is used to trim the flow ratio rather than for positioning the diluent valve, the response time of the viscosity measurement becomes considerably less important, to the point where the blending control system has a good chance of success.

Another common example is shown in Fig. 14.7. Owing to the swell effect, direct measurement of the level in a boiler drum is not reliable enough for direct level control. To maintain the correct inventory of water in the boiler system, the approach usually taken is to material balance the boiler. Each kilogram of steam that comes out of the boiler must be replaced by a kilogram of water that goes in. Therefore, the steam flow out of the boiler is measured, and the measurement sets up the flow of boiler feed water to the boiler using a ratio control system.

This would ensure the right amount of water in the boiler, if it were not for the fact that the measurements of the steam flow and the feed water flow may not be entirely accurate. In addition, the boiler can lose inventory in ways other than producing steam, such as, for example, when it is blown down. To compensate for these secondary effects, a level controller is installed on the boiler drum, and its output is used to trim the flow ratio.

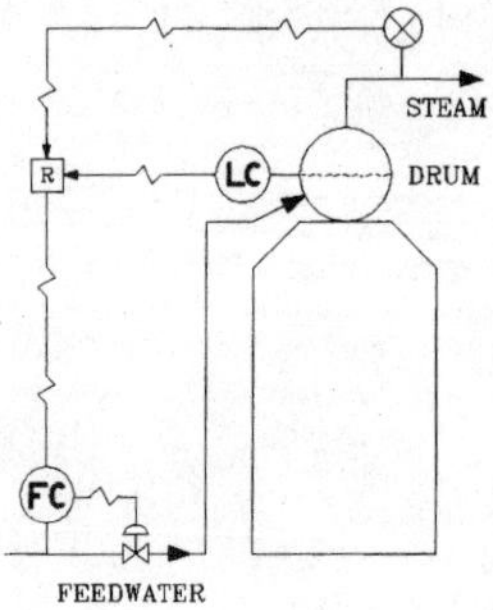

Figure 14.7 A three-element boiler feedwater control system is an example of ratio control with adjusted ratio.

Since trimming is all that is needed, the level controller operates at a relatively low gain and reset rate. At the conservative controller settings that are used, the swell effect in the boiler drum, which tends to foul up level measurement, is not a hindrance to meeting the real control objective, which is to maintain the proper water inventory in the boiler. This control system is standard on boilers, and it has come to be called a *three-element boiler feed water control system.* The three elements are the steam flow, the feed water flow, and the drum level.

A third example of ratio control, with the ratio adjusted by a product controller, is in the proportioning of airflow to fuel flow when firing a boiler or a large fired heater. In these ratio systems, the ratio controller is on the airflow, which is ratioed to the fuel flow so that the correct proportions of air to fuel are maintained.

The real objective, however, is to end up with the correct amount of excess air in the flue gas after the fuel has been burned. Excess air can be determined by measuring the percentage of oxygen remaining in the flue gas. Here again, the measurement is not particularly good dynamically, so attempting to control excess air using a percentage oxygen controller whose output goes to the control damper on the airflow would have little hope of success. Instead, the output of the oxygen controller should trim the ratio of airflow to fuel flow.

A fringe benefit of using this control scheme is that if the oxygen analyzer has to be taken out of service for maintenance, the ratio control system will still be operational. The only loss is that the ratio must be set manually instead of automatically until the oxygen controller is back on line.

In a ratio control system in which the ratio is adjusted by a control signal rather than set manually, the ratio module is actually a multiplication module that can generate the function $S_c = R \times X$, where X is the independent measurement, and R is the trim signal from the product controller. There is no bias in this function, and in fact, there should not be, since in all three examples just described, the flow of the ratio-controlled stream should be zero when the independent flow rate is zero.

An analog multiplication module is a tricky item to make, and not every instrument manufacturer has one in its line. This sometimes leads to an addition module being substituted for a multiplication unit. The difference in their actions is that a multiplication module will rotate the ratio line around the zero point, which is the theoretically correct thing to do, whereas the addition module moves the ratio line to a parallel position. The gain change and bias graphs in Fig. 14.1 show these two effects. Using an addition module can give acceptable results provided that only small adjustments to the ratio are required.

14.2 Cascade Control

A variation of the conventional feedback control system that is sure to be encountered sooner or later is the arrangement in which the output of one automatic controller is sent not to a control valve but to the setpoint of another controller. This arrangement is called a *cascade control system.*

It is incorrect to describe a cascade control system as a control system having two controllers but only one control valve. A selective control system fits the same description. In cascade control, one controller sets the setpoint of another controller, while the second controller positions the valve. It is this arrangement that distinguishes cascade control.

The most common example of cascade control is the use of a valve positioner on a control valve. The valve positioner is actually a valve-position controller. Its setpoint, to which it controls the stroking of the valve, is determined by the output of the level, pressure, flow, or temperature controller to which it is linked.

A variety of names have been used to identify the two controllers in a cascade system. The following are some examples:

Controller whose output determines the setpoint of the second controller	Controller whose setpoint is set by the first controller
Primary	Secondary
Outer	Inner
Master	Slave

For the purposes of the discussion here, the terms *primary* and *secondary* will be used.

A valid question, and one that has to be answered, is, "Why would anyone want to use this control arrangement?" When we get right down to it, only one variable, the primary variable, is being controlled. The secondary variable is technically under control, but it is being controlled at the value that will hold the primary variable at its setpoint. The setpoint of the secondary control loop is not chosen independently.

While using two controllers instead of one to control one process variable certainly has appeal to the supplier who sells the instruments, the cost of the instrumentation and the added complexity have to be justified by some tangible improvement in control. There are, incidentally, plenty of cascade control systems in the field that could not justify their existence on this basis. There seems to be a feeling among some designers that if a particular process variable is going to be difficult to control, then using a cascade control system will somehow make it easier. What usually happens is that the difficulties that would have defeated a conventional feedback control system will end up making cascade system an exercise in futility as well.

All this points to one conclusion: If cascade control is to be used instead of single-loop control, then there has to be a legitimate reason. The single most profitable reason for using cascade control is *to transfer a disturbance out of a control loop that recovers slowly into another loop that recovers quickly.*

In the temperature-control system shown in Fig. 14.8*a,* oil is being heated in a gas-fired heater. The outlet temperature of the oil is being controlled by varying the flow of fuel gas to the heater. One of the disturbances to this system is the pressure of the gas at the main. If it changes, the temperature of the oil will change, and the temperature controller will have to find a new valve position that will bring the temperature back on control. Since the heater has some very large thermal lags, it could take almost an hour to bring the oil temperature back to its setpoint and level it out.

Now consider the cascade control system that would do the same job. In Fig. 14.8*b,* the gas pressure at the inlet to the burners (the firing pressure) is being controlled by a pressure controller (PC). This is a tight control loop; the recovery time is 15 to 20 seconds. The setpoint of the PC is being set by the output of the temperature controller (TC) on the heated oil. Through this arrangement, the TC actually sets the gas firing pressure at the correct value for holding the oil temperature at its setpoint.

If a change in gas main pressure occurs, it shows up immediately as a change in the gas firing pressure controlled by the PC. The required corrective action is taken by the PC, and in 20 seconds or less the firing pressure is back to where it should be. By using cascade control, a disturbance (gas pressure) has been transferred out of

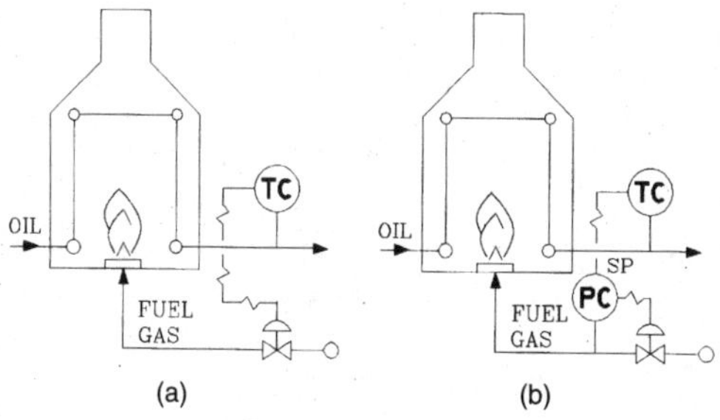

Figure 14.8 In a fired heater outlet temperature control system, cascading the temperature controller to a pressure controller eliminates the effects of gas supply pressure variations.

the temperature loop, with a recovery time on the order of 1 hour, into the pressure loop, which has a recovery time of 20 seconds. In the real plant situation, the TC would never see the effect of the change in gas pressure.

14.2.1 Cascade control design requirements

If any cascade control system is to be successful, two conditions must be met. In fact, the success of the cascade system will depend on these conditions.

1. Cascade control will prove ineffective if the secondary controller cannot control the secondary variable to the setpoint commands sent by the primary controller. This fact should be self-evident, but it seems to get overlooked. Specifically, the period of oscillation in the secondary loop should be no greater than one-fifth the period of oscillation in the primary loop. Another way of putting this is that the secondary loop must be capable of recovering from a disturbance in no more than one-fifth the recovery time of the primary loop.
2. If the measurement in the secondary loop is nonlinear, as it would be in the case of an orifice meter, it should be linearized by adding a suitable function module or by software instructions. Otherwise, the primary controller will have a problem with improper gain settings as the load in the primary control system changes.

When the secondary control loop is a flow-control loop that has an orifice meter as the measurement device, the output signal from the primary controller will not be setting flow rate. It will be setting what is actually being controlled in the secondary loop, the differential pressure across an orifice plate. Since flow rate will be proportional to the square root of the differential pressure, the secondary loop will have a high gain at low flow rates and decreasing gain as the flow rate increases. One has only to look at a square root scale for an orifice meter to see that the same amount of setpoint movement at the low end of the scale represents a great deal more in flow rate than it does at the high end of the scale. If the flow measurement is not linearized, then it may be necessary to lower manually the gain of the primary controller when it is calling for a low flow rate or to raise the gain of the primary controller when it is calling for a high flow rate in order to have acceptable stability in the primary loop.

This problem with controller gain is in the primary loop. There is also a problem in the secondary loop. For an orifice meter, the input is flow rate, while the output is differential pressure. Owing to the square law relationship between differential pressure and flow rate, the static characteristic has a low gain at low flow rates, increasing as the flow rate increases. Thus, if there is an appreciable requirement for flow-rate changes in the secondary loop, the gain of the secondary controller may have to be increased at low flow rates or decreased at high flow rates in order to retain the correct stability in the secondary loop. Notice that this is just the opposite from the controller gain adjustment that may be necessary in the primary loop.

14.2.2 Secondary benefits of cascade control

While transferring a disturbance from a loop that has a long recovery time to another loop that has a relatively short recovery time is probably the most significant benefit to be gained from cascade control, there are others that are worth consideration.

- Whenever a cascade control system can be used, there is usually a net improvement in recovery time for the primary variable. Taking whatever fraction of the dynamics that can be extracted from the process and incorporating them into a closed loop turns out to be a better dynamic arrangement than having the same dynamic elements strung out in a chain in a conventional feedback control loop. The result is that the period of oscillation and the recovery time of the primary loop are reduced.

The following computer-simulated case was run to test this statement. The test process had two essential time constants, 3 and 0.5 minutes, along with 2.0 minute of dead time. In a standard feedback control loop, these dynamic elements would result in a period of oscillation of 7.9 minutes.

If a cascade control system could be set up so that the 0.5-minute time constant and 0.33 minute of the dead time were in the secondary loop, with the remaining 3.0-minute time constant and 1.67 minutes of dead time in the primary loop, then the period of oscillation of the primary loop, which would include the closed secondary loop, would be reduced to 6.8 minutes.

The period of oscillation in the secondary loop would be 1.1 minutes, which conforms with the cascade control design criterion requiring that the period of oscillation in the secondary loop should be less than one-fifth the period in the primary loop.

The recovery time for the primary process variable would be reduced from about 20 minutes to about 17 minutes. This improvement, while tangible, is hardly phenomenal, which is why this benefit is relegated to the secondary category. The degree of improvement will depend on the fraction of the dynamic elements that would ordinarily be in the primary loop which can be separated out and placed in a closed secondary loop. Thus this potential benefit may be worth investigating.

- If the secondary controller is a flow controller, then the primary controller will be calling for an actual flow rate of material or energy. This is in contrast to a single-loop system, in which the controller is calling only for a valve position. If valve position alone is ordered, the material or energy flow through the control valve will be determined by pressure conditions in the fluid system, as well as by the setting of the control valve.
- By setting up a cascade control system, transferring from control of one process variable to another process variable may be made easier for operators of the plant.

This last assertion requires explanation. Figure 14.9 shows four sketches all pertaining to the same process situation. Note that in each of the four cases, the use of a pressure transmitter and a flow transmitter is implied, but Fig. 14.9*b* is the only one in which the transmitters are actually shown.

In Fig. 14.9*a,* purge gas is being drawn off a process in which a chemical reaction is being carried out. The purge gas line has a control valve so that the flow of purge gas can be manipulated to achieve either of two control goals. Depending on circumstances, the flow of purge gas may be set to hold the backpressure on the process at a constant value, or it may be required to remove the purge gas at a constant rate, i.e., on flow control.

To be able to switch from pressure control to flow control and back, there must be a switch somewhere that can be thrown to change the

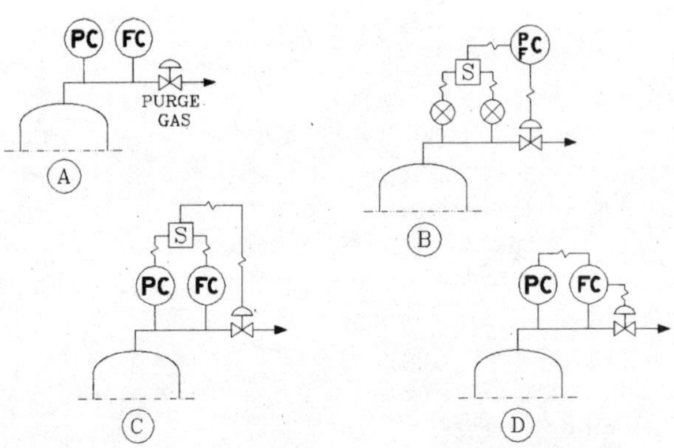

Figure 14.9 Possible schemes for selecting either of two process variables to be controlled.

variable controlled. In addition, and perhaps more important, is the fact that this switching operation should be designed so that any plant operator can do the switching by himself or herself, without requiring an instrument technician to be present and without jarring the process or causing other control problems.

Figure 14.9*b* through *d* shows the available ways of resolving the switching problem. Figure 14.9*b* is the cheapest method in hardware, since only one automatic controller is needed. The selector switch *S* receives the measurement signals from both the pressure and the flow transmitters and, depending on how the switch is positioned, sends one or other of the measurement signals to the single automatic controller. The controller output goes to the control valve.

This method has at least three problems. First, the controller is sometimes a pressure controller and sometimes a flow controller. Consequently, it will require an elaborate dual pressure and flow scale. Second, when the switch is thrown, the chances are minimal that the signal levels from the two transmitters will be the same. Therefore, the controller will have to correct for a bump that will reflect back on the process. Third, the optimum controller settings will be different for flow control than they will be for pressure control, so in addition to changing the switch, it will be necessary, in principle anyway, to alter the controller gain and reset settings.

Owing to the process and control difficulties which it presents, Fig. 14.9*b* is accordingly unacceptable.

Figure 14.9*c* has two controllers, one for each of the controlled variables, pressure and flow. The selector switch receives the output signals from the two controllers and sends the selected signal to the control valve. With two controllers, the problem of different controller settings is resolved, since each controller can be tuned properly for the control job it is required to do.

The method in Fig. 14.9*c* does not eliminate the problem of the process bump, however. As a matter of fact, it is likely to be worse than in the case of Fig. 14.9*b* because it is customary for flow and pressure controllers to have automatic reset. Suppose that the switch is positioned so that the pressure controller is connected to the control valve. The flow controller does not know that its output is going nowhere, but it has a live power supply and a measurement signal coming in. Consequently, it will still attempt to bring the controlled variable, flow, which it can no longer affect, to its setpoint.

Even though the flow controller changes its output in the right direction, the control error does not get any smaller. Eventually, the automatic reset in the controller will drift the controller output all the way up to maximum or all the way down to zero, and reset windup has set in. When the selector switch is thrown to change over control, the controller that has been off line will take over the control valve with an output signal that is either zero or maximum. The control valve will be moved, initially, either to fully closed or fully open. The bump under these circumstances may be severe. Thus the method in Fig. 14.9*c* is not acceptable either.

Finally, Fig. 14.9*d* shows a cascade control system in which the pressure controller is the primary controller, and the flow controller is the secondary. Both controllers can be tuned correctly. The required selector switch is the remote-local switch in the secondary controller. The plant operators will know how to handle this switch as a result of their job training.

If the switch is in the local position, the pressure controller is shut out, and the system will be on flow control. If the switch is in the remote position, the output from the pressure controller sets the set-point of the flow controller, in cascade control style, to control the pressure. This meets both requirements for a bumpless transfer and for the operator to be able to deal with the switching without requiring assistance from an instrument technician.

14.2.3 Tuning the controllers in a cascade control system

A question that occasionally comes up is, "How does one go about tuning the two controllers that make up a cascade control system?" The procedure is quite straightforward.

1. Leave the secondary controller on remote. Put the primary controller on manual at some appropriate value. This will fix the set-point of the secondary controller.
2. Tune the secondary controller using any suitable procedure. If the cascade control system has been well designed and meets the requirement that the period of oscillation in the secondary loop is no greater than one-fifth the period of oscillation in the primary, then it should be possible to tune the secondary controller using the period-of-oscillation method.
3. When the optimum settings for the secondary controller have been determined, dial them in, and switch the secondary controller to automatic.
4. With the secondary controller on automatic, tune the primary controller using an appropriate method, most likely by making a process reaction curve. Do not be concerned that the output of the primary controller is going to another controller instead of to a control valve. The output step change will still produce a process reaction curve, and the optimum settings calculated from the reaction curve data will be the best values for the primary controller.

14.3 Selective Control

There is a fundamental law of process control that states that in any process situation, the number of variables that can be controlled is limited to the number of variables that can be manipulated. Much as people would like to get around this law, it is a control fact of life, and it sometimes creates a problem. The classic example is shown in Fig. 14.10*a*.

In this case, a feed surge drum collects liquid streams from two or more sources at varying rates. The different streams are blended in the drum and then are passed on to an upgrading operation. There are two process variables that are of concern.

1. The flow rate of the blended stream to the upgrading plant should be kept constant for the sake of stable operating conditions in that plant. This implies that the stream should be on flow control.
2. The level in the feed surge drum should be held within limits, since overflowing the drum or allowing it to run dry would create awkward problems in themselves. This implies that the flow out of the feed surge drum should be on level control.

There is, unfortunately, only one variable that can manipulated, the flow rate of the stream to the upgrading plant. Therefore, which of the two possible controllers should be connected to the one available control valve? One suggestion that may come up is to connect the flow controller to the control valve and have the level controller set the setpoint of the flow controller—in other words, set up a cascade control system.

Since both controllers would be functional at the same time, this might give the appearance that both level and flow are being controlled. The fact is, however, that the rate of flow of the feed stream would not be controlled to suit the downstream plant. It would be controlled at whatever value is required to maintain the level in the drum. Connecting the output of the level controller directly to the control valve would produce the same result.

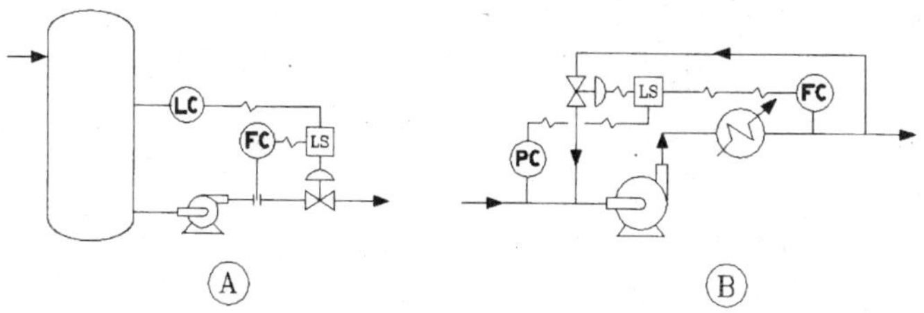

Figure 14.10 Selective control for (*a*) a feed surge drum and (*b*) a centrifugal compressor.

A compromise solution to this problem is to use a selective control system. In selective control, the output signals from the two controllers are connected into a selector module. The selector module then decides which of the two controllers should be connected to the control valve, based on the logic that has been programmed into it.

In the example, from the operations point of view, the preferred method of control is to flow control the stream to the upgrading plant. By using a selective control system, flow control can have priority, as long as the level in the drum is above the safe low limit. This can be achieved by having a control valve that opens with an increasing control signal and a selector module (LS) that selects the lower value of the two signals it is receiving.

If the level in the drum is above the low-limit setpoint, the output of the level controller will be comparatively high, since it will be trying to open the control valve. The output from the flow controller will be lower than the level controller's output; consequently, the selector module will connect the flow controller's output to the control valve. Flow control will be in effect provided that the level in the drum does not drop below the low-limit setpoint. If this happens, the level controller will decrease its output.

When the output of the level controller falls to the point where it is less than the output from the flow controller, the selector module switches the output of the level controller to the control valve and shuts out the flow controller. If, and when, the level in the vessel builds back up again, the output signal from the level controller will increase, and when it becomes greater than the output of the flow controller, the selector module will switch back to flow control.

Another way of viewing the function of the selector module is that it connects the control valve to the controller that needs the lower flow rate through the control valve.

Figure 14.10*b* shows another application of selective control. This application involves a gas compressor that is moving gas formed in a reaction process. The preferred method of control, from the operations standpoint, is to control the pressure at the suction of the compressor, since this is actually the pressure of the reaction process. The compressor has a greater capacity for moving gas than the process has for making it; therefore, the pressure is controlled by recycling gas back from the discharge of the compressor to its suction.

However, the compressor has requirements of its own, the most important one being that the flow of gas through the compressor should not drop below a safe limit. Consequently, the flow of gas from the compressor is also measured and controlled. If the gas flow rate drops below the safe limit, then the control valve must be transferred to the output of the flow controller to ensure than the minimum flow is maintained.

This application requires a control valve that closes on an increasing control signal. The selector module (LS) selects the lower of the two control signals. In other words, it selects the controller that requires the higher gas flow rate through the control valve.

The actual controllers in a selective control system are the standard type; they do not need the facility of having the setpoint adjusted by an external signal. It also would be possible to have more than two controllers involved, although cases where this happens are infrequent.

While the explanation of how selective control works seems straightforward enough, there are some fine points that have to be considered before selective control is decided on. First, the application in Fig. 14.10*a* allows the level controller to take over only if the level in the drum gets too low. What if there is also a requirement to switch to level control if the level gets too high?

To meet this additional specification, the drum will require a second level controller with its setpoint set at the safe high-level limit on the drum. A second selector module, which selects the higher of two control signals that it receives, is also required. The high selector module is installed in series with the low selector so that the output signal from the flow controller must pass through both of them to reach the control valve.

Using this arrangement, the control valve is connected to the output of the flow controller only if the flow controller's output signal is lower than the control signal from the low-limit level controller and higher than the control signal from the high-limit level controller. This will be the case if the level in the drum is in between the setpoints of the two level controllers.

A second consideration is the problem that arises when automatic reset is used in the controllers. In the case of Fig. 14.10*b*, reset will be required in both controllers. In Fig. 14.10*a*, it will be required in the flow controller and probably in the level controller as well. In these selective control systems, both controllers will be functional, but only one will be connected to the control valve. The other controller will not be aware that it can do nothing to affect the variable that it is supposed to be controlling, and with automatic reset active, the output of the controller will wind up.

Then, if the variable controlled by the off-line controller crosses its setpoint, the off-line controller will not take over the control valve right away. There will be a delay, due to the time required for the automatic reset in the controller to unwind the controller output from maximum or minimum to the point where the selector module trips.

If the selection process involves two analog controllers, pneumatic or electronic, it is possible to circumvent the reset windup problem by

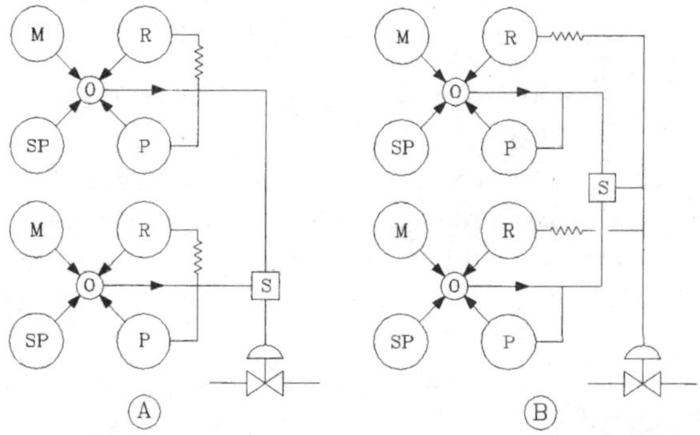

Figure 14.11 Modifying the controller circuitry eliminates reset windup.

using the special hookup shown in Fig. 14.11. The explanation requires a brief look at the circuitry of analog controllers generally.

The output signal generator of the controller, shown as the smaller center circle with the letter O, can be affected by any of the four working sections of the controller. These are shown by the larger circles and are, respectively, the measurement section M, the setpoint section SP, the proportional section P, and the reset section R. A change in the signal level in any of these four sections can change the output from the controller.

The functions of the measurement and the setpoint sections of the controller are self-explanatory. The proportional section creates the reduction in controller gain and also has the refinements necessary to make the gain adjustable. The reset section cancels out the effect of the proportional section, but only after a time delay caused by the restriction in the path to the reset section.

When the controller eventually comes to a steady state following a disturbance, the measurement and setpoint signals will be equal, which is the desired state of affairs. The proportional and reset section signal levels also will be equal, but not necessarily at the same value as the measurement and setpoint.

Figure 14.11*a* shows what would be the normally expected hookup, with the output of both controllers going to the selector module S, which picks one of the outputs to be sent to the control valve. The other output is dead-ended, and the controller responsible for that output signal winds up due to reset action.

Figure 14.11*b* shows how the circuitry can be modified to cure this problem. It is necessary to break into the controller circuit and isolate the connection to the reset section. Then the reset section of both controllers can be connected to the signal coming out of the selector module and going on to the control valve. The result is that no matter which controller is connected to the control valve, the reset section of the other controller is balanced on the signal that is positioning the control valve. If control is switched to the controller that has been off line, the upset is minimal because the reset section of the controller that comes on line has been balanced at the correct value.

A third point is that although examples of selective control applications usually involve selecting the right controller to position the control valve, selective control also may involve the selection of measurements. One possibility is a fixed-bed reactor in which the temperature is measured at a number of places in the bed. The requirement will be to control the highest temperature measured, wherever that might be in the reactor. This would require a selector module whose inputs were all the measured temperatures. The duty of the selector would be to pick the highest temperature and switch it in as the measurement for the one temperature controller.

14.3.1 Design procedure

The design procedure for a selective control system should include these steps:

1. Decide on the best failure position for the control valve (closed or open).
2. Establish the correct action (direct or reverse) for the two automatic controllers.
3. Establish whether a low-signal selector or a high-signal selector is required for the selective control system.

14.4 Multivariable Control

In most industrial operations, a plant will have several, perhaps dozens, of control loops for holding liquid levels, pressures, flow rates, and temperatures at their correct values. Problems will sometimes occur in plants with many automatic control loops because some of the control loops interact with each other. Figure 14.12 shows a level-control loop and a temperature-control loop that are coupled together through the plant. In this application, a liquid stream comes off a feed surge drum on level control and then passes through a heat exchanger where it is heated. The temperature controller manipulates the flow of the hot medium to the exchanger so as to hold the temperature of the feed stream coming off the exchanger at the right value.

Opening the temperature-control valve will not affect the level. On the other hand, if the level controller moves the level-control valve, this will create an upset in the temperature-control loop which the

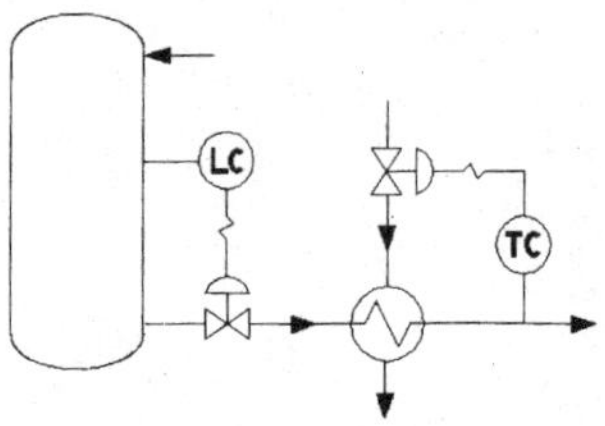

Figure 14.12 The TC is coupled to the LC through the plant. Corrective action by the LC will upset the TC.

temperature controller will have to correct for on whatever recovery time happens to be present in the temperature-control loop.

Suppose that in a particular plant two of the control loops are *A* and *B*. The situation is not good if corrective action (moving the control valve) by controller *A* causes a disturbance in control loop *B*. Controller *B* may have enough disturbances to deal with without having to correct for control action taken by controller *A*. The situation is far worse, however, if the subsequent corrective action taken by controller *B* causes an upset in control loop *A*. Then each control loop can upset the other. When this happens, it is unlikely that either loop will ever settle down on automatic control.

In order to run the plant, it may be necessary to leave one of the loops on manual control. Interaction between control loops occurs because there is a coupling path between them through the plant which results in control action initiated by one controller creating an upset in another control loop. The objective of studying multivariable control systems is to determine how much automatic control is possible under these circumstances.

Specifically, multivariable control deals with three questions:

1. Is it possible to put a number on the extent of the interaction between two interacting control loops?
2. Can both interacting loops be operated on automatic control?
3. Would it be possible to decouple the control loops and eliminate the interaction so that both control loops can operate normally on automatic?

The last question leads to a method by which interacting loops can sometimes be decoupled. While it is not possible to eliminate the coupling path through the plant, it may be possible to create an equal and opposite effect that will cancel out the interaction between the control loops. Since this is a general approach, it could be argued that *multivariable control* might better be called *decoupling control*.

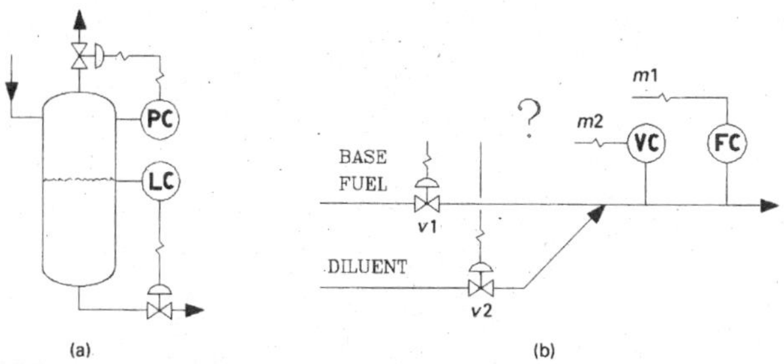

Figure 14.13 Interaction extremes: (*a*) zero interaction; (*b*) interaction to the extent that it may not be possible for both controllers to operate on automatic.

14.4.1 Interaction extremes

The two sketches in Fig. 14.13 show the two extremes that can exist in interaction. In Fig. 14.13*a,* there are a pressure-control loop and a level-control loop on the same process vessel. In this case, moving the pressure-control valve has no effect on the level. Similarly, moving the level-control valve does not affect the pressure. In this application there is no problem with interaction, and it is obvious which control valve should be connected to which controller.

Figure 14.13*b* is the other extreme. The flow controller (FC) controls the overall flow rate of the blended fuel stream. The viscosity controller (VC) controls the viscosity of the blended stream by manipulating the amount of diluent that is blended in. Moving either control valve will upset both the controlled variables. This is interaction at its worst. It may be difficult to establish how the control valves and the controllers should be paired or even if it is possible to have both variables on automatic control at the same time.

14.4.2 Open-loop matrix approach

Possible answers to the questions raised about interacting control loops may be found if the relevant information is presented in matrix form. Referring again to Fig. 14.13*b,* both control loops are transferred to manual control (i.e., open loop). Then the control valves are opened one at a time by increasing the signals m_1 and m_2 to observe the effect on the variables, flow rate and viscosity, which are to be controlled. This produces the following data:

- Opening either valve v_1 or valve v_2 increases flow.
- Opening valve v_1 increases the viscosity.
- Opening valve v_2 decreases the viscosity.

Let us put this information into a matrix, in which a plus sign represents "increases" and a negative sign represents "decreases," as shown below.

	Flow	Viscosity
v_1	+	+
v_2	+	−

F. G. Shinskey has proposed the following theory regarding matrices of this type. This grouping is called a *positive matrix* because it has three positives and one negative. A matrix that has three negatives and one positive is also a positive matrix. A positive matrix is good news because it means that stable control should be possible with both loops on automatic, provided that the controllers and control valves are correctly paired.

If the matrix has four signs all the same, or if it has two positives and two negatives, then the matrix is a *negative matrix*. Under these circumstances, the interaction is such that it will not be possible to have both control loops on automatic, at their optimum settings, no matter how the controllers and the valves are paired.

Figure 14.14 is an application that has a negative matrix involving two temperature-control loops on a fractionating column. For fractionation to function properly, a temperature gradient must be established up and down the column. The top of the column is cooled by pumping back reflux, which is condensed overhead vapors. The bottom of the column is heated by boiling up part of the bottom liquid, and injecting the vapor back into the column. For consistent fractionation, both column temperatures should be held at their correct values.

Opening the valve on the reflux causes both temperatures to decrease. Opening the valve on the heat to the reboiler causes both temperatures to increase. Consequently, the matrix will have two positives and two negatives, which makes it a negative matrix. It will not be possible to get both temperature controllers to function on automatic, at their optimum settings, using the arrangement shown in Fig. 14.14.

14.4.3 Proper pairing

The point has been made that if the matrix is positive, then it is possible to have stable operation with both loops on automatic, *provided that the controllers and the control valves are paired correctly.* The logical question, therefore, is, "How can the correct pairing be determined?"

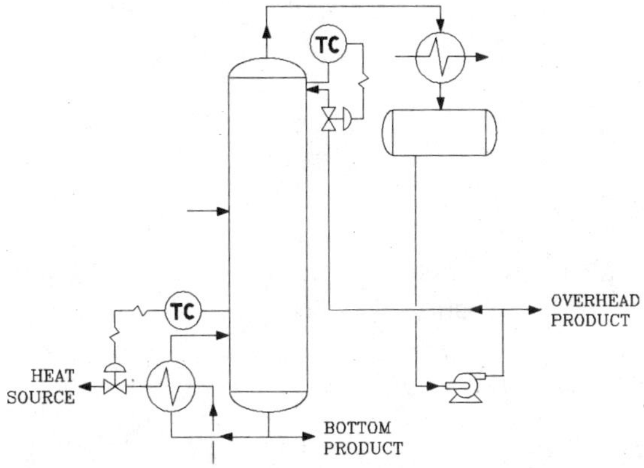

Figure 14.14 A fractionator control application with a negative matrix.

Let us rerun the test on the application shown in Fig. 14.13*b,* only this time we will record not only the direction which the controlled variable changes but also how much it changes. Control valve opening is measured in terms of control valve stroke, so percent valve opening and percent controller output signal are equivalent. This yields the following data:

- With v_2 fixed, a 5 percent increase in v_1 opening caused the flow F to increase by 8 percent and the viscosity V to increase by 3 percent.
- With v_1 fixed, a 10 percent increase in v_2 opening caused the flow to increase by 5 percent and the viscosity to decrease by 12 percent.

Gain is the change in output divided by the change in input generally, so instead of the signs only, the gain value, with the proper sign, is placed in the matrix. Since a 5 percent increase in m_1 produces an 8 percent *increase* in flow, the gain resulting from this pairing is +1.6. The other values are determined in the same way and plugged into the matrix. The matrix formed in this way is the *open-loop matrix.*

	Flow	Viscosity
v_1	+1.6	+0.6
v_2	+0.5	−1.2

At this point, some manipulation of the matrix is required in order to reduce it to a form from which valid conclusions can be drawn. This is not a math textbook; therefore, the steps for reducing the matrix will be shown, but without any attention to the mathematical theory underlying the manipulation of matrices. In any case, the manipulation process in this example is relatively straightforward, since the matrix is the smallest possible, a 2×2.

Step 1: Replace the upper-left value and the lower-right value with the product of these two values. *Retain* the positive or negative sign that results from multiplying the two numbers.

Step 2: Replace the upper-right value and the lower-left value with the product of these two values. *Reverse* the sign which results from multiplying the two numbers. The result of steps 1 and 2 is shown in the following matrix.

	Flow	Viscosity
v_1	−1.92	−0.30
v_2	−0.30	−1.92

Step 3: Add the two numbers that are in any row or in any column. (The result is the same, whichever pair is chosen.) In the case of the preceding matrix, the sum of the two numbers is −2.22.

Step 4: Divide each of the numbers in the matrix from steps 1 and 2 by the sum determined in step 3. This will result in a new matrix, shown below. This new matrix is called the *relative-gain matrix.* Notice that in the relative-gain matrix the numbers in each row and in each column add up to 1 and that they are all positive if the math has been done correctly.

	Flow	Viscosity
v_1	0.86	0.14
v_2	0.14	0.86

When the relative-gain matrix has been obtained, the reduction of the matrix is complete. The values in the relative-gain matrix provide two items of information that are useful. First of all, the correct way to pair the controllers and control valves is on the higher number, in this case on the 0.86. This means that the flow controller should be connected to control valve v_1 and the viscosity controller to control valve v_2.

Second, the gap between the higher and lower numbers in the relative-gain matrix is an indication of how much interaction the two control loops are going to have to cope with and thus how well they are likely to perform on automatic control. The difference between 0.86 and 0.14 is favorable and indicates that stable operation should be attainable in both loops with both controllers functioning at a decent gain setting.

The ultimate best values to end up with in the relative-gain matrix would be 1.00 and 0.00. As the gap between the numbers narrows, however, it will become more and more difficult to achieve satisfactory control in the face of the interaction that is going on. Lower and lower controller gain settings will be needed, to the point where one has to wonder whether any control action is taking place at all. If the numbers become 0.50 and 0.50, then all hope is gone of having both loops on automatic.

14.4.4 Possible remedies for interacting control systems

In general, problems with interacting control systems can be approached in any of three ways. There are, unfortunately, situations in which none of the three procedures will achieve the goal desired, i.e., to have both control loops operating on automatic control. If it comes to this, then what remains is to put the lower priority loop on manual.

Method 1: Detune the controllers. The obvious question is, "Which one, and how much?" The ultimate in detuning would be, of course, to put one controller on manual. The chances of arriving at the best pattern for detuning the controllers will be improved if the data required for setting up a relative-gain matrix can be obtained. The spread of the numbers in the relative-gain matrix will then be a useful guideline for the detuning procedure.

Generally speaking, detuning the controllers means lowering the gain setting on either or both of the controllers, since it is primarily the amount of proportional response that will affect the stability of the control loop, as well as any other loop with which it may interact. It should not be necessary to alter the reset rate or the derivative time.

If the ratio of the numbers in the relative-gain matrix has a good spread, say, 2:1 or better, then the controller gain in the faster-recovering loop should be left at its optimum value and the desired stability attained by lowering the gain in the controller on the slower loop.

If the ratio of the numbers in the relative-gain matrix is less than 2:1, then detuning should be carried out by lowering the gain in both controllers in equal fractions until the desired stability is reached.

Method 2: A decoupling system. As the name implies, the intent is to decouple the two interacting loops so that they act independently, with the automatic controllers tuned to their optimum settings. All this sounds highly commendable, but the fact remains that there is no way to remove the interaction paths between the control loops by way of the plant.

What sometimes can be done, however, is to design the control systems so that when one controller changes its output, a fragment of the output signal, modified so as to have an equal and opposite effect to the interaction effect, is combined with the output of the second controller. The net result, hopefully, is that the second controlled variable will then be affected only by the output signals from its own controller. A second decoupling path is included, from the output of the second controller to the output of the first, so that any interaction between them through this path will likewise be canceled out.

Figure 14.15 shows a possible decoupling control scheme in block form. It appears involved at first, but the explanation is relatively straightforward. The circular symbols with the crosses are simple addition modules. The square symbols represent gain modules. The upper loop (loop 1) is controlling flow. The lower loop (loop 2) is controlling viscosity. For both loops, the process block contains two paths: the normal path that establishes the value of the variable which that loop controls, and a second path that causes interference with the other controlled variable.

Considering first loop 2, which is controlling the viscosity, the output of the viscosity controller m_2 would normally proceed along the straight-line path through the process at g_{22} and produce the appropriate value of the viscosity V. Unfortunately, V also can be affected by interaction that proceeds from the output of the flow controller m_1 through the g_{21} element of process 1.

The function of the decoupler d_1 is to create a signal that is equal and opposite to the interaction effect so that the net effect on V is negligible. This is done by splitting off the m_1 output signal, modifying it through a proper value of the gain d_1, and subtracting the resulting signal from the m_2 signal.

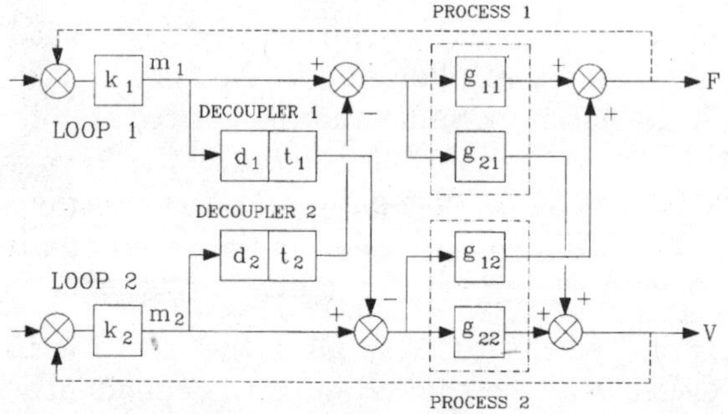

Figure 14.15 A decoupling arrangement for two control loops.

The values of the g elements are those which would be recorded for making an open-loop matrix. The value that should be used for d_1 can be calculated based on the appropriate g values, along with the following line of reasoning. If the interaction path from the loop 1 controller to the controlled viscosity is traced out, it will be seen that the m_1 signal passes first through a summation module, which has a gain of 1, then through the g_{21} module, and finally arrives at a second summation module, which also has a gain of 1. The magnitude of the m_1 signal at this stage will be therefore $(m_1 \times g_{21})$.

If the m_1 signal is traced through the decoupling route, it passes first through the decoupler, which has a gain of d_1, and arrives at a summation module in the m_2 path, where it subtracted from the m_2 signal. The gain of this summation module is also 1. The value of the signal that will be subtracted from the m_2 signal will be $(m_1 \times d_1)$. However, this subtracted component of the m_1 signal then proceeds on through the g_{22} element in the process so that its ultimate effect on the controlled viscosity is $(m_1 \times d_1 \times g_{22})$.

Thus, for effective compensation, $m_1 g_{21} - m_1 d_1 g_{22}$ must equal 0, and

$$d_1 = \frac{g_{21}}{g_{22}}$$

Following the same line of reasoning, the value of d_2 for decoupler 2 should be

$$d_2 = \frac{g_{12}}{g_{11}}$$

Suppose that the values taken to form the open-loop matrix for the flow and viscosity control loops were used to design a decoupling system of the type shown in Fig. 14.15. The original data were as follows: With m_2 constant, a 5 percent increase in m_1 caused the flow to increase by 8 percent and the viscosity to increase by 3 percent. This puts the value of g_{11} at 1.6 and the value of g_{21} at 0.6. With m_1 constant, a 10 percent increase in m_2 caused the flow to increase by 5 percent and the viscosity to decrease by 12 percent. The values of g_{12} and g_{22} should be 0.5 and -1.2, respectively. Therefore,

$$d_1 = \frac{0.6}{-1.2} = -0.5 \qquad \text{and} \qquad d_2 = \frac{0.5}{1.6} = 0.313$$

The significance of the negative sign for the value of d_1 is that the output of the d_1 module should decrease when its input increases.

The remaining item in the Fig. 14.15 decoupling scheme is the two modules that are marked t_1 and t_2. The entire development of the decoupling system so far has been based on gain values, which, as we know, exist in the steady-state world. However, feedback control systems are made up of dynamic elements as well as elements that have static characteristics. The t_1 module is included to compensate for any differences in the reaction times that are in the interaction and the decoupling paths.

With t_1 set at the right value, the decoupling signal from d_1 will affect the controlled viscosity through the g_{22} path at essentially the same time as the interference signal affects the viscosity through the g_{21} path. Effective compensation requires that the compensating signal not only be of the correct magnitude but also that it arrive at the correct moment.

The system could be tested in the following manner. Set loop 2, the viscosity control loop, on manual control, and allow the viscosity to settle at a steady value. Any disturbances to this loop will now show up as a change in viscosity. Then make a setpoint change to the flow controller. If the viscosity remains constant, the decoupling is accurate. If the viscosity drifts away to a new value, the d_1 value is not correct. If the viscosity drifts away from its steady value and then comes back again, the d_1 value is correct, but the dynamic compensation t_1 is not set properly. The test should then be repeated with the flow loop on manual and a change in setpoint in the viscosity loop.

The decoupling system requires two gain modules, two summation modules, and two delay modules to assemble it. These modules can be individual analog components. However, it will clearly be neater job if the decoupling system can be configured in software.

It should be understood that the decoupling scheme shown in Fig. 14.15, along with the accompanying explanation, has been presented for the purpose of illustrating the principles that are involved in decoupling. It should not be assumed that anyone could put together the control scheme in Fig. 14.15 from a kit of parts and get it to work. There are other factors involved, including the problem of getting the correct initial values for the outputs of the two controllers so that they will start up harmoniously. Another problem is how to tune the controllers, in view of the fact that the output of each controller affects the process through two paths: the normal direct path and a parallel path through the decoupler, down the output path of the second controller, and back through the interaction path.

Setting up a decoupling system is no job for a neophyte in control. It is best left to an expert. A more detailed explanation of decoupling systems is contained in F. G. Shinskey's book, *Process-Control Systems*.

Method 3: Rearrange the control loops so as to change the matrix to a positive matrix. This will reduce the effect of interaction. The words *if possible* apply here because in many cases such rearrangement is simply not practical. It is possible, however, in the case of the fractionation col-

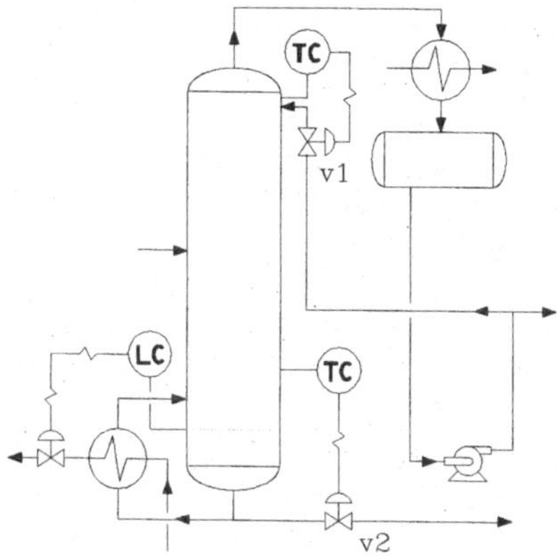

Figure 14.16 Rearranging the controls creates a positive matrix.

umn example in Fig. 14.14. The new control scheme is shown in Fig. 14.16.

An understanding of the overall control method in Fig. 14.16 depends on understanding the function of the level controller (LC). This level controller is controlling the level of the liquid in the bottom of the column by manipulating the quantity of liquid that is boiled off. If the level rises, the level controller brings in more heat to vaporize the liquid at a faster rate and bring the level down again. The matrix for this case can now be determined.

If the reflux control valve v_1 opens wider, the temperature at the top of the column will decrease (−), the same as in Fig. 14.14. However, the additional liquid reflux that flows down the column will cause the level in the bottom of the column to rise. The level controller will open the heat-control valve to compensate for this, and the temperature at the bottom of the column will rise (+).

If the bottom liquid draw-off control valve v_2 opens wider, it will cause the level in the bottom of the column to drop. The bottom level controller will cut back on the heat to compensate, which will cause the temperatures at both the bottom and the top of the column to decrease (−). Thus the matrix will appear as the one below. This is a positive matrix, which indicates that conditions are satisfactory for both temperature controllers to operate together on automatic.

	Top Temperature	Bottom Temperature
v_1	−	+
v_2	−	−

Chapter

15

Control Valves

Every text on process control needs a section on control valves. The control valve is, after all, an important component of the control system. The question is, "What needs to be said about control valves that has not been said numerous times already in other texts?"

It is not the purpose of this text to recapitulate already published material. Consequently, there will be no discussion of valve end connections, bonnets, packing, metallurgy, or valve noise. Persons who require information on these topics will find their local control valve representative ready to oblige. If the fact that what used to be called a *leak* is now being called a *fugitive emission* is any indication, we can look forward to many profound articles on the philosophy of the design and use of various types of stuffing boxes and packings.

What really makes a control valve different from any ordinary plant valve? Answer: two things. First of all, a control valve must have the ability to provide smooth, consistent increases in flow rate as the inner valve (plug, piston, vane, or whatever) is moved from the closed to the open position. The same applies , in reverse, as the inner valve is moved in the opposite direction. This is called having *good throttling characteristics.*

The second is that the control valve has to be positioned, for the correct flow rate, by a pneumatic or electronic signal originating from an automatic controller. Therefore, the control valve cannot be set with a handwheel or a lever. It requires a special type of actuator to do the positioning.

The discussion in this chapter will be configured around these two considerations.

15.1 Throttling Characteristics

The tap on the outside of one's house, to which the garden hose is attached, does not have very good throttling characteristics. About 80 percent of the total available flow is obtained with the first half turn of the handle. The other two turns get only the last 20 percent. Throttling characteristics of this type would not work very well in process control systems.

The throttling characteristics for a control valve are revealed in a graph of the control valve's *static* characteristics. Whatever throttling characteristics the control valve may have will be built into the valve by the manufacturer. Currently, only two separate characteristics are available for process control. These are *linear* characteristics and *equal-percentage* characteristics. Authorities on this subject may contend that there is a third characteristic, *quick opening*. However, the quick-opening characteristic is properly used only in control systems in which the controller operates in on-off control. Since stability is a requirement of nearly all process control systems, this means that the control valve must be able to function in a throttling position, something that a quick-opening control valve is not designed to do. Therefore, quick-opening characteristics will not be considered here.

In Fig. 15.1, the solid-line graphs are the theoretical static characteristics for two different control valves, one with a linear characteristic and the other with an equal-percentage characteristic. The static characteristic for any control valve is a graph of valve opening versus valve stroke. These two terms may appear to refer to the same thing, but strictly speaking, they are not the same. *Valve stroke* refers to the amount that the inner valve is moved by the actuator. *Valve opening* refers to the amount of area, inside of the valve body, that is open to flow. Valve stroke is the input; valve area is the output.

To complicate matters further, *valve opening* does not necessarily mean the same as *flow rate* through the valve. The reason for this is that the fluid flow rate created at the control valve depends on two factors: the valve opening, as before, and the difference between the

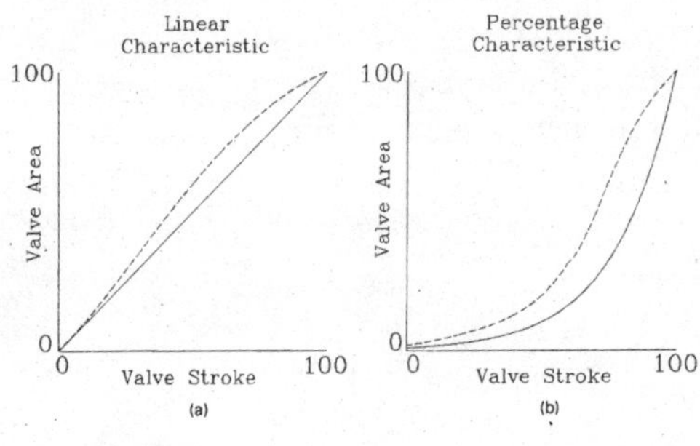

Figure 15.1 Linear and equal-percentage valve characteristics.

pressures at the inlet and the outlet of the valve body. It is this pressure difference, or *pressure drop,* as it is usually called, that provides the driving force that moves the fluid through the valve. In most cases, increasing the pressure drop, and hence the driving force, will increase the flow though the valve, even though the valve opening has not been changed.

The linear characteristic is based on the rule that a change in stroke produces a proportional change in valve area, valve area being equivalent to valve opening. In mathematical terms, this means

$$\Delta A = k\Delta x$$

where A and x are the area and the stroke, respectively, and k is the proportional constant. When this differential equation is solved, it becomes

$$A = C + kx$$

C being another constant. If the characteristic is plotted on a graph that has 0 to 100 percent scales for both the valve stroke and the valve area, then the value of the constant k will be 1, and C will be zero. The graph is an obvious straight line, and the valve characteristic is called *linear.*

The so-called equal-percentage characteristic valve follows the rule that a change in stroke produces a proportional *fractional* change in valve area. Written as a differential equation, it looks like

$$\frac{\Delta A}{A} = k\Delta x$$

Notice here that ΔA is the change in area, so $\Delta A/A$ will be the fractional change. If this differential equation is worked out to a conclusion, it will become,

$$A = C\varepsilon^{kx}$$

In the percentage characteristic graph in Fig. 15.1, the values of C and k are 0.10 and 0.461. For this type of valve characteristic, *equal fraction* would have been a more descriptive name. However, *equal percentage* is the name that was originally chosen, and for all intents and purposes, it is now cast in concrete.

The graphs in Fig. 15.1 show something else that is of interest, namely, how closely the characteristics of an actual control valve match the theoretical characteristics. The broken-line graphs are taken from the published data of one of the leading control valve manufacturers for a linear valve and for an equal-percentage valve, respectively. Yet another question that might be asked is, "Just how closely will the characteristics of two supposedly identical control valves match each other?"

15.1.1 So what if control valves can be characterized?

The answer to this question depends on the circumstances. We need to keep two facts in mind. The first is that control valve characteristics are *static* (steady-state) characteristics, and static characteristics affect the stability of the control system, not its period of oscillation and recovery time. This is so because the significance of a nonlinear characteristic in any control system component is that the gain of the component will change when its operating point changes. If the gain of any component in the control system changes, the gain of the whole system changes, and so does its stability. The system becomes either more or less stable.

The second fact is that what we ultimately want is a control system that will recover with an amplitude ratio of about 1/4 over the expected range of operations of the control system without having to change the gain setting in the controller to compensate for the different system gains at different operating points.

In the 1990s, much of what is being done in process control is being accomplished with digital computer-based control systems. One feature of all digital systems is that it is easy to generate steady-state functions. Thus, if any component in the control system, such as the measuring element, or the process, or the control valve, turns out to have a nonlinear static characteristic, then it is a simple matter to write a compensating function in the system software that is equal and opposite to that of the nonlinear component so that the combination of the two characteristics is essentially a linear characteristic.

Plants that are controlled by analog measurement and control hardware are a different matter. It is not that easy to incorporate oddball static characteristics into analog systems to compensate for apparent nonlinearities in particular components. There are analog function generators available, to be sure, but they do not have the flexibility to produce any desired function, as do the digital systems. Furthermore, analog function generator modules require maintenance. An analog function generator can lose its calibration just like any measuring element. It must be tuned up from time to time by someone who understands not only how the function generator works but also how it fits into the system as a whole.

Another consideration is the range of operation, or loads, through which the control valve will be expected to function. If the difference between the highest and lowest operating points is relatively small, say, 20 percent or less, then even if the overall characteristic is a nonlinear curve, the segment of the curve between the highest and lowest operating points will be almost a straight line. Oil refineries tend to run in this manner. They are brought up to their normal throughput, remain at this level for a year or more, and then are shut down. It is for this reason that in plants such as these, problems with variable stability due to changes in operating level hardly ever come up. Hence discussions about whether linear or nonlinear characteristics should be used are mainly academic.

There are certain real-life processes that have a nonlinear static characteristic in which the gain falls off as the process is pushed to higher and higher operating levels. A possible example might be a fired heater, in which the outlet temperature of the heated liquid is being controlled by the quantity of gas fired in the heater, as in Fig. 14.8. In this application it may be found that as the outlet temperature is pushed to a higher and higher value, it becomes more and more difficult to raise the temperature, owing to the fact that a greater fraction of the energy input is going into losses. Specifically, it may require a larger increase in gas firing rate to change the temperature from 280 to 285°C than it does to raise the temperature from 150 to 155°C.

Another case, which is fairly common, is that of a control valve located in a process system in which the beginning and ending pressures of the system remain relative constant irrespective of the flow rate through the system. Along with the control valve, the system will contain piping and probably miscellaneous process equipment such as heat exchangers. It is an unfortunate law of nature that pressure drop is required to drive the fluid through any of these devices. Furthermore, the pressure drop increases as the *square* of the ratio of the new and old flow rates.

If the flow rate in this system is to be increased, where will the additional pressure that nature requires come from? The overall system pressure drop does not increase to help the situation; consequently, the only place the extra pressure drop can come from is the control valve. The outcome is that as the flow rate in the system goes up, the pressure drop at the control valve goes down.

It is the pressure drop at the control valve that provides the driving force for moving the fluid through the valve. At a lower pressure drop, the same change in valve position produces a smaller change in flow rate. Therefore, as the flow rate increases (another way of saying that the control system goes to a higher load), the control valve gain falls off.

The control valve with the equal-percentage characteristic was intended to compensate for control system components that suffer a reduction in gain as the system moves to higher loads. The increasing gain contributed by the percentage characteristic compensates for the reduction in gain in the process, the control valve, or the measuring system. The compensation is never perfect, but the equal-percentage valve in combination with the nonlinear component provides a more linear characteristic overall than if a control valve with a linear characteristic were used.

The point is that the equal-percentage characteristic in the control valve is supposed to be used in combination with a component for which the gain decreases at higher system loads. Thus the combination of the two is more linear, over the range of system loads encountered, than either component is by itself. Yet there are "rule-of-thumbers" who insist on specifying equal-percentage characteristics for all control valves, on the mistaken belief that percentage characteristics always give better control.

Some poor souls actually believe that the equal-percentage characteristic in the control valve compensates for the nonlinear characteristic in an orifice meter. For an orifice meter, the input is flow, and the output is differential pressure. Thus the orifice meter characteristic is a square law, not a square root law. Combining an equal-percentage characteristic with a component that follows a square law could have catastrophic results. However, people do this, and get away with it, simply because once the process is up an running, the changes in load are minor. And once again, if two points on a curve are close enough together, then the segment between them is almost a straight line.

15.1.2 How does one decide on the correct characteristic?

The choices for the control valve characteristic are linear or equal percentage. Which one should be chosen, and on what basis?

From the discussion so far, it would seem that it should only be a matter of estimating whether the static characteristics of the process, and the measuring system, are linear or nonlinear. Then the best characteristic for the valve would be the one which, when combined with the static characteristics of the other components, would give the closest approach to linearity overall. If it were only that simple.

Consider the temperature-control system shown in Fig. 15.2. In this system, glycol solution is being heated in a steam-heated

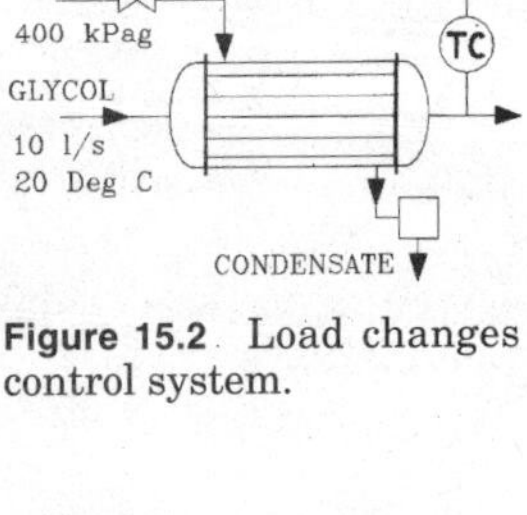

Figure 15.2 Load changes in a steam heater control system.

exchanger. The required temperature of the glycol solution coming off the heater is 50°C. The solution usually flows through the heater at a rate of 10 l/s, and its temperature entering the heater is usually 20°C.

By definition, a load change in the control system is any change that requires the control valve to move to a different position in order to keep the controlled variable on control. Let's consider the load changes that can affect this temperature-control system.

- The inlet temperature of the glycol solution may vary from 5 to 30°C.
- The flow rate of the solution may vary from 4 to 16 l/s.
- The steam pressure at the inlet to the control valve may vary from 400 to 100 kPag.

Notice that each of these cases represents an increase in load because the control valve will have to open wider to compensate for the load change to maintain the outlet temperature at the desired value, 50°C. The setpoint for the outlet temperature is always 50°C; therefore, it is not a potential load change in this application.

Let us assume that the efficiency of the heat exchanger remains relatively constant irrespective of the load on the exchanger. That is, the fraction of the input heat that is lost somewhere and does not go into heating the glycol solution remains about the same no matter how great the demand for heat may be. It is true that changes in temperature difference across the exchanger tubes and changes in the flow rate of the glycol solution through the tubes will change the *rate* at which heat is transferred, in accordance with the familiar exchanger rule $Q = km\ \Delta T$. However, what we are concerned with here is the amount of heat that is transferred, not how quickly it is transferred.

If the efficiency of the exchanger is constant over its range of operating loads, then the same amount of heat will be required to raise the temperature of the glycol solution 1°C, irrespective of the incoming temperature of the glycol solution or its rate of flow. In both cases, the heat to the glycol solution (the output) will be proportional to the heat provided by the steam (the input), and the static characteristic of the exchanger will plot as a straight line, indicating that the gain of the exchanger is not changing with load.

Under these circumstances, it seems clear that for the combination of the control valve and the process to have linear characteristics overall, then the control valve should have linear characteristics. Using an equal-percentage valve in this case would present a nonlinear system to the automatic controller.

Now consider the case where the load change is a decrease in the steam pressure at the inlet of the control valve. If this pressure falls off, so does the pressure drop across the control valve, along with the ability of the control valve to create steam flow to the heater. The capacity of the control valve, i.e., the rate of flow of steam when the valve is wide open, will be less if the pressure drop is less. The consequence of this is that at a lower steam inlet pressure the same movement of the control valve creates less of a change in steam flow, which is another way of saying that the gain of the control valve is falling off. Quantitatively, the valve gain at 100 kPag inlet pressure will only be about 40 percent of the gain at 400 kPag.

To compensate for the decrease in gain caused by decreasing pressure drop, the control valve can be provided with an equal-percentage characteristic, which produces an increase in valve gain as the valve is opened wider. The combination of these two effects will not produce perfect linearity but will be better overall than if a control valve with a linear characteristic were used.

What this points out is that it is not always a simple matter to decide which characteristic, linear or equal percentage, the control valve should have. In the example, two of the possible load changes would require a valve with linear characteristics, while equal-percentage characteristics are needed to cope with the third load change. This implies that it may be necessary to analyze the process to determine which of the potential load changes is going to predominate. In some cases, the only option is to make the best guess as to whether linear or equal percentage should be used. If the guess proves to be incorrect, then order a new set of trim for the valve.

In any event, it is worth remembering that control valve characteristics are an exercise in static characteristics, and as such, they affect the gain of the control system, though not its period of oscillation. Therefore, they may be a factor in problems with varying stability at different operating points.

Even if the process characteristic is nonlinear, if the highest and lowest operating points are not far apart, the locus between them will be almost a straight segment. *Not far apart,* in the control system sense, means that the ratio of the maximum load to the minimum load should not be greater than 2:1. If this is the case, either valve characteristic will probably work acceptably. The same applies if the process and control valve gains are such that the controller operates at a relatively high gain, say, 3 or more.

The possibility of process gain changes at different loads can be investigated by running process reaction tests. Chapter 13, and especially Fig. 13.2, elaborates on the fact that the point where the tangent to the process reaction curve intersects the vertical ($t = 0$) axis has a bearing on the gain that should be set on the automatic controller. Expanding on this idea, if two process reaction curves were run, one at minimum load and the other at maximum load, and the controller gain that resulted was the same in both cases, then the conclusion that follows would be that the gain of the process was not changing with load, at least not with the type of load change that was used in the test.

It would be a mistake, however, to extrapolate this conclusion to the belief that the process will react linearly to all types of load changes.

15.2 Spring-Diaphragm Actuators

15.2.1 Forces and sizing

First of all, why should anyone be concerned about these matters, when the supplier of the control valve will take care of sizing the actuator as part of his or her service? Many control valve users will agree with this notion, while, on the other hand, they are totally committed to sizing control valves themselves, supported in most cases by an elaborate computer program. This practice is followed even though the valve supplier also will size the control valve if he or she is given this responsibility.

The control valve will fail to perform its function if it is not sized properly and if it fails to throttle the flowing fluid correctly because the actuator cannot cope with the forces created inside the valve body. An understanding of actuator sizing may well be as important to successful automatic control as an understanding of valve body sizing, which is dealt with in Chap. 17.

Figure 15.3 presents sketches of the basic parts of a control valve that opens as the air pressure on the top of the diaphragm is increased. When the valve is closed (Fig 15.3*a*), the difference between the inlet and outlet pressures at the valve body causes a downward force on the plug, which tends to push the valve open. The magnitude of this force will be equal to the pressure drop across the

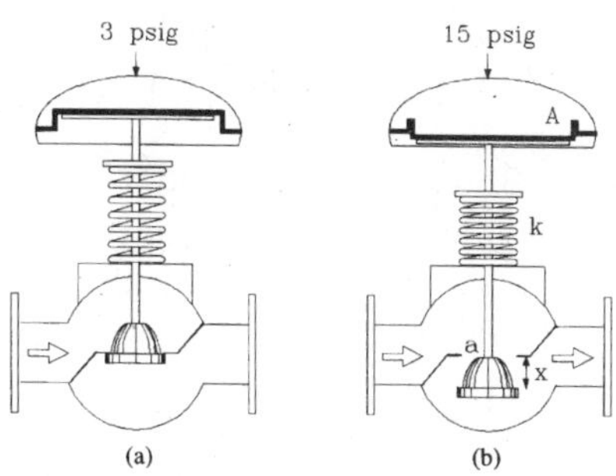

Figure 15.3 Force factors in a working control valve.

valve body (inlet pressure minus outlet pressure) multiplied by the area of the valve plug that is exposed to this difference in pressure.

When the valve is wide open (Fig. 15.3*b*), the plug is essentially out of the flow path, so any *vertical* force which the flowing fluid exerts on the plug is negligible. Using the valve-closed and the valve-open situations, the size of the spring and diaphragm of the actuator can be worked out. The calculations are based on balancing the net upward and downward forces, much like the problems which students encounter in Applied Science I.

Since the most popular operating range for control valves is 3 to 15 psig, the fps system of units will be used here. Forces will be in pounds, distances in inches, areas in square inches, and pressures in pounds per square inch. The following variables need to be defined:

- A is the effective area of the diaphragm in the valve top (in^2).
- y is the length of the full valve stroke, from the closed to the open position (in).
- k is the spring constant of the valve spring (lb/in).
- a is the area of the plug on which the difference in inlet and outlet pressures acts when the valve is closed (in^2).
- ΔP is the difference between the inlet and outlet pressures at the valve body (psi).
- S_i is the initial compression of the valve spring, i.e., the amount that the spring is compressed during calibration to cause the valve to begin to open at an air loading pressure of 3 psig.

Consider first the forces that are acting when the valve is closed. The only upward force is the initial compression of the spring S_i. There are two downward forces: the pressure drop acting on the plug and the 3 psig acting on top of the diaphragm. Therefore,

$$S_i = \Delta P \times a + 3A \tag{15.1}$$

When the valve is open, the upward force will be the sum of the initial compression of the spring plus the additional spring force that results from compressing the spring through a distance equal to the stroke of the valve. There is only one downward force, that contributed by the 15 psig acting on top of the diaphragm. Equating the upward and downward forces gives

$$S_i + ky = 15A \tag{15.2}$$

If Eqs. (15.1) and (15.2) are combined so as to eliminate S_i and then rearranged to obtain an expression for k, the result is

$$k = \frac{12A - \Delta Pa}{y}$$

This relation contains a rather important message. The size of the diaphragm A will be fixed for any given actuator. The A term actually represents the nominal size of the actuator. As the unbalanced force from within the valve body (ΔPa) becomes greater, the proper value for the spring constant k becomes smaller. This is not a desirable development because the lighter the spring, the more susceptible the valve is to buffeting from the action of the flowing fluid, and it may be more difficult to position the valve consistently at the right spot for good control. Directionally, the stiffer the spring, the better.

By completing this calculation, the correct k value for the spring can be worked out. The missing element, however, is the size of the actuator, or more specifically, the value of A. So far it has been assumed that this is known. The control valve supplier will have a whole range of actuators from small to large, each with its own size of the diaphragm. Which one should the supplier provide for a particular valve he or she is bidding on, keeping in mind that the larger the actuator, the more money the supplier will have to ask for in his or her quotation? From the user's point of view, is it possible to verify that the actuator recommended will be large enough to handle the fluid forces and still position accurately?

A useful guide to answering the second question is the size of a factor that has come to be called the *bench set.* When a control valve is calibrated for correct travel, the calibration is done in the shop of either the valve supplier or the user. It is never done in the line with the process fluid flowing through it. Inasmuch as there is no force at all coming from the inside of the valve body when the valve is in the shop, a 3- to 15-psig calibration done in the shop will not turn out to be a 3- to 15-psig calibration when the valve is in service.

For a control valve that opens with increasing pressure on the diaphragm, as in Fig. 15.3, the valve would be wide open at 15 psig, since the vertical forces on the valve are negligible when the valve is open. The same would not apply to the closed position, because there is a force at work inside the valve body tending to push the valve open. Thus it might require a pressure of only 1 psig, rather than 3 psig, to start the valve opening when it is in service.

The *bench set* is the amount that the normal 3- to 15-psig operating range has to be modified in a shop calibration to compensate for the force from inside the valve body so that in service the valve will actually have a range of 3 to 15 psig.

The vertical force from inside the valve body is equal to ΔPa. The air pressure on the diaphragm required to overcome this force is equal to

$$\frac{\Delta Pa}{A}$$

If the valve in Fig. 15.3 is to have an operating range of 3 to 15 psig when it is in service, then in the shop it will have to be calibrated for a range of

$$\left(3 + \frac{\Delta Pa}{A}\right) \text{ to 15 psig}$$

Suppose that this turned out to require the control valve to have a shop calibration of 5 to 15 psig in order to have an operating range of 3 to 15 psig in service. The difference between the 5 psig and the nominal 3 psig, or 2 psi, is the *bench set.*

Since the bench set is equal to $\Delta Pa/A$, we can see that the bench set increases with the pressure drop at the valve and with the unbalanced area inside of the valve body on which the pressure drop can act. The bench set decreases with the size of the actuator, as evidenced by the diaphragm area A term in the denominator.

The ideal situation for any spring-diaphragm actuator is that the forces developed by the diaphragm and the spring are of such magnitude that they completely swamp out any force that originates from inside the valve body. This will virtually ensure accurate positioning. An increasing bench set is an indication that the force which the actuator can develop is losing its dominance over the force which comes from within the valve body. If this happens, the force created by the flowing fluid may become a factor in how well the valve is able to throttle the fluid stream at the right value for control. Any bench set that is greater than 2 psi should be cause for concern that the actuator is not large enough to do the job. The value of the bench set is usually stamped on the data plate of the control valve.

Furthermore, if a control valve has an operating range of 3 to 15 psig, it will move through its full travel, closed to open, with a pressure change of 12 psi on the diaphragm. Any bench set that is required will be subtracted from the nominal 12-psi span. A control valve that needs a bench set of 2 psi requires that the valve move through its full travel with a change in air loading pressure of 10 psi. To accomplish this, the valve actuator has to be equipped with a lighter spring, i.e., a spring with a smaller k value. This in itself is not conducive to accurate positioning.

In the case of a control valve that opens as the air pressure on the diaphragm increases, the unbalanced force inside of the valve body works with the pressure on the diaphragm and against the spring. If a bench set is required, the calibration range for the valve will be shorter than 3 to 15 psig. For a control valve that closes as the pressure on the diaphragm increases, the unbalanced force works with the spring and against the pressure on the diaphragm. Then, if any bench set is required, the calibration range will be greater than 3 to 15 psig. A required bench set of 2 psi would make the calibration range 3 to 17 psig. The bench-set adjustment always affects the air loading pressure at which the valve should be closed.

A calculation of the value of the bench set requires a value for the pressure drop ΔP. Bearing in mind that this should be the pressure drop present when the valve is closed, what value should be used? In some circles, ΔP is taken to be that which would be present with the nominal line pressure at the inlet of the valve and *zero* pressure at the outlet. This is a conservative, and therefore safer, approach. However, if there actually is a positive pressure at the valve outlet when the valve is closed and the calculation is based on zero pressure instead, then the bench-set value will be incorrect, and the operating range of the control valve, in the line, will not be 3 to 15 psig.

At this point, the reaction of many persons will be, "So what?" It all depends on the importance that one attaches to having the standard 3- to 15-psig operating range maintained. Plant operators sometimes need to know when a control valve is closed or when it is open and how much. Each control loop is provided with a meter or a display that shows the operator the controller output, and this is supposed to correlate 1:1 with valve position.

Valve stroke transmitters that measure and transmit the stroke of a control valve are available, but they are rarely used. If the standard operating range for the control valve is not adhered to, then it is possible that the operator will be lead to believe, based on what he or she sees, that a valve is closed when in fact it is not. The point is that the accuracy with which a meter that displays controller output will reliably read out valve stroke depends on the valve having the standard 3- to 15-psig range.

The most obvious solution to eliminating, or at least minimizing, the bench set is to specify a larger actuator, one with a bigger diaphragm and a stiffer spring, but there is another alternative worth considering. This is to use an operating range of 6 to 30 psig instead of 3 to 15 psig. By so doing, the same actuator will develop twice the force, and any bench set necessary will be cut in half.

Whether or not the use of a different operating range will cause other problems for the user is, of course, the user's decision. However, if the suggestion is turned down, it should not be because the hardware required to go to a 6- to 30-psig range is not readily available. Current-to-air converters, valve positioners, and volume boosters with a 6- to 30-psig output are all available as standard equipment from the manufacturers of control valves.

15.2.2 Merits of spring-diaphragm actuators

The merits of spring-diaphragm actuators should not be discounted. They include the following:

- Simplicity. They are easy to understand and easy to repair. There are fewer things to go wrong. Therefore, they are reliable.
- They cost less than other types of actuators.
- The spring gives the actuator inherent proportional valve travel as the air pressure on the diaphragm is increased. A separate positioning device is not required for this purpose.
- They have a fail-safe position if air pressure is lost.

These are very strong arguments in favor of the spring-diaphragm actuator. The conclusion to be drawn from this is that for *any* process control valve, the spring-diaphragm actuator should be the logical first choice.

On the negative side, spring-diaphragm actuators require that the plant produce and distribute instrument-quality air, which is admittedly an expensive utility. It is for this reason that attempt after attempt has been made to devise an airless control valve actuator. Unfortunately, any airless actuator that has come on the market up to 1990 has not come close to matching the positive features of the spring-diaphragm actuator listed above.

Other weak points of the spring-diaphragm actuator are as follows:

- On some applications, they do not position the valve quickly enough to ensure good control.
- Much of the force developed by the air pressure on top of the diaphragm goes into compressing the spring. When the force from within the valve body is very great, due to either a large valve size or a high pressure drop, it may be necessary to dispense with the spring in order to develop sufficient actuator force to position the valve correctly.
- They operate on low-pressure air only.

These considerations sometimes render it necessary to specify a piston actuator rather than a spring-diaphragm unit.

15.3 Piston Actuators

Reliable as it is, at some point a spring-diaphragm actuator will run out of force or speed or both. Much of the force developed by a spring-diaphragm actuator goes into compressing the spring rather than positioning the inner valve against the fluid forces that develop inside of the valve body. The positioning speed is limited also by the relatively low air pressures used and the size of the pneumatic pilot relay that loads and unloads the air pressure to the diaphragm head.

There are many applications in which the inner valve has to be positioned accurately despite large forces acting on it. And there are going to be processes that react very quickly to input changes. Obviously, if the process reacts faster than the valve can be moved, it will be difficult to maintain control over the controlled variable.

When extra force or speed is mandatory, the spring-diaphragm actuator has to give way to the piston actuator. A piston actuator will develop greater force because

- Higher air pressures, 100 psig or more, can be used.
- A conventional piston actuator does not have a spring; consequently, all the available force goes into positioning the valve.

Once again, a hardware component that has been selected to circumvent the problems of an earlier component turns out to be capable of resolving those problems. However, as always happens, the new component brings along with it a set of problems of its own. For the piston actuator, these include the following:

- Piston actuators cost more money.
- Since a pure piston actuator does not have a spring, it lacks inherent proportional positioning. Therefore, a piston actuator must have a valve positioner. There is no choice in this matter.
- Since a piston actuator has no spring, it does not have a predictable fail-safe position. If it loses its air supply, the valve may go open, go closed, slowly or quickly, or it may remain in its last position. What it does will depend on the forces exerted from inside the valve body, the weight of the valve parts that can move, and possibly other factors.

15.4 Positioning Problems and Solutions

It goes without saying that accurate and consistent throttling of the control valve is a requirement for good automatic control. Without accurate positioning, the correct flow of fluid to the process, which is the process input, could not be achieved, nor could the balanced effect be established that is necessary if the controlled process variable is to be held at the desired value.

Two factors tend to work against accurate positioning. The first is the pressure drop between the inlet and outlet of the valve body acting on any unbalanced area of the inner valve. This causes a vertical force that may hinder the actuator from positioning the inner valve at the right place in its stroke.

The second is the valve stem and the stuffing box. The valve stem, which transmits the motion initiated by the actuator to the inner valve, must by encased in some type of sealant or packing that will prevent leakage of the fluid inside the valve body to the outside along the stem. To make the seal effective, the stem must be highly polished so that that the packing fits smoothly against it. The polished condition is also important so that as the stem moves back and forth through the packing, the packing will not be scuffed or otherwise damaged.

In addition, the packing must be of the correct type for the application, and it must be kept under the right degree of pressure. The nature of the fluid which the valve is handling, as well as its temperature and pressure at the valve body, will generally dictate the type of packing that should be used. Maintenance preferences also may play a part in the choice of valve packing materials.

In a well-designed stuffing box, the correct packing pressure will be maintained by a spring inside the stuffing box that applies a constant pressure of the right amount to the packing. In simpler stuffing box designs, the pressure is set by tightening two packing nuts that bear down on a packing gland, which in turns bears down on the packing. There is definitely a correct amount of packing pressure. Too little pressure will allow the fluid inside the valve to leak out along the valve stem. Too much pressure will cause the packing to bind on the valve stem and prevent the valve from stroking smoothly.

When a control valve starts out new, or just overhauled, it will usually work smoothly and without any leakage problem. Unfortunately, over a period of time, the packing begins to wear. The time period will depend on what the conditions are like inside the valve body and the amount that the valve is moving, hour in and hour out, to keep the controlled variable on control. Inevitably, however, some leakage will develop.

In certain cases, such as low-pressure steam, water, or air service, the leakage may simply be tolerated, but in most cases, it cannot. When the leak is great enough that it poses a potential hazard and action is required, the correct procedure is to take the valve out of service, replace the packing, and refinish or replace the valve stem. Unfortunately, a production-minded operator may elect to fix the leak with the valve in service by tightening down farther, sometimes much farther, on the stuffing box nuts. The additional pressure this puts on the packing usually corrects the leak for only a short time. The harm that results from this expedient is that it clamps the packing more tightly onto the valve stem so that instead of stroking smoothly, the valve moves in a series of jerky steps.

15.4.1 Balanced control valves

Manufacturers of control valves are well aware of the positioning problem created by the pressure-drop force that comes from inside the valve body. Fortunately, not all control valves are constructed as shown in the simplified sketches in Fig. 15.3. This design maximizes the problem, since the pressure difference acts on an area that is almost as large as that of a circle with a diameter equal to the nominal size of the valve.

In the plug and port style control valves that preceded the cage type, the balanced effect was achieved by having two plugs and seats instead of one, with the stream directed in between them. Then the force created by the pressure difference acted upward on one of the plugs and downward on the other.

As is often the case, this solved one problem but created another. In the closed position, it was virtually impossible to get both plugs to seat together. The plug that could not close tightly, because the other plug did, caused the valve to leak. To many control valve users, this was considered to be an acceptable price to pay for having both the force effect minimized and better positioning. If a dead-tight shutoff were required, a second valve, possibly a gate valve or a cock, would be installed in the line in series with the control valve.

Contemporary cage-type control valves also come in balanced and unbalanced forms. Figure 15.4 illustrates this. Without the small passage through the top of the piston, there would be an upward force on the piston equal to the pressure drop across the valve body multiplied by the area of the piston. This passage through the piston equalizes the

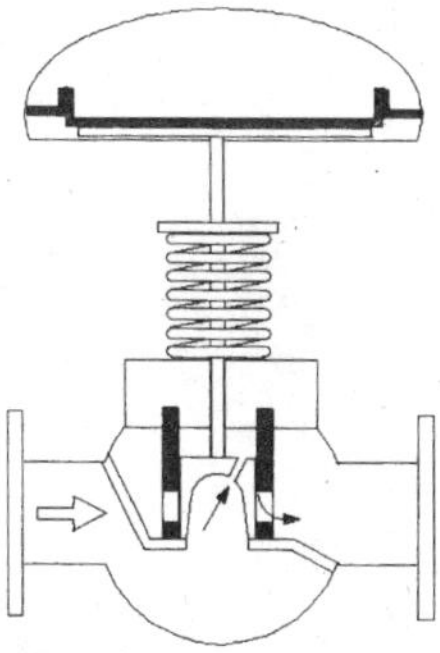

Figure 15.4 A balanced-cage-type control valve.

pressures above and below the piston and reduces the force to the product of the pressure drop and the cross-sectional area of the valve stem.

Once again, balancing out the force in this way creates a leakage path that did not exist before. The leakage path, shown by the two small arrows, goes up through the passage through the piston, downward between the outside of the piston and the inside wall of the cage, and out though the normal ports in the cage. This leakage path does not exist in an unbalanced cage valve, which can consequently shut off much better. The seating point in an unbalanced cage valve is the annular surface at the bottom of the piston.

The fact that leakage may occur in the closed position is not a reason for avoiding the use of balanced control valves. Control valves are, after all, primarily intended to provide smooth throttling, not tight shutoff. For most applications, where either the valve size or the pressure drop is not small, the best choice is to use the balanced control valve. If a tight shutoff is required, then a second valve, which is designed for shutoff rather than throttling, should be installed in series with the control valve.

15.4.2 Valve positioners

Most instrument hardware specialists know that if a control valve is likely to have positioning problems due to either of the causes noted, then the solution is to equip the control valve with a valve positioner. A *valve positioner* is really a valve-position controller. It operates in a regular feedback control loop. The positioner senses the stroke of the valve by means of a mechanical link that connects the valve stem to the input of the positioner. The setpoint of the valve-position controller is the control signal sent out from the automatic controller that is controlling the process variable. The pneumatic output of the valve positioner goes on top of the valve diaphragm.

If the control signal were 60 percent, for example, this would require that the valve be stroked through 60 percent of its travel. The positioner checks the position of the valve stem by means of the linkage provided, and if the stem position does not correspond to 60 percent, the positioner loads or unloads the pressure on the diaphragm to bring the valve travel to 60 percent.

The pressures which the positioner can apply to the top of the diaphragm or bleed off from the diaphragm are not strictly limited to 3 to 15 psig. The positioner can put out any pressure from zero to almost its full air supply pressure, which could be as high as 50 psig, to attain the length of stroke called for by the automatic controller.

These revelations would tend to the conclusion that valve positioners are a good thing, so why not put a valve positioner on every control valve that has a spring-diaphragm actuator? Some people actually do this as a matter of policy. The belief is that the valve positioner cannot do any harm and may actually improve control in those cases where precise positioning is required. This arbitrary practice is not recommended, however, for the following reasons:

- Equipping a control valve with a valve positioner when the positioner is not contributing anything only benefits the manufacturer of the valve positioner.
- Specifying a valve positioner when one is not needed runs contrary to the philosophy that simplicity begets reliability. A valve positioner, like most instrument components, requires periodic servicing, including calibration and possible replacement of parts. If it is not contributing anything worthwhile, a valve positioner is just one more thing that can go wrong.
- It is most important to recognize that when a control valve is provided with a valve positioner (i.e., a valve-position controller), then a cascade control system is created. The primary controller is the automatic controller that is controlling the process variable. The primary controller sends its output signal to the setpoint of the secondary (valve-position) controller.

This cascade control system will be subject to the rules of proper system design that apply to all cascade control systems, in particular the rule that states that the recovery time in the secondary loop should be no greater than one-fifth the recovery time in the primary loop. In a flow-control loop, especially when using electronic transmission and control, the transmitter, controller, and process all react very quickly. The component with the biggest dynamics is the control valve. Thus, if a valve positioner is used, the major dynamic element, the control valve, will be in the secondary loop, and the recovery time in the secondary loop will actually be greater than the recovery time in the primary loop. Under these circumstances, it may be difficult to get the primary loop to stabilize. The gain of the primary controller may have to be set so low that it becomes questionable whether there is any control action taking place at all.

The bottom line, therefore, is that there really are times when using a valve positioner can do more harm than good.

Finally, the purpose of a valve positioner is to cause accurate positioning of the control valve. Its purpose is *not to get more valve stroking speed.* While it is possible that some control valves will stroke faster with a valve positioner, this is merely a fringe benefit that should not receive any special emphasis. If a faster-moving control valve is what the control loop needs, then there is a better way to achieve this.

15.4.3 Hysteresis and limit cycling

Hysteresis is one of the assorted positioning problems that can happen to control valves. Hysteresis may be caused by packing that has been tightened too much or by a poorly designed pilot valve. In any case, the net result is that while the valve may respond to an acceptably small change in input signal when it is required to move in the *same direction* as it did previously, it requires a much larger change in input to get the valve to reverse its direction.

If the static characteristics of a control valve with hysteresis were plotted, they would show that for the same input values, the output, i.e., valve position, would increase along one path and decrease along another, as shown in Fig. 15.6. If it happened that a control valve with hysteresis were used on an integrating process (one without self-regulation) and with an automatic controller that had automatic reset, then conditions would be right for a phenomena called *limit cycling* to occur.

Suppose that the process in this case is a liquid storage vessel that is being pumped down at a constant rate. The vessel is under level control with the control valve on the inflow, as in Fig. 15.5. The control valve opens as the control signal to it increases.

This process has no self-regulation. The level in the vessel will be the integral, over time, of any difference between the inflow and the outflow. Suppose that at some moment the operating point is point A in Fig. 15.6, the inflow and the outflow are balanced, and the level is constant at the setpoint. Then there is a short-term loss of pressure on the supply side of the control valve that temporarily unbalances the system and causes the level to fall. This will probably trigger a rather bizarre chain of events.

First, the level controller will respond to the difference between the vessel level and the setpoint, and automatic reset in the controller will begin to increase the output to open the control valve and bring the control error back to zero. Since the level is too low, the controller must temporarily unbalance the system, with inflow greater than outflow, in order to restore the level to the right value. Thus the output of the controller will go to point B.

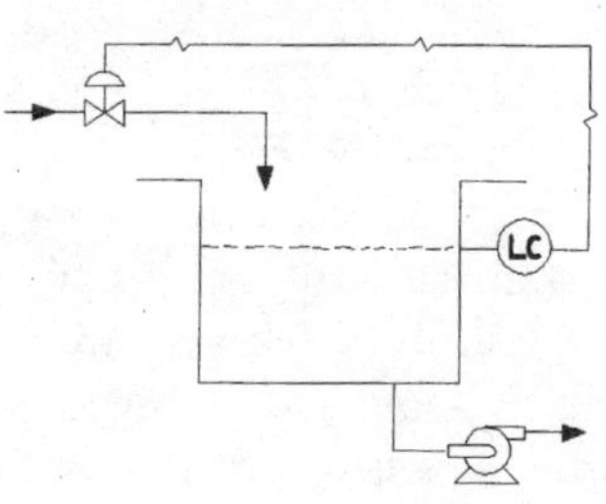

Figure 15.5 Level control on a storage tank: a possible case for limit cycling.

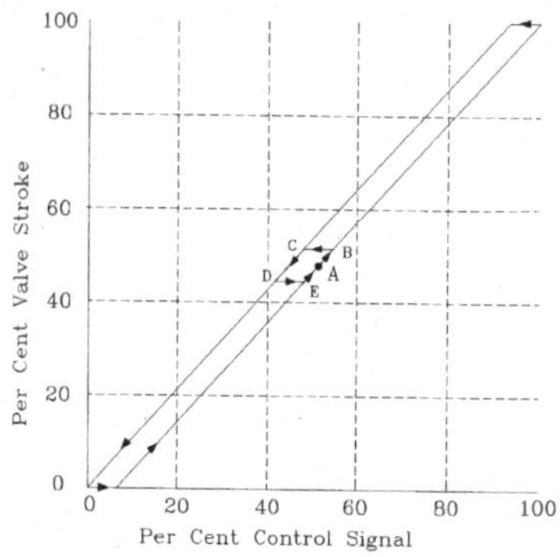

Figure 15.6 Limit cycling in a level-control loop.

When the level is back up to the setpoint, the controller must reduce its output to close the control valve slightly and bring the inflow and outflow into balance. Unfortunately, owing to hysteresis in the valve, the controller output has to cross over to the parallel operating line, which defines valve positions as the valve receives a decreasing input signal. Since this happens through automatic reset, it takes a certain amount of time to get there (point *C*). While this is going on, the process itself is integrating the difference between the inflow and the outflow, and the level in the vessel is still increasing.

By the time the controller output reaches point *C,* the level will be too high, which will require a correction in the opposite direction. To bring the level back down again to the correct value, the controller has to unbalance the system, with the inflow less than the outflow. Thus the automatic reset drifts the controller output down the operating line to point *D.*

When the level is correct and the control error is zero once again, the controller has to increase its output to open the control valve slightly and get the inflow and the outflow back in balance. This means that it must cross over to the other operating line, the one that defines valve positions as the input to the valve is increasing. Once again, this takes time, and while this is going on, the process is integrating the difference between the inflow and the outflow so that by the time the controller output reaches point *E,* the level is too low again.

The ultimate result is that the system cycles around the *BCDE* loop. This is *limit cycling,* and as already stated, it is caused by the combination of hysteresis in the control valve, an integrating process, and integration (automatic reset) in the automatic controller. It is important to understand that limit cycling is not caused by too much gain in the control system. This being the case, lowering the gain in the controller will not remove the cycling. It will only reduce the fre-

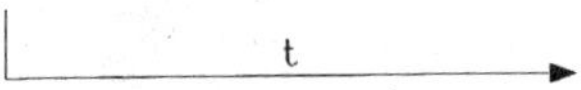

Figure 15.7 The pattern of level variations due to limit cycling.

quency at which the cycling takes place. The cycles do not have the sine-wave form but appear as in Fig. 15.7.

Confusion will result if it is assumed that limit cycling is natural oscillation due to high gain and that the period at which the cycling takes place is the true period of oscillation of the control system. Using the period of the limit cycle to determine the optimum value for the automatic reset in the controller is a trap that awaits the unwary.

Since it takes a combination of three factors to create limit cycling, eliminating any one of them will correct the difficulty. The process is what it is, so nothing can be done to change it. Consequently, the alternatives are to remove the automatic reset from the controller and operate with proportional action alone or to eliminate the hysteresis in the control valve by installing a good valve positioner.

15.5 The Stroking Speed Problem

In order to position a control valve with a spring-diaphragm actuator, air has to flow into or out of the diaphragm chamber. The speed with which the valve strokes depends on how quickly the air moves in and out. Pneumatic controllers, current-to-air converters, and valve positioners all have a pneumatic three-way pilot valve whose function is to load or unload air to the top of the diaphragm as may be required. The limiting factor in how quickly the valve will move will generally be the size of the ports in the pilot valve. The larger the ports, the faster the air can be moved into and out of the actuator, and the quicker the valve will be positioned.

To a lesser extent, the maximum value of the air supply pressure is also a factor. A higher air supply pressure, if it can be used, usually improves valve speed. Most pilot valves that are used in conjunction with spring-diaphragm actuators have a limit of about 50 psig on their air supply.

If a control valve appears to be positioning too slowly, the solution is *not* to install a valve positioner. The function of a valve positioner is to achieve *more accurate positioning*. It is *not* to get faster stroking. For improving speed, the device that is required is a volume booster.

A *volume booster* is a pneumatic 1:1 relay. It duplicates, at its output, the same pressure that is applied to its input. However, it can be provided with an air supply on the order of 50 psig, and by virtue of its larger ports, it can load and unload a spring-diaphragm actuator faster than other devices that have smaller pilot valves.

A question that is bound to arise is, "Does this mean that it would be possible to install a volume booster between the valve positioner and the actuator so as to achieve both better positioning and better speed?" The answer, unsatisfactory as it may appear, is, "It may work, and it may not." In this arrangement, there will not be any problem with the volume booster, but there may be a problem with the valve positioner.

The reason is that some valve positioners need a definite amount of volume on their downstream side. If the required volume is not there, the positioner may become unstable. Connecting the output of the positioner to a volume booster, which has very little input-side volume, may tip the positioner over the edge.

15.6 Control Valve Failure Mode

During the era when pneumatic control systems were the only choice, the matter of the control valve failure mode was relatively straightforward. The pneumatic control signal from the automatic controller to the control valve was what positioned the control valve. Accordingly, the only concern was what the control valve should do if the control signal went to zero.

Modern process control systems do not work this way any more. In today's control world, the control signal does not provide the power that positions the control valve. Control valves come equipped with ancillary devices such as I/P converters and valve positioners. These devices need an air supply so that they can generate a pneumatic output for positioning the control valve, even though the control signal from the automatic controller may be electronic.

So now there are two considerations where there used to be one:

- What should the control valve do if the control signal goes to zero?
- What should the control valve do if the air supply to the I/P converter or the valve positioner goes to zero?

There are three things the control valve can be made to do in the event of control signal or air supply failure:

1. It can go to the closed position.
2. It can go to the open position.
3. It can lock in the position it was in when the failure occurred.

This subject is relevant because process designers do not realize that all these possibilities exist, nor do they realize that it is necessary to specify what the control valve should do on failure of the air supply, as well as on failure of the control signal. Process designers tend to go blithely onward with their process flow diagrams, marking *FC* for fail closed or *FO* for fail open beside each control valve. This amount of information is not adequate any longer.

In a great many cases, the control valve should do the same thing if either the control signal or the air supply go to zero, but this should not be accepted as a general rule. It is quite possible, for example, that for the well-being of the process, a control valve should close on loss of the control signal but remain in its last position in the event of loss of air supply. Having the control valve meet a specified failure-mode requirement can be easy or difficult depending on whether the valve has a spring-diaphragm actuator or a piston actuator.

15.6.1 Failure modes using a spring-diaphragm actuator

The spring in a spring-diaphragm actuator ensures that the control valve has an inherent fail-safe position. If the air supply fails, the spring will position the valve at one end of the valve stroke. This can be specified as the valve-open (FO) or valve-closed (FC) position at the discretion of the designer. The specification is met by the supplier through the orientation given to the inner valve or, failing that, by selecting a direct- or a reverse-acting actuator.

If the control valve is required to lock in its last position on air supply failure, an air-failure lock is required. This is a small air-operated shutoff valve installed in the air line to the actuator. The air supply pressure is sensed by the failure lock at its input port. If the air supply pressure falls to an unsafe low value, the lock snaps shut and traps the prevailing pressure in the top of the actuator. The control valve remains locked in its last position until the air supply is restored, when the lock reopens.

An air-failure lock is a simple, inexpensive accessory available from the control valve supplier. Having a control valve with a spring-diaphragm actuator lock in its last position on air failure is only slightly more complicated than having it fail open or fail closed.

15.6.2 Failure modes using a piston actuator

Inasmuch as a piston actuator has no spring, it lacks an inherent fail-safe position. If its air supply fails, the valve may open or close, quickly or slowly, or remain in its last position.

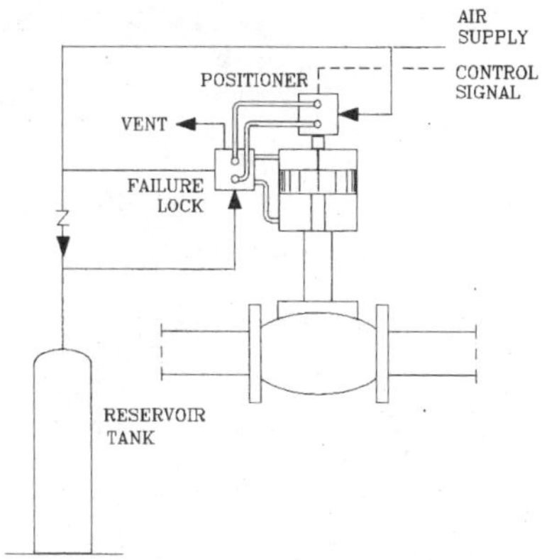

Figure 15.8 Piping requirements for fail open or closed when using a piston actuator.

It is a relatively complicated matter to have a control valve with a piston actuator fail open or closed. Inasmuch as the actuator has no spring, a reservoir tank containing enough air at supply pressure to stroke the valve once is required. The tank will probably store air at 100 psig or more. It may be classified as a pressure vessel and, as such, may have to be inspected and certified by some official body.

In addition, a pneumatic relay valve that loads one side of the piston from the reservoir tank and vents the air from the other side of the piston will be required in the event that the air supply pressure drops below the unsafe value. Figure 15.8 illustrates what has to be done to meet the fail-open or fail-closed requirement.

It is much easier to have a control valve with a piston actuator lock in its last position on air supply failure. This would require only an air-failure lock, similar to the unit that would be used for a spring-diaphragm actuator, except that it is designed to block two air lines to either side of the piston instead of one.

Because of the complexity and cost involved in having a control valve with a piston actuator fail open or closed, the first choice is to have it fail in its last position. Most process designers do not know this. If they insist on fail open or fail closed, then they should be made aware of the complications this creates.

15.7 Manipulating the Process Without a Control Valve

Conserving energy is an important matter in just about every industry. One facet of automatic process control that conservation of energy touches on is the pressure drop consumed by control valves. In some circles, this is considered to be an energy waste that should be eliminated or at least minimized. Since a control valve must have pressure drop in order to be able to vary the flow of material or energy to the process, a moot question is, "Is it possible to manipulate the process without using a control valve?"

If a control valve cannot be used, then the only recourse is to vary the output of the pump, compressor, or other device that is providing the driving force behind the fluid. This can be done in either of two ways:

- By varying the speed of the driver that turns the fluid mover.
- By installing a variable-speed coupling between the driver and the fluid mover that permits the driver to run at its design speed but enables the speed of the fluid mover to be changed.

Before discussing the merits of these alternatives, there is a term that needs to be defined because it is relevant in this matter. The term is *turndown*. The turndown of a variable-throughput device is the ratio of its maximum practical output to its minimum practical output. Turndown is expressed as a ratio. If, for example, the turndown of device is said to be 5:1, this means than the useful operating range of the device is from 100 percent of its output down to 20 percent of its output. Below 20 percent, it is implied that the performance of the device will not be dependable.

The reason that turndown is important is that for process control, any variable-output device that is used in place of a control valve should have at least the turndown capabilities that a control valve has. When control valves are the issue, the word *turndown* is usually replaced by the term *rangeability*. Rangeability and turndown are the same thing, except that rangeability is used in reference to control valves, while turndown is a more general term.

The rangeability of a good-quality control valve is on the order of 10:1. This implies that a control valve's useful range is from 100 percent of its flow capacity down to 10 percent of its capacity. Attempting to throttle the flow of fluid at flow rates below 10 percent of capacity in most cases will give indifferent results.

15.7.1 Variable-speed drives

Of the types of drivers used in industrial processing operations, electric motor drives and steam turbine drives predominate. Considering electric motor drives first, literature that has been appearing recently suggests that considerable work has been done by manufacturers of heavy electrical equipment on devices that make it possible for all types of electric motors to run at variable speed. This development should be promising for automatic process control. Some manufacturers are even aware that an important requirement is to be able to set the speed of the motor using a control signal with a 4- to 20-mA range.

The key questions to be answered are, "How much is a variable-speed controller going to cost?" and "What turndown capability will it have?" The answer to the first question establishes which is the better economic alternative—buying and operating the variable-speed controller or expending the energy at the control valve. The answer to the second question will determine whether the range of speeds at which the driver can function will be wide enough to accommodate the various control system loads.

Another consideration is how quickly the speed of the motor can be changed using the variable-speed device. Once again, if the process can react faster than the flow of material or energy to the process can be altered, then automatic control is in doubt.

Varying the flow of gas or liquid to the process by varying the speed of a steam turbine that drives a pump or a compressor has been done successfully for many years. The control system designer needs to be aware that a steam turbine has a critical-speed point at which the frequency of rotation exactly matches the natural vibration frequency of the turbine. The turbine must not be allowed to operate at this point; otherwise, the resulting vibration would destroy it. Since the critical-speed point must be outside the design operating range of the turbine, this may limit the turndown of the turbine to a relatively low value, perhaps 2:1 or 3:1.

If a decision is made to manipulate the process by varying the speed of a steam turbine driver, the *output of the automatic controller must never go to a control valve in the steam line to the turbine.* Many steam turbines can change speed with alarming quickness, faster than most control valves can be moved to hold the turbine in check. The correct method is to have the turbine manufacturer provide the turbine with a governor whose speed setting can be changed by a standard control signal. Control is obtained by having the output of the automatic controller set the setpoint of the governor in cascade control fashion. This will ensure that the speed of the turbine is kept safely under control. In addition, if the turbine has speed limits, these can be maintained by means of mechanical stops on the governor.

15.7.2 Variable-speed couplings

A variable-speed coupling is a machinery component whose input and output are both rotary motion. The input turns at a constant rate, as set by the electric motor or other driver that powers the coupling. The output speed can be varied to suit the process application. Changing from one speed to another may be done either mechanically or hydraulically. For automatic control, the manual-speed change lever can be replaced by an actuator that will accept a control signal.

It is generally true that where there is an established need to make something better and there is a profit to be earned by so doing, then sooner or later industry will come up with the better product. However, impressions of the variable-speed couplings that have been available up to 1990, *as a component for process control,* are that their performance is abysmal. Slow response, limited turndown, and high cost are just part of the story. The primary problem is hysteresis, which has been so severe in some brands that it would be virtually impossible to control anything if a variable-speed coupling were part of the control loop. The conclusion that follows has to be that when variable-speed couplings were originally designed, automatic control was never really taken into consideration.

Someday, hopefully soon, some manufacturer of variable-speed couplings will take notice of the automatic process control field and will build a unit that has the features which the control field requires. Until that time, no one who values his or her reputation could consider them to be a viable alternative.

15.7.3 Fans

Fans are used to move large quantities of gas, often air, against low pressures. To manipulate the amount of gas that is moved, there are actually five possible options:

1. Vary the speed of the fan.
2. Instead of one large fan, use several small ones, and turn on only the number needed to attain the required gas flow.
3. Operate the fan at constant speed, and vary the pitch of the fan blades.
4. Operate the fan at constant speed, and throttle the gas flow with louvers.
5. Operate the fan at constant speed, and manipulate the gas flow by venting off some of the gas which the fan has discharged.

From a strictly control point of view, varying the speed of the fan to get the proper gas flow is an excellent method. The relationship between gas flow rate off the fan and the speed of the fan is almost linear. The turndown of the fan itself is good. The problem with this method is the cost of the variable-speed driver. Specifications on turndown, speed of response, and hysteresis for the variable-speed driver should be investigated as well.

Using several smaller fans instead of one large one has the merit of being a simple control approach. It is therefore a reliable method. The difficulty is the obvious one. The gas flow rate will be manipulated in step fashion as the fans are turned on and off. This does not conform with the concept of smooth throttling characteristics. On some applications, however, this will prove satisfactory. Varying the airflow in cooling towers is often handled by this means.

While variable-pitch propellers work for airplanes and ships, for process control applications they are risky. There are two potential problems. The first is hysteresis in the rotation of the blades, which has been observed to be severe in some cases. The second is that the relation between gas flow and blade rotation is often nonlinear. At its worst, there can be a section in the center of the static characteristic where rotating the blades causes very little change in gas discharge rate. Little or no change in output in response to an input change is probably the worst fault that any control system component can have.

Manipulating the gas flow using louvers is often used in air-cooled heat exchangers and in the air and flue-gas ducts of boilers. The method generally works well.

Louvers have the appearance of venetian blinds and, in fact, work exactly like them. The blades in the louvers are connected by a linkage that causes them to open and close together like the slats in a venetian blind. A single actuator is used to position the whole set of blades, and this is where the potential trouble lies.

The louvers will usually come as part of an exchanger or boiler package, and the manufacturer of the louvers will be expected to supply the actuator, all mounted and ready to operate. The louver manufacturer will probably obtain the actuator from a manufacturer of control valves.

Since the louver supplier is competing with other suppliers of the same type of operating equipment, he or she will naturally avoid doing anything that will put the equipment at a price disadvantage in terms of competitors. Consequently, the actuator that will be provided for stroking the louvers will be the smallest and least expensive one the louver manufacturer believes will do the job.

If, prior to shipment, the purchaser of the louver set sends an inspector to the manufacturer's plant for an acceptance test, it will no doubt be demonstrated to the inspector's satisfaction that the actuator positions the louvers as it should. The point that is often missed, however, is that this test is done when the louvers are brand new and well lubricated and with everything indoors at room temperature. The question of how well the louvers will work after they have been outdoors exposed to the weather for a year and when the temperature is −20°C is overlooked.

Even though the attitude may seem somewhat arbitrary, smart prospective owners will notify the louver manufacturer that they are willing to pay the extra cost for an actuator that is one size larger than the unit which the manufacturer would ordinarily supply.

Finally, controlling the gas discharge from a fan by venting off what is not needed will work in principle, although energy conservationists

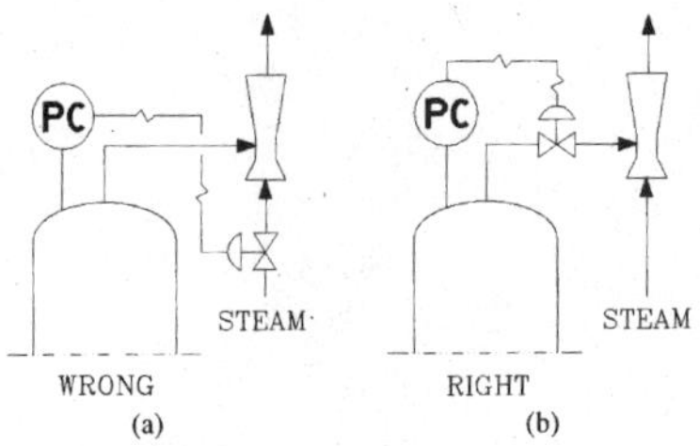

Figure 15.9 Two methods of controlling pressure using a steam-jet eductor.

will not like the idea. The method is limited to nonhazardous gases. It does not appear to be widely used, and it is mentioned here mainly for the sake of the completeness of the list of available methods.

15.7.4 Steam-jet eductors

Certain processing operations, such as drying and fractionation, are conducted at very low, subatmospheric pressures. A device that is often used to create the low pressures required is a steam-jet eductor. If the eductor is capable of creating a higher vacuum than is required by the process, how should the process pressure be controlled? Figure 15.9 shows two possible methods, one wrong and the other right.

Figure 15.9 is also a good example of the deceptiveness of single-line sketches. The steam supply line to the eductor will be fairly small, perhaps 1 or 2 in. On the other hand, owing to the very low pressure, the exhaust line between the vessel and the eductor will be much bigger, perhaps as large as 6 or 8 in. Large pipes mean large control valves, which will naturally cost more money. This is argument 1 in favor of locating the control valve in the steam supply line instead of in the exhaust line.

Argument 2 is the conservation argument. It is irresponsible to use steam to create a high vacuum and then waste part of the vacuum across a control valve. Therefore, use only as much steam as is necessary.

As valid as these two arguments appear, they do not stand up against the counterargument that if the flow of steam to the eductor is restricted by a control valve, then the eductor will not work properly. That is where the matter has to rest.

Chapter

16

Realistic Control Valve Pressure Drops

Even in the most amicable of engineering departments, one of the most controversial subjects is the matter of how much pressure drop should be assigned to control valves. The confrontation is generally between the process designers, who realize that pressure drop consumes energy and therefore should be minimized, and the instrumentation engineers, who know that it is the pressure drop that provides the driving force which moves the fluid through the control valve. Therefore, the greater the pressure drop, the greater is the ability of the automatic controller to affect the process and change the value of the controlled variable.

In the extreme, if the pressure drop at the control valve were zero, the controller would have zero ability to change the value of the controlled variable, and there could be no automatic control. It is most curious that despite the obvious importance of this aspect of automatic process control, there is so little published in the way of specific guidelines as to how to determine the proper amount of control valve pressure drop. By *proper* is meant that amount which will ensure successful functioning of the control system yet avoid unnecessary waste of energy.

Fisher Controls, in its otherwise informative *Control Valve Handbook,* says nothing about how to establish control valve pressure drop. The *Masoneilan Handbook for Control Valve Sizing* suggests that the control valve should have one-third the total friction drop in the system. This could be helpful, except that a few sentences later the authors state that "under certain circumstances" 25 percent might be satisfactory or possibly even 10 percent. Designing on the basis of such vacillating guidelines obviously will be difficult.

The most questionable statement of all appears in volume 2 of Liptak's *Instrument Engineers' Handbook* (p. 412), to the effect that the amount of control valve pressure drop that should be specifically assigned for control is "none at all." The inference is that process designers, in their usual charitable manner, can be depended on to design "fat" into the system in the form of excess pressure drop. This fat will end up in the control valve and will provide all that is necessary for automatic control. This reasoning would be laughable were it not for the fact that the Liptak volumes are used by a great many people as their number 1 reference for process measurement and control.

Arguing over the arbitrarily chosen percentages of system pressure drop that should be given to a control valve is at best a waste of time, because there is no real engineering basis for the percentage numbers. What is needed is a method based on logical, credible principles. Some authors are on record as stating that no such method exists. There is a method, though, and it will be described here.

A typical situation is shown in Fig. 16.1. We will use this diagram not only to identify the problem but also to establish some basic truths and to define some terms. Assume that a process design engineer is working on the charge circuit for a hydrocarbon feed stream that passes through some preheat exchangers, then through a fired heater, and finally ends up in a fractionator. The fractionator operates at 20 psig, and 15 psi is necessary to lift the liquid from grade up to the feed nozzle. At what is considered to be the design flow rate in this circuit (call this flow rate Q_d), there will be a 32-psi pressure drop in the preheat exchangers, a 60-psi drop in the fired heater, and a 30-psi drop in the piping. Since the intent is to have this stream on flow control, there also will be a 2-psi drop in the orifice meter.

Let us call the pressure drop taken by the piping and all the process equipment in the system, including the orifice but *excluding* the control valve, the *friction pressure drop* and designate it by the letter F. In this example, F = 124 psi.

Having reached this point in the development of the design, the design engineer now approaches the instrument engineer with the

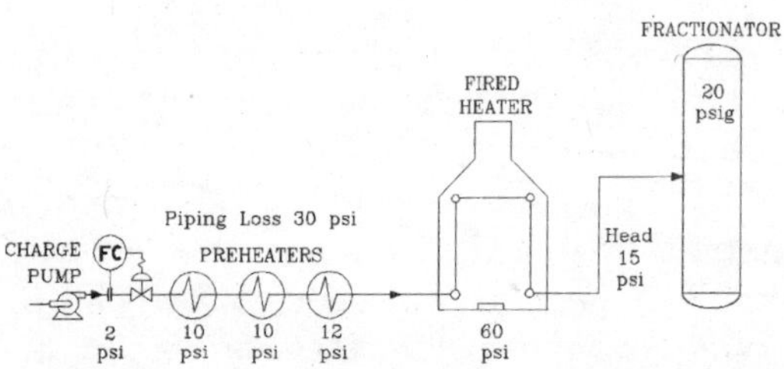

Figure 16.1 A fractionator charge circuit application.

question, "How much pressure drop is required for the control valve?" Once this figure is established, the design engineer can complete the pressure balance, determine how much pressure is needed at the start of the circuit, and draw up the specifications for a suitable pump.

To answer the question, the instrument engineer falls back on the popular rule that one-third of the system drop should be in the control valve. This would put the control valve pressure drop at one-half of F, or 62 psi. The design engineer takes this number, calculates that the pump discharge pressure would have to be 221 psig, and says, "No way." This would necessitate a big pump with a bigger driver that would consume far too much energy. Why does the pressure drop at the valve have to be so high?

The question puts the instrument engineer on the defensive. His or her response may be, "Well, how much pressure drop can you allow us?" The process designer's answer, "How about 10 psi?" The instrument engineer knows intuitively that the control system cannot work over any significant range of loads if the control valve has only 10 psi, but he or she lacks any soundly based argument to prove the point. Eventually, the two engineers settle for 25 psi, and the instrument engineer buys the coffee and hopes for the best. Surely, there has to be a better way.

The friction pressure drop F has been defined as the pressure drop taken, at design flow rate Q_d, by the process equipment and the piping. Bear in mind that F must change if Q_d changes. Notice also that this system, like most systems, starts and ends at two points of relatively constant pressure. Let us call the starting and ending pressures for the system P_s and P_e, respectively. In Fig. 16.1, P_e is the fractionator pressure plus the static head, i.e., 35 psig. P_s will be the pump discharge pressure, which would be 184 psig if the control valve were to be assigned a pressure drop of 25 psi. The phrase *relatively constant pressure* means that the pressures P_s and P_e do not change to any appreciable extent when the flow rate in the system changes. The difference between P_s and P_e provides the driving force for moving the fluid through the whole system from beginning to end. The pressure-balance equation is

$$P_s = P_e + F + \Delta P \tag{16.1}$$

where ΔP is the control valve pressure drop. Rearranging this gives

$$\Delta P = (P_s - P_e) - F \tag{16.2}$$

We should note carefully what Eq. 16.2 is telling us. One of the most aggravating questions process designers can put to instrument engineers is, "Why can't you put in a larger control valve and take less pressure drop?" This is an out-and-out trap, and instrument engineers owe it to themselves to know how to avoid being drawn into it. The answer is obvious from Eq. 16.2: *The pressure drop at the control valve can only be what is left of the overall system pressure drop after the friction pressure drop has been deducted.*

Suppose that in the Fig. 16.1 example, the control valve size turned out to be 2 in with a pressure drop of 25 psi assigned. Then the 2-in control valve is taken out and a 3-in control valve is installed in its place. Would the pressure drop at the control valve change? No, it would not. The control valve pressure drop must be what is left of the overall system drop ($P_s - P_e$) after F is deducted, so it is still 25 psi. The 3-in control valve would be pinched down in its stroke farther than the 2-in valve, but the pressure drop at a flow rate of Q_d would still be 25 psi. Put in a 4-in control valve and the pressure drop would still be 25 psi at the design flow rate Q_d, although the 4-in valve would be pinched down even further.

The bottom line is that the valve size has no effect on the pressure drop, although the opposite is certainly true. Control valve pressure drop is determined *only* by the pressure-balance equation (16.2). Period.

Figure 16.1 reveals still one more fundamental concept. Suppose that for operating purposes it is necessary to increase the flow rate to a value of Q_d +10 percent. The laws of nature say that pressure drop will vary as the square of flow rate. Consequently, if the flow rate increases to 110 percent of Q_d, the friction pressure drop will increase to 121 percent of F, or 150 psi. This is an increase of 150−124, or 26 psi. Now then, where is this additional 26 psi going to come from? Since P_s and P_e do not change significantly with flow rate, the additional 26 psi is not going to be made up by any increase in the overall system pressure drop. The inescapable conclusion is that it has to come out of the control valve.

Thus evolves the important concept of the control valve as a *pressure-drop bank*. It is a bank on which the devices that make up the friction drop draw on for more pressure drop when it is necessary to increase the flow rate in the system. It also points out the direction which the *engineering* of the control valve pressure drop should take.

The underlying principle is that at the design stage, the control valve must be provided with enough pressure drop so that when the system flow rate is increased to the anticipated maximum rate, the control valve can give up the additional pressure drop demanded by the friction drop and still have the minimum of pressure drop left that will permit the automatic controller to keep the controlled variable on control.

In the Fig. 16.1 example, we calculated that an increase in flow rate to 110 percent of Q_d would increase the friction drop F by 26 psi. This would have to come out of the control valve. If the control valve were assigned only 25 psi in the first place, then the flow-control loop would be out of control at this higher flow rate. The control valve would go wide open, and the flow rate would be determined by the vagaries of the system.

In certain cases there will be no need to perform calculations to determine control valve pressure drop. One example is a steam pressure letdown station. The valve inlet and outlet pressures are the higher and lower steam pressures, respectively, and the pressure drop is their difference. Another example is a hydrocarbon-water separator drum from which the water is being dumped to an open sewer on interface level control. The drum pressure is the valve inlet pressure, and the sewer pressure (zero) is the valve outlet pressure. Unfortunately, these simple cases are usually in the minority.

In general, the pressure drop ultimately assigned to a control valve must take into account the following factors:

1. How much pressure drop will the control valve have to give up to the system friction drop if the flow rate increases from the design value Q_d to the anticipated maximum flow rate (call this Q_m)?
2. How much allowance should be made for a possible falloff in the overall system pressure drop ($P_s - P_e$) created by an increase in system flow rate?
3. Even in the wide-open position, a control valve will impose some resistance to flow; therefore, it will create some pressure drop. This will be a function of the type of control valve used (plug, cage, ball, butterfly, etc.).

What we now need are methods for putting numbers on each of the contributing factors listed above.

16.1 Allowance for an Increase in Friction Drop

To calculate this, we need to know the ratio of the maximum anticipated flow rate Q_m to the design flow rate Q_d, in other words, Q_m / Q_d. Note that *the actual flow rate values need not be specified, only the ratio.* Usually, the process designer will be willing to make an estimate of Q_m / Q_d based on his or her knowledge of how the plant will be run or the upper-limit values of such factors as pump discharge rates. If the process designer does not want to commit on an estimate of the Q_m / Q_d ratio, then the instrument engineer will have to make his or her own best estimate based on whatever transients in flow rate may be expected as the control loop recovers in its typical cyclic fashion from an upset. Some suggested values for Q_m/Q_d are given at the end of this chapter.

At the design flow rate Q_d, the friction drop is F psi. At the maximum flow rate Q_m, the friction drop will become

$$F \times \left(\frac{Q_m}{Q_d}\right)^2$$

The *increase* in pressure drop, consequently, will be

$$F \times \left(\frac{Q_m}{Q_d}\right)^2 - F \qquad \text{or} \qquad \left[\left(\frac{Q_m}{Q_d}\right)^2 - 1\right] \times F$$

The value of F, as determined by process design, is really the best estimate. At this point in time, the plant will exist only on paper, and it will not be possible to get an accurate fix on pipe lengths, numbers and types of fittings, and velocity and physical properties of the stream. Therefore, it is probably prudent to apply a small factor of safety, say, 10 percent, to the preceding calculation. As a result, the allowance for the increase in friction drop will be

$$1.1\left[\left(\frac{Q_m}{Q_d}\right)^2 - 1\right]F \tag{16.3}$$

16.2 Allowance for a Possible Falloff in Overall System Pressure Drop

The overall system pressure drop is $P_s - P_e$. A reasonable allowance, considering that P_s and P_e are both supposed to be points of relatively constant pressure in the system, would appear to be 10 percent of $P_s - P_e$. However, when this figure is used, it will be found to be practical in some cases but not in others.

Consider the situation in which fluid is flowing through a control valve from one vessel that operates at 300 psig to a second vessel that operates at 250 psig. The overall system pressure drop would be 50 psi. If the allowance for the loss of system driving force were 10 percent of this value, i.e., 5 psi, then the pressure in either of the vessels would only have to change 2 percent for the allowance to be wiped out. What this tells us is that this allowance should take into account not only the system pressure drop but also the pressure level at which the system operates.

The rationale, therefore, will be to use the system end pressure P_e as an indicator of the pressure level of the overall system and divide the 10 percent allowance equally (5 percent and 5 percent), between the system pressure drop $P_s - P_e$ and the system pressure level P_e. The allowance consequently becomes

$$0.05 \times (P_s - P_e) + 0.05 \times P_e \tag{16.4}$$

which simplifies to $0.05P_s$.

16.3 Allowance for the Control Valve Itself

Even in its wide-open position, the control valve will impose some resistance to flow and will require some pressure drop. The value will vary with the design of control valve used and probably with the application. This factor in the required control valve pressure drop will be identified by the letter *B*, for *base pressure drop.* By working with average line velocities, assuming that in the majority of cases the control valve will be one pipe size smaller than the line, and by averaging calculated values of *B* over a range of valve sizes, the following values of *B* were arrived at:

Control valve type	*B* (psi)
Single plug	11
Double plug	7
Cage valve (unbalanced)	4
Cage valve (balanced)	4
Butterfly valve	0.2
V-ball	1

Taking into account the pressure drop allowances that should be designed into the control valve for all three reasons listed earlier, the aggregate required control valve pressure drop therefore is

$$\Delta P = 0.05P_s + 1.1\left[\left(\frac{Q_m}{Q_d}\right)^2 - 1\right]F + B \tag{16.5}$$

A look at some examples may help to clear up any skepticism. Figure 16.2 shows a temperature-control valve in the fuel gas line to a fired heater. The system clearly starts at the fuel gas header, where the pressure is 35 psig, but where does the system end? At first glance, it may appear to end at the inlet to the gas burners. However, the requirement for both the start and end points of the system is that they be points at which the pressure remains relatively constant despite changes in flow rate in the system. This would not apply to the burner inlet because the pressure here varies considerably with

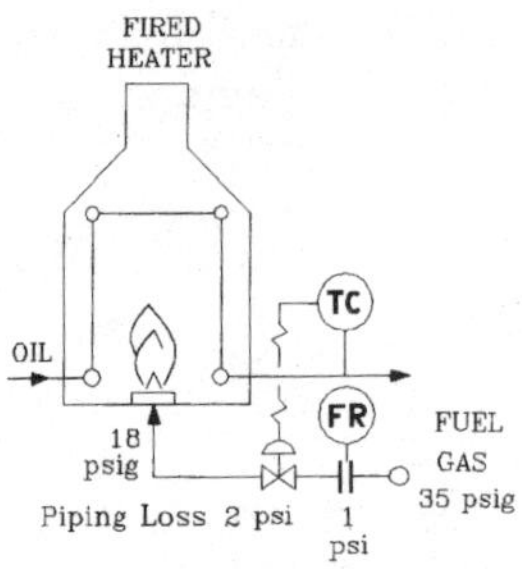

Figure 16.2 A fired heater outlet temperature-control application.

changes in gas firing rate. The real end of this system is in the firebox, where the pressure is 0 psig.

Based on the anticipated variations in heating load on the heater, the value of Q_m was estimated to be 140 percent of Q_d. The control valve selected is a balanced cage type.

The friction drop F at Q_d will be in the orifice plate (1 psi), the piping (2 psi), and the burner (18 psi). Thus $F = 21$ psi. The available pressure drop is calculated from the pressure-balance equation (16.2)

$$\textit{Available } \Delta P = (P_s - P_e) - F = (35 - 0) - 21 = 14 \text{ psi}$$

The pressure drop required to make the specified Q_m/Q_d ratio of 1.4 possible is calculated from Eq. 16.5. The value of B for the cage control valve in Fig. 16.2 is 4 psi. Therefore,

$$\textit{Required } \Delta P = 0.05 \times 35 + 1.1 \times (1.4^2 - 1) \times 21 + 4 = 28 \text{ psi}$$

These two calculations show that 28 psi is required for the control valve, but only 14 psi is available. Furthermore, the required value, 28 psi, was not determined from a crystal ball percentage of the overall system drop. It was based on the shifts in pressure drop that the laws of nature require. Thus the instrument engineer at last has a well-founded answer to the question, "So what if we can't give you 28 psi?" The answer is, "Unless the control valve has 28 psi pressure drop at the design flow rate Q_d, it will not be possible to increase the flow rate to anywhere near $1.4 \times Q_d$ ($= Q_m$) before the control valve is robbed of all its pressure drop and the temperature-control loop goes out of control."

This example was an actual oil refinery case. Engineering the required control valve pressure drop using Eq. 16.5 not only puts the comparison of required versus available pressure drop on a sound basis, but it also has a valuable side benefit. If the requirement for pressure drop turns out to be that much greater than what is actually available, as in this instance, an examination of the numbers that went into the Eq. 16.5 calculation will sometimes show which one is creating the need for so much ΔP. In this particular calculation, the big factor is the 18 psi pressure drop at the gas burners.

When this became apparent, the combustion engineer was contacted and was asked if the firing pressure could be lowered without adversely affecting the performance of the burners. His opinion was that the spuds could be reamed out a few thousandths, which would still give acceptable burner efficiency but would lower the firing pressure to 10 psi. At this new value, F becomes 13 psi. Then

$$\textit{Available } \Delta P = 35 - 0 - 13 = 22 \text{ psi}$$

$$\textit{Required } \Delta P = 0.05 \times 35 + 1.1 \times (1.4^2 - 1) \times 13 + 4 = 19.5 \text{ psi}$$

By making this modification to the burners, the available control valve pressure drop was increased to the point where the temperature-control valve still had enough pressure drop left to keep the temperature loop on control even if the gas flow rate increased to 140 percent of the design rate.

Example 2 (Fig. 16.3) also was an actual case. It concerned the revamp of a fractionating column. The process designers were not keen on purchasing a new distillate pump even though they knew they needed higher reflux rates. The question was brought back to the instrument engineering group, "Will the control valve have enough pressure drop?"

In this example, the system in which the reflux control valve was located started at the pump discharge, where the pressure was 150 psig. This was the value of P_s. The end of the system was in the fractionation column, which was to operate at 100 psig, but 19 psi of static head was required to lift the reflux from grade up to the reflux nozzle, and static head is not changed by flow rate. Consequently, P_e was 119 psig.

The friction drop F was 18 psi; 2 psi contributed by the orifice plate and the rest by the piping. The available pressure drop for the cage-type control valve, from the pressure-balance equation was

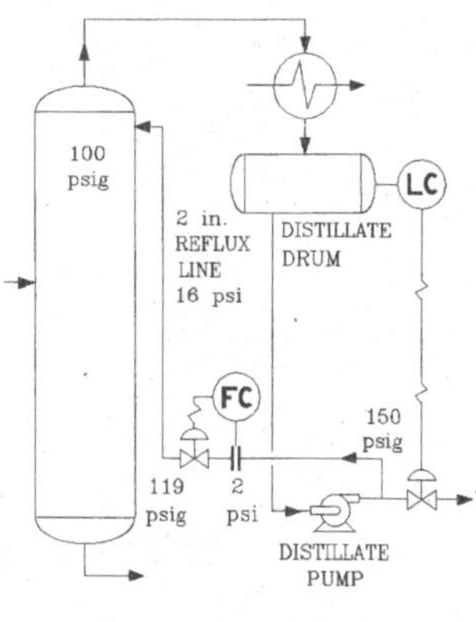

Figure 16.3 A distillation column reflux line application.

$$\Delta P = (150 - 119) - 18 = 13 \text{ psi}$$

The designers put the most likely value of Q_m / Q_d at 1.25. Calculating out the required pressure drop, therefore, gave

$$Required\ \Delta P = 0.05 \times 150 + 1.1 \times (1.25^2 - 1) \times 18 + 4 = 22.6 \text{ psi}$$

There was obviously not enough pressure drop to do the job. However, in examining the numbers that went into the required pressure-drop equation, it was clear that the biggest contributor was the piping drop, 16 psi. This seemed like an unusually large number, and the process designer was questioned about it. It turned out that he already had misgivings about the size of the reflux line, which was only 2 in. The question raised by the instrument engineer simply confirmed his suspicions.

A decision was consequently made to increase the reflux line size to 3 in. This lowered the piping pressure drop to 4 psi and the friction drop F to 6 psi. The second round of calculations therefore produced an available pressure drop of 25 psi and a required pressure drop of 15.2 psi. Once again, calculating the required control valve pressure drop using this method showed not only that the situation as it stood was not feasible but also how to correct the problem.

For the final example we should go back to the case with which we began, Fig. 16.1. This was for a new plant, and the deliberations relating to the correct amount of pressure drop to be provided at the control valve would have been more appropriate for an Arabian marketplace than for an engineering office. Determining the right pressure drop in this case is slightly more complex, but only slightly.

The difficulty is that in order to use the required pressure-drop equation (16.5), one has to know the start-of-system pressure P_s. However, P_s cannot be determined from the pressure balance until the control valve pressure drop is known. Nevertheless, what is known is that in this application the system end pressure P_e is 35 psig and the system friction drop F is 124 psi.

Furthermore, the process designer is willing to commit on a value of Q_m = 120 percent of Q_d, so Q_m / Q_d is 1.2. The most suitable control valve type will be a cage valve for which the value of B is 4 psi. From the pressure-balance equation,

$$Available\ \Delta P = P_s - 35 - 124 = P_s - 159$$

From the required pressure-drop equation,

$$\begin{aligned} Required\ \Delta P &= 0.05 \times P_s + 1.1 \times (1.2^2 - 1) \times 124 + 4 \\ &= 0.05P_s + 64 \end{aligned}$$

Equating these two expressions for the control valve ΔP produces a system start pressure P_s of 235 psig. Using $P_s = 235$ in either equation gives a control valve pressure drop of 76 psi.

It is worth noting that this number is even greater than the pressure drop that would be arrived at by assigning one-third of the system pressure drop to the control valve. Yet 76 psi is the pressure drop that must be in the control valve at the design flow rate for this system if it is to be possible to increase the flow rate to 120 percent of Q_d and still have the control loop remain on control.

Remember that a pressure drop of 25 psi was the process designer's best offer and that it was reluctantly accepted by the instrument engineer. If 25 psi were actually selected for the control valve, then P_s becomes 184 psig. Using these two numbers in the required pressure-drop equation, the value of Q_m / Q_d can be back-calculated. It turns out to be 1.04! In other words, with a 25-psi pressure drop at design flow rate, the flow rate in this system could only be increased 4 percent over Q_d before the control valve went wide open for lack of pressure drop and the control loop went out of control.

This essentially completes the description of the method, but there are some loose ends that should be dealt with. First, process designers usually will calculate the value of the friction pressure drop F at the design flow rate Q_d. Occasionally, however, they will calculate F at Q_m. This does not happen often, but it happens often enough that one should always check with them to verify the flow rate at which F was calculated. If F was calculated at Q_m, it should be scaled down by the factor $(Q_d / Q_m)^2$.

Second, if no commitment can be obtained from the process designer regarding the Q_m / Q_d ratio, then the suggested values for calculating the required control valve pressure drop are 1.1 for flow-control valves and 1.25 for level-, pressure-, and temperature-control valves. These numbers are based purely on anticipated flow-rate transients at the control valve as the control loop recovers from a disturbance.

Third, if the required-pressure-drop method described here is used and the control valve size turns out to be line size, this is no cause for concern. It may be an indication that the piping is on the small side, but the control loop can still do the job that is expected of it.

Last, Eq. 16.5 for calculating the required control valve pressure drop will work equally well whether the numbers inserted are in pounds per square inch or kilopascals, as long as consistent units are used.

Chapter 17

Sizing Control Valves

Current methods for sizing control valves use formulas based on determining a valve capacity coefficient C_v, which is defined by the relation

$$Q = C_v(\Delta P/G)^{1/2}$$

where Q = liquid flow rate through the valve (U.S. gpm)
ΔP = pressure drop between the valve inlet and outlet (psi)
G = specific gravity of the liquid at its flowing temperature

Although the capacity coefficient C_v is defined in terms of liquid flow, it also stands for the capacity limits of control valves in steam and gas service.

17.1 Sizing Control Valves Using SI Units

The coefficient C_v has units, and the units are in the fps system, not the SI system. Consequently, to retain the established control valve sizing formulas while using SI units, a possible tactic is to introduce an SI valve capacity coefficient, which will be called the C_{vm} (for *metric* C_v).

If the C_{vm} concept is used, the sizing equation for control valves in liquid flow service becomes

$$Q = C_{vm}(\Delta P/G)^{1/2} \tag{17.1}$$

where Q = liquid flow rate through the control valve (dm³/s or l/s)
ΔP = pressure drop, valve inlet to valve outlet (kPa)
G = specific gravity of the liquid at its flowing temperature

In Eq. 17.1, dm^3/s (or l/s) and kPa replace gpm and psi, which were the units in the original sizing equation.

At this time, most control valve capacity data published by manufacturers is in terms of the C_v, not the SI C_{vm}. The relation between these two coefficients is

$$C_{vm} = C_v \times 0.024 \tag{17.2}$$

In other words, the SI C_{vm} is about 1/40 of C_v.

Equation 17.1 can be used to calculate the rate of flow through a control valve of specified size. It also can be rearranged so that C_{vm} can be calculated, given the required flow rate and available pressure drop.

17.2 Sizing Considerations

Even though the calculations for determining the optimal size for a control valve are relatively straightforward, there are some complicating factors that must be considered:

- For what rate of flow should the control valve be sized?
- For liquid flow, will the liquid tend to vaporize as it flows through the control valve?
- For steam or gas flow, is the vapor velocity through the valve body in the sonic velocity region?

These factors must be known if proper sizing is to be ensured. An important basic fact is that it is *volume flow at line pressure and temperature* that determines the size of a control valve, not mass flow.

17.3 Excess Capacity

Excess capacity may be built into a control valve in either of two ways:

1. Control valves come only in discrete sizes. After calculating the required C_v (or C_{vm}), the designer will generally select the control valve with the smallest C_v that is at least as large as the calculated required C_v. The C_v of the valve chosen usually will be somewhat larger than the required C_v. Consequently, excess capacity will be available.
2. The designer may deliberately add excess capacity to the control valve to ensure satisfactory control.

It is excess capacity that is really the dominant factor in control valve sizing, to the point where small correction factors (e.g., for viscosity or piping geometry) are of questionable importance.

The control valve is one of the major components of the control loop. When the loop is disturbed, proper recovery will be in doubt if any component in the loop runs into a limit, or saturates, while the loop is going through its transient oscillations. A control valve that is sized so that it is at 80 to 90 percent of its capacity when the controlled process variable is on control will not be able to cope with much in the way of control-loop transients before it is running up against its wide-open limit.

Theoretically, the best policy from the viewpoint of automatic control would be to size the control valve so that when the flow rate is its normally expected value, the valve would be operating at 50 percent capacity. Such a policy would result in larger control valves generally and would be considered extravagant, and therefore unacceptable, by those whose responsibility is to monitor costs.

The decision as to how much excess capacity is realistic is consequently a compromise between control-loop performance and cost. The following is a set of guidelines that have been used successfully.

1. *For flow-control loops.* Size the control valve for 150 percent of normal flow rate at the normal flow pressure drop or 120 percent of maximum flow rate at the maximum flow pressure drop, whichever results in the greater C_v (or C_{vm}).
2. *For level-, pressure-, and temperature-control loops.* Size for 180 percent of normal flow at the normal flow pressure drop or 120 percent of maximum flow at the maximum flow pressure drop, whichever results in the greater C_v (or C_{vm}).

17.4 Sizing for Liquid Flow-Control Valves

17.4.1 Effective pressure drop

The design of modern control valves is such that in liquid flow situations, jet flow may occur inside the valve body. This is really a vena contracta effect in which velocity energy is created at the expense of pressure energy. It may actually cause the pressure inside the valve body to drop to a value less than the pressure at the valve outlet.

If the pressure inside the valve body is reduced below the vapor pressure of the liquid at its prevailing temperature, the liquid will attempt to boil, and vapor will form. When this happens, opening the valve farther only creates more vapor and greater volume flow, while the mass flow increases only marginally.

This condition is called *choke flow.* It is more likely to be a problem with pure liquids than with mixtures, since pure liquids, such as water, will have a clearly defined relationship between vapor pressure and temperature.

The choke-flow effect is taken into account in liquid flow valve sizing by calculating the effective pressure drop ΔP_e. The effective pressure drop is that part of the apparent pressure drop (inlet pressure minus outlet pressure) which is actually effective in producing flow through the control valve. The equation is

$$\Delta P_e \text{ (kPa)} = K_m (P_i - R_c P_v) \tag{17.3}$$

where K_m = valve cavitation index
P_i = pressure at the valve inlet (kPaa)
R_c = critical pressure ratio
P_v = vapor pressure of the liquid at the valve inlet temperature (kPaa)

The valve cavitation index K_m (also called the *valve recovery coefficient*) depends on the design of the control valve. Some typical values are given in Table 17.1.

The critical pressure ratio R_c is a function of the ratio of the vapor pressure of the liquid at the valve inlet temperature to its critical pressure. Values of R_c for water are given in Table 17.2 and for other liquids in Table 17.3.

If the calculated value of ΔP_e turns out to be less than the apparent pressure drop (inlet pressure minus outlet pressure), then the ΔP_e value should be used in the control valve sizing equation, not the apparent pressure drop.

TABLE 17.1 Typical Values for K_m and C_1

Control valve type	K_m	C_1
Cage, balanced		
Linear	0.71	33.6
Equal percentage	0.74	31.5
Cage, unbalanced		
Linear	0.81	35.1
Equal percentage	0.83	35.3
Globe valves		
Single-ported	0.76	33.7
Double-ported	0.73	34.1
Butterfly valves	0.45	20.8
V-ball valves	0.60	26.8

Note: The values given in this table for K_m and C_1 are averaged values. If the values of ΔP_e and C_{vm} must be calculated with maximum accuracy, a second calculation should be made, using the K_m or C_1 value given in the manufacturer's data for the control valve type and size which will actually be used.

TABLE 17.2 Values of R_c for Water

°C	0	10	20	30	40	50	60	70	80	90
0	0.950	0.950	0.950	0.950	0.949	0.949	0.949	0. 948	0.948	0.947
100	0.945	0.943	0.941	0.937	0.933	0.929	0.923	0.915	0.908	0.898
200	0.888	0.877	0.865	0.852	0.840	0.827	0.815	0.804	0.795	0.787
300	0.780	0.773	0.765	0.755	0.743	0.726				

TABLE 17.3 Critical Pressure Ratio R_c (All Liquids)

P_v/P_c	0.00	0.01	0.02	0.03	0.04	0.05	0.06	0.07	0.08	0.09
0.0	0.950	0.937	0.925	0.915	0.905	0.896	0.888	0.880	0.874	0.867
0.1	0.862	0.857	0.852	0.848	0.844	0.841	0.837	0.834	0.831	0.829
0.2	0.826	0.824	0.822	0.819	0.817	0.815	0.813	0.811	0.808	0.806
0.3	0.804	0.802	0.800	0.797	0.795	0.793	0.790	0.788	0.786	0.784
0.4	0.782	0.780	0.778	0.776	0.774	0.773	0.772	0.771	0.771	0.771
0.5	0.771	0.771	0.770	0.770	0.769	0.768	0.767	0.767	0.766	0.765
0.6	0.764	0.763	0.762	0.760	0.759	0.758	0.757	0.756	0.755	0.754
0.7	0.753	0.752	0.751	0.750	0.749	0.748	0.747	0.746	0.745	0.744
0.8	0.743	0.741	0.740	0.739	0.738	0.736	0.735	0.733	0.731	0.729
0.9	0.727	0.725	0.722	0.719	0.716	0.713	0.710	0.706	0.701	0.697

Apart from the problem of sizing the control valve correctly, a calculated value of ΔP_e that is less than the apparent pressure drop is an indication that cavitation can occur. Cavitation has been known to cause damage to the control valve body and trim, as well as to the downstream piping. The usual corrective measure is to select a control valve that has a higher K_m. This usually means a more expensive valve, possibly in two ways. First, an anticavitation inner valve in itself is expensive. Second, a valve with an anticavitation inner valve will generally have a lower C_v than a valve of the same size with a regular inner valve. Consequently, the valve size may have to be increased to get the equivalent flow capacity.

17.4.2 Effect of viscosity

In theory, a high liquid viscosity will reduce the capacity of a control valve. This leads to the conclusion that viscosity should be taken into account when calculating the required C_v (or C_{vm}). In reality, a capacity correction for liquid viscosity is not necessary unless the Reynolds number is less than 10,000. Even at a Reynolds number of 4000 (the lower limit for turbulent flow), the loss in capacity is only 10 percent.

If a very low Reynolds number is encountered, the manufacturer's data should be consulted.

17.4.3 Accuracy

Control valve capacity coefficients have to be determined through laboratory research, so their accuracy will be affected by the quality of the manufacturer's laboratory facilities, the workers, and the procedures used. Not all manufacturers are equal in these respects. Thus the accuracy of published C_v data should be considered to be limited to the third significant figure at the very most.

Furthermore, any discussion of the correctness of determining the specific gravity of a liquid by dividing its density by the density of water at 0, 40, or 15°C is not relevant. The density of water can be considered to be 1000 kg/m^3 for control valve sizing calculations.

17.5 Sizing for Steam or Gas Flow Control Valves

Steam and gas control valve capacity equations are more complicated than the liquid flow equation due to the compressible nature of these fluids. A steam or gas stream may undergo density changes as it passes through the valve, even though the mass flow rate remains constant, as the laws of nature require. Since control valve capacity depends on volumes, not mass, changes in density can affect sizing.

There is another natural phenomenon that can have a bearing on the steam or gas discharge rate through a control valve. When the velocity of the stream is created by pressure difference, the velocity will have a natural upper limit which is the velocity of sound in the fluid at the valve inlet temperature and pressure. The fluid velocity will approach the sonic limit when the pressure drop becomes critical. This occurs if the downstream pressure is reduced to about one-half the inlet pressure (both pressures in absolute units).

When the pressure drop reaches the critical value, lowering the downstream pressure further will not produce any increase in fluid velocity, even though the control valve pressure drop is being increased. Consequently, sizing equations for compressible fluids must have a built-in check that accounts for whether or not the pressure drop is greater than the critical value.

The capacity equations below are basically those developed by Fisher Controls converted for SI units.

For steam:

$$W\,(\text{kg/s}) = (529 \times 10^{-6}) \times (P \times \rho)^{1/2} \times C_1 \times C_{vm} \times K_c$$

where P is the valve inlet pressure (kPaa) and ρ is the density of the steam at the valve inlet (kg/m^3). If the steam density is given in terms of specific volume in cubic feet per pound, then

$$\rho = \frac{16.02}{\text{specific volume}}$$

For gases and vapors:

$$W\ (\text{kg/s}) = (184 \times 10^{-6}) \times P \times (M/T)^{1/2} \times C_1 \times C_{vm} \times K_c$$

where P is the valve inlet pressure (kPaa), T is the valve inlet temperature (degrees kelvin), and M is the molecular weight of the gas.

For purposes of calculating the mass flow rate, the density of a gas can be estimated with adequate accuracy using the universal gas law:

$$\text{Density of the gas (kg/m}^3) = 0.1203 \times \frac{MP}{T}$$

(with P in kPaa and T in degrees kelvin). The equivalent equations for steam and gas flow, in fps units, are included in the Appendix.

17.5.1 K_c factor

The term K_c is a valve capacity correction factor for steam or gas flow (only) when the pressure drop is less than the critical value. Fisher Controls has observed that the relationship between the ratio of the downstream pressure to the inlet pressure and the capacity correction closely approximates a sine function. Consequently,

$$K_c = \sin\left(\frac{3417}{C_1} \times \sqrt{\frac{\Delta P}{P}}\right)$$

The expression inside the parentheses is in degrees, not radians. When it becomes 90 degrees, the stream velocity is at the sonic value, and the value of K_c becomes 1.00. If the expression inside the brackets gives a value greater than 90 degrees, then the pressure drop is greater than critical, and the value of K_c should be set at 1.00.

17.5.2 C_1 factor

C_1 is a factor specific to the type of control valve chosen for the application. Typical values of C_1 for different styles of control valves are given in Table 17.1.

Chapter

18

Perspective on Process Computer Control

It was about 1950 when the first digital computer made its appearance in a process control application. The earlier installations were proclaimed as the forerunner of a virtual revolution in control technology and process plant operations. Those who were skeptical about this were considered to be negative thinkers, perhaps even technologically obsolescent.

And yet, very few new developments in process instrumentation and control have really revolutionized continuous processing operations. New technology is usually developed to rid us of the problems associated with the old way of doing things. It often accomplishes this. However, the new technology will inevitably bring with it its own set of peculiar problems. The question this leaves us with is, "Are we better off now, all things considered, than we were before we made the change?" This was the question we had to ask when we changed from local to centralized control, from pneumatic to electronic, from steam tracing to electric, from instrument panels to consoles, from analog to digital, and from monolithic digital systems to distributed computer systems.

Intelligent process control designers know that this question, "Will we be better off with the new method and its problems than we were with the old way and its problems, which we nevertheless understood and knew how to live with?" has to be reckoned with. They also know that generalized answers are usually not trustworthy. The fact that someone else ended up better off as a result of changing to a new method does not mean that the same thing will happen in our plant. The circumstances have significant influence.

This is why it is worthwhile to stand back and take a perspective view of computer process control. What are the valid arguments for it? At the outset, there are two arguments that are *not* valid. The first, which might be called the "state-of-the-art magazine argument," is that if you do not now have an on-line computer controlling your operations or at least have definite plans to install one, then you are in the backwash of technology, and you can expect to lose out to your competition.

Any reasoning person would see that this argument is nonsense. On-line computers are expensive to obtain and expensive to own. Process computer installations run into the millions of dollars. They need a crew of highly paid people to maintain and upgrade them. And a computer does not necessarily do a better job of monitoring, evaluating, or controlling than a human being. The objective is to operate in the most profitable manner, not to have the leading business and technical journals publish articles about your new automated systems.

The second argument is much more subtle but still invalid. It goes like this. A digital computer can do *any* control job; therefore, digital computers should do *all* control jobs. The first part of this statement is 99 percent true. The second part is wrong. It behooves every conscientious process control designer to pick out the best hardware or system for the job.

The field of process control seems to attract rules of thumb. They are used extensively by persons who are uninformed or who are not inclined to spend the effort or the money to do the job right. All rules of thumb should be treated with suspicion; all except one, which has been proven true over and over. It says, *"Simplicity and reliability go hand in hand."* Thus the best design decision is to opt for the simplest way to achieve the required results in measurement or in control. Digital computers are not simple devices.

Authorities on process computer control may point out that process computers operate in real time. "What other kind of time is there?" is the reaction of many of us who were brought up by way of the measurement and control route. This is just one example of computer jargon that has characterized the arrival of digital computers in industrial processing, and that is now around for us to stumble over. I am personally not sure that I understand what "real time" means. However, my impression is that a system that functions in real time processes its input information as quickly as the input information enters the system. Recalling that process computers have inputs that are process measurements, and which regularly change their values, the computer system acts on a new value of any input immediately. It does not put the new value into storage somewhere until it can get around to dealing with it.

One might well ask why there is any need to have a perspective on process computer control, inasmuch as the chances that the plant where one works will ever have a process computer are practically nil. Make no mistake, if management becomes intrigued by the more glamorous aspects of computer control, then justification for the project by those who want it often becomes a simple matter, especially if the plant has a high-volume output.

Consider a plant that produces 10 million liters of product each day, with a value of 20 cents per liter. The value of the product turned out by this plant is $2 million per day. The proponents of the computer project can state, correctly, that if through the optimization routines which the computer can run the output of the plant will be increased by a modest 1 percent, then the computer will earn $20,000 per day. At this rate, the procurement costs of the computer, say $5 million, will be recovered in 250 days.

Management will rarely turn down an investment opportunity with this kind of return, so the computer project will become a reality. The trouble is that after the computer is finally in and has been debugged, several months late, no one checks to see if the promised 1 percent increase in production was ever realized. Very likely this could not be done anyway, since production in most high-output plants is far more sensitive to mishaps such as mechanical failures and operating mistakes than it is to any optimizing that a computer might do.

There are some real benefits to be obtained from process computers when they are applied in the right circumstances. The successful applications, which were no doubt publicized, as well as others, for which the owners had to settle for less than expected or which were eventually abandoned because they did not pay back the cost of keeping them on line and getting the results intended, reveal that real benefits can be obtained in at least three ways.

18.1 Logic

One of the more significant changes that has taken place in process control is the increased use of programmed logic. The term *logic* here means

- *On-line mathematics:* addition, multiplication (scaling), squaring and square rooting, and function generation for the purpose of linearizing nonlinear relationships. The ability to perform on-line math has a significant benefit in that it makes possible monitoring, and even controlling, process variables that cannot be measured directly but which can be computed from other measurable variables. Examples are catalyst circulation rates and boiler and fired heater efficiencies.

- *Setting limits on variables and the signals that represent them.* This means measurements, controller outputs, setpoints, and differences between the actual value of a process variable and its setpoint.
- *Selecting variables.* Selecting the highest, the lowest, or the middle value, or the average value, of more than one related signal for display, alarm, or control.
- *Conditional moves.* If a particular variable is greater than or less than another variable, then a switch is to be made to a different mode of measurement or control.

One needs only to look at a control schematic diagram for a modern boiler control system to see how much of the control system is really logic and how little of it is PID control. Logic is not done at all well by chains of analog modules. One module per function to be performed is required. This means that someone has to be available who understands not only the specific function of the module but also how that module fits into the whole control scheme. Furthermore, analog modules can lose their calibration over time and have to be removed and tuned up if the system is to work properly.

Logic is far easier to set up, and to change, when it is done in software. Changes to logic systems that are made up of analog modules require the addition of still more modules and their associated hookup. Furthermore, and perhaps more important, logic in a digital system is more reliable because it never loses its accuracy.

Since PID control can be done digitally, this leads to the conclusion that if logic predominates over analog control in the overall plant control scheme, then an all-digital system is probably the best choice.

18.2 Advanced Control

In Chap. 11 the point was made that a feedback control system cannot really hold the controlled variable at the setpoint. It can only keep bringing it back to the setpoint after the control system is disturbed. Consequently, a feedback control system will give satisfactory results if

- The disturbances are minimal in size and number or if they spread themselves over an extended period it time. This is another way of saying that the disturbances are not large or sudden.
- The control system has a short period of oscillation and therefore a short recovery time. Under these circumstances, the disturbances are not really a problem because the control system can quickly compensate for them.

However, lacking at least one of these conditions, feedback control systems perform rather poorly. Control of the product off a fractionating column is an example. In order to maintain decent control when disturbances are not minor and when the recovery time is long, some more elaborate control scheme will be required. These more elaborate schemes, i.e., more elaborate than conventional feedback control, are usually lumped together under the heading *advanced control.*

A characteristic of all advanced control schemes is that they require the capability to compute on line. It can be reliably stated, therefore, that the advent of on-line digital computing has made it possible to use control methods that were not hitherto available. Provided that these schemes are properly engineered and tuned, considerable improvement in the performance of control systems on hard-to-control processes can be expected.

One useful control strategy, especially in the case of control of fractionators, is to use the computer to material balance or energy balance the process. When this method is used, the computer is provided with a model of the process, from which the computer can calculate how much the controlled variable will be changed by a known change in any of the other variables (manipulated variable and disturbance variables) that can affect it. The disturbance variables and the manipulated variable are all measured and become inputs to the computer.

Whenever any disturbance variable changes, instead of waiting for the effect to work its way through the process and cause the controlled variable to deviate from the setpoint, the computer notes the extent of the change and calculates how much this would cause the controlled variable to change. Then it calculates how much the manipulated variable would have to be adjusted to create an equal and opposite effect on the controlled variable. This corrective action is fed into the process, via the manipulated variable, in phase with the disturbance so that the net effect on the value of the controlled variable is zero.

Control schemes that function in this way have come to be classified as *feedforward control. Material balance control* or *energy balance control* would have been better names but presumably would not have been as catchy. The nature of product quality-control systems is such that it may be necessary to develop a material balance. Some temperature-control systems are difficult to control well using feedback control. For these systems, an energy balance is required rather than material balance.

Figure 18.1 shows a fired heater control system for which the requirement is to maintain the oil temperature constant at the heater outlet. The manipulated variable is the flow of fuel gas to the heater. The potential disturbance variables are the inlet temperature of the

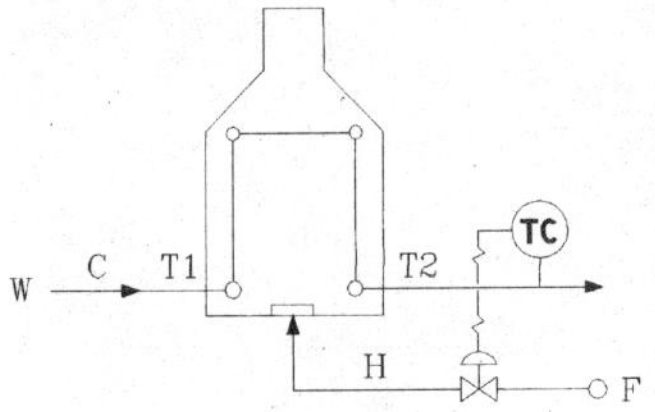

Figure 18.1 A fired heater outlet temperature-control system.

oil, the oil flow rate, and possibly, changes to the setpoint. A computer model can be developed around an energy balance in the following way. The relevant variables are

Oil flow rate = W kg/s

Oil temperature at the heater inlet = T_1 (°C)

Fuel gas flow rate (the manipulated variable) = F sm^3/s (standard cubic meters per second)

Specific heat of the oil = C kJ/kg•°C

Calorific value of the fuel gas = H kJ/sm^3

Oil temperature at the heat outlet = T_2 (°C)

The disturbances in this control system are W, T_1, and T_2 (the setpoint).

For an energy balance, the heat supplied equals heat consumed. The heat supplied comes from the fuel gas. It will be

$$F \times H$$

The heat is consumed by heating the oil and by losses, which are assumed to be 30 percent of the heat fired. Therefore, the heat consumed will be

$$W \times (T_2 - T_1) \times C + 0.3 \times F \times H$$

Equating the heat supplied to the heat consumed gives

$$0.7FH = WC(T_2 - T_1)$$

Rearranging this in terms of the manipulated variable F gives

$$F = WC\,\frac{(T_2 - T_1)}{0.7H}$$

This equation is the model of the process. W and T_1 are the disturbance variables. Both of these must be measured independently.

Their measurement signals become inputs to the control computer. T_2, the setpoint, is a manual input to the computer. C, H, and the 0.7 term are parameters that are programmed into the model. Their values may have to be adjusted from time to time.

This is basically the way in which a feedforward control system is designed. However, there are two other factors that must be dealt with before the design can be considered to be workable.

The first is that the corrective action (the change in the manipulated variable) must be applied to the process over the same time interval as the disturbance. Corrective action that comes too soon results in overcorrection, which will drive the controlled variable away from the setpoint in the opposite direction. Consequently, corrective action must be applied with the proper amount of delay, depending on how quickly or how slowly the disturbance variable affects the controlled variable. In practice, it is usually not possible to predetermine the right amount of delay. The delay effect must be made adjustable, and the correct amount must be found on the operating plant by experimentation.

The second is that the model of the process will never be perfect. There will be disturbances other than the ones specifically singled out for inclusion in the model. The measurements will never be entirely accurate. And the values of the parameters in the system may change. The end result is that the corrective action taken through feedforward control may not hold the control variable exactly at the setpoint.

To get around this problem, feedforward control systems usually have a feedback control loop overlaid on them. The function of the feedback system is to correct for any residual control error after the feedforward system has done its job. This feedback loop probably will operate at a relatively low gain and reset rate. It will not have derivative, since its function is not to stabilize the control system but to eliminate any uncorrected control error caused by deficiencies in the model and the measurements.

In a well-designed feedforward control system, one would expect that when the control system is disturbed, about 95 percent of the required corrective action would be provided by feedforward control and the remaining 5 percent by the feedback control loop.

Figure 18.2 shows the control computer and the various measurement signals that are its inputs. This computer control system actually has two feedback control loops in it: the feedback system on the outlet temperature described already and a feedback loop on the flow of fuel gas. The second feedback loop is part of a cascade system in which gas flow control is the secondary loop. The setpoint of the gas flow-control loop is the gas flow rate calculated by the computer to be correct for

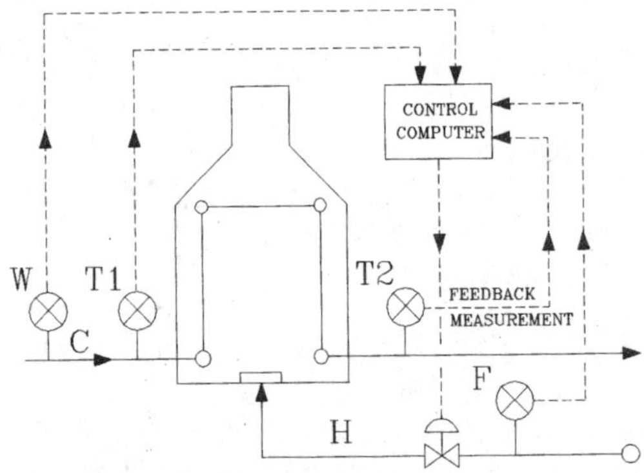

Figure 18.2 A feedforward temperature-control system for a fired heater.

compensating for disturbances. The linearizing of the flow measurement signal, the proportional plus reset response, and any characterization of the control valve are all done digitally in the computer.

Although feedforward control is currently a big item in the advanced process control repertoire, it is not by any means the only one. In plants that require large numbers of control loops, there may be interaction between two or more of the loops. What happens is that corrective action in one feedback control loop causes an upset in another control loop. If the corrective action subsequently taken in the second loop reacts back and upsets the first control loop, the two loops get into conflict with each other, and neither will settle down, at least not at the controller settings that are determined to be optimum for each one.

In the absence of any other remedy, both loops may have to be operated at much lower than optimum values of gain, reset, and derivative, or one of the loops may have to be operated on manual control. However, if on-line computing is available, it may be possible to develop a control algorithm using matrix algebra that will effectively decouple the two control loops and allow them to operate independently at close to their optimum P, I, and D settings. The technique is described under "Multivariable Control" in Chap. 14.

In the section entitled "Ratio Control" in Chap. 14, there is a description of a three-element boiler feed water (BFW) control system which is an example of a ratio control system with feedback trim. If the features of this control system are examined, it appears to have all the characteristics of a feedforward control system, complete with feedback trim. Is a three-element BFW control system a ratio control system, or is it a feedforward control system?

The answer is that it is both. A ratio control system is really a special case of a feedforward control system, in particular, one in which only a single disturbance is measured. Feedforward control encompasses control systems in which two or more disturbances are measured and fed into the model. In the control of light ends fractionators, the objective is to control the quality of the product stream, while the principal disturbances are the flow rate and the quality of the feed stream. Accordingly, both these variables have to be measured and fed into the computer. The computer's model of the fractionator is based on the simple and reliable hypothesis that you can only take out of a fractionating column what you put into it. Using this principle, the computer sets the withdrawal rate of the primary component at the rate at which that component is being fed into the column. By so doing, it material balances the column.

18.3 High-Speed Monitoring

For profitability reasons, the optimum size of plants has been increasing and, along with it, the size and the cost of the machines that go into such plants. Continuous production of great value often depends on the performance of large, expensive pumps, compressors, motors, conveyers, or other continuously operating machinery. Other process equipment may not move but may consume huge amounts of costly energy. Some machines are so costly that it is not practical to have a standby spare.

In order to avoid breakdowns and the resulting production losses, operating equipment that is this important must be monitored closely. Can any human being be expected to do this? Checking gauges for a whole shift is a menial, dull, and onerous task that requires little discretion or judgment. No one could reasonably be expected to devote his or her full time to it. Yet the critical nature of the equipment, as well as the consequences of having it down unscheduled, is such that this level of attention is necessary.

A high-speed digital computer-based monitoring system is a practical solution. Important variables can be monitored on a second-to-second basis. Their values can be compared with programmed safe limits and an alarm given if any condition becomes abnormal. Furthermore, past values can be stored, from which a trend can be computed as a guide to the seriousness of a developing bad situation. Finally, if an accident does occur, the computer will contain data that may lead to determining the real cause.

18.4 Computer-Supervised Startup and Shutdown

While the optimization capabilities of on-line supervisory computers have been tried out on a grand scale, very little attention seems to have been given to their possible use in supervising plant startups and shutdowns. This seems curious, especially in view of the extent to which computer monitoring is used in the startup and launching of space vehicles. Oil refineries and petrochemical plants still seem to start up and shut down their process units relying only on the knowledge and skill of their process operators, assisted by thick manuals of instructions that are stashed away in the operator's desk drawer.

Visualize instead a startup procedure in which the whole startup sequence is programmed, step by step, into an on-line supervisory computer. The computer displays the steps to be taken, in their proper sequence, one step at a time. No step is displayed until the preceding step has been carried out properly and verified by a manual input from the operator in charge or by a signal which the computer receives directly from the process.

In addition, the computer constantly checks the process by monitoring not only temperatures, pressures, flow rates, and levels in vessels but also whether or not critical valves are in the correct open or closed position and whether or not key rotating equipment is on and operating properly. In the event that anything does not happen as it should, the supervisory computer alerts the operator or may even abort the startup.

When the startup operation has been completed and the unit is on line, the computer will contain complete documentation of what happened. One could conclude that a computer-assisted procedure such as this would lead to safer startups and shutdowns. These are, after all, the times when the likelihood that something will go wrong is greatest. Despite this, not much seems to have been done yet toward using on-line computers in this case.

18.5 Evolution of Process Computer Control

The earliest uses of digital computers for controlling operating plants in real time were naive attempts to move computers designed for business offices into an industrial environment. The efforts were not very successful. Computers that were used to having young women with a gentle touch typing on their keyboards in a nice clean office environment were not happy when stalwart plant operators pounded the keyboards with big fingers inside of bigger mitts or spilled coffee or pop on the keys while saying nasty things to the computer because it did not do what they expected it to do. These computers also did not like being subjected to the dust or to the traces of hydrogen sulfide, chlorine, or other contaminants that were present in the atmosphere. As a result, process computer control got off to a rather slow start.

This text will spare the reader from the tedium of having to peruse diagrams depicting computer architecture. Suffice it to say that the evolution of process control computers went through three recognizable phases. In the first phase, the plant retained all its individual, panel-mounted analog instruments. The measurement signals on which these instruments relied were branched off to the computer through an interface unit, one of whose functions was to convert the individual measurement signals from an analog to a digital form.

Thus, having access to all the plant measurement information, the computer went through its optimization routines and worked out the optimum setpoints for the control loops that had the greatest impact on the profitability of the plant. These setpoints were then relayed back to the appropriate panel-mounted controllers via the interface unit, which converted the digital setpoint signals back to analog.

Although phase 1 was eventually abandoned, it had three virtues that are worth mentioning. First of all, if the computer quit, which, as we know, does happen, the plant still carried on under control by the analog controlling instruments. Thus a computer outage was not a catastrophe. It merely meant that the optimization routines were temporarily out of service.

Second, the ultimate decision as to whether the computer was doing the right thing by altering the setpoint of any control loop was left up to the operator. By means of a switch on each of the panel-mounted controllers, the operator could switch out the setpoint generated by the computer and set his or her own setpoint.

Third, by using the phase 1 method, it was possible to overlay supervisory computer control on a plant that was equipped with pneumatic instruments. It simply required some additional components to make the conversions from pneumatic to electronic, and vice versa. It is interesting to note that these advantages would still be considered valid today, even in the sophisticated control environment that currently prevails.

Considering its merits, why was phase 1 discarded? Mainly for two reasons. The first was that successful operation of the whole computer system depended on the performance of the interface unit. At that point in time, instruments were obtained from an instrument manufacturer, whereas computers were obtained from a computer maker. Each apparently considered that the other should accept the responsibility for designing and building the interface. As a result, the interface units that were produced were generally of questionable quality, and this affected the reliability of the system as a whole.

The second reason was that there was a developing philosophy to the effect that operating process units from instrument panels was inefficient. The right way was to have the operator seated in front of a console, with all the relevant information available to him or her on the screen, instead of having the operator pace up and down in front of an instrument panel that might be 50 ft long. Accordingly, phase 1 was not acceptable because of instrument panels.

The objective of phase 2, therefore, was to streamline the system to get rid of the panel instruments and the interface unit. This objective was achieved by bringing all the measurement signals from all the field-mounted transmitters directly into the computer. The required analog-to-digital conversions were now a function of the computer itself. Having the process measurements in digital form, the computer accepted the setpoints as manual inputs via a keyboard and calculated digitally the PID control response for each control loop using the algorithms that were programmed into it.

The computer also generated graphics that were displayed on the screen, along with any process information which the operator called out via the keyboard. The output signals that were to go to the various control valves were converted from digital to analog by the computer also. Thus the panel instruments and the interface were gone, and just about everything of importance operated in the digital mode. This arrangement came to be called *direct digital control,* or *DDC.*

One needs only a cursory look at the DDC strategy to realize that there is a significant potential problem. Since the computer is doing the control job for all the control loops, then a computer outage would cause every control loop in the plant to switch to manual control at the same time. In addition, there would be no measurement information available to the operator.

Nevertheless, in some notable cases the proponents of DDC were determined enough that a second computer was justified and installed to back up the first computer. One of the ploys that was used was to have the 20 percent most important control loops assigned to both computers so that if either computer failed, the other would immediately take over control of these critical loops. In other cases, critical loops were backed up by individual analog controllers mounted on an auxiliary instrument panel that was located in an unobtrusive corner of the operations center.

It eventually dawned on designers that for the sake of having everything done in the digital mode via DDC, a great deal of expense, due to the backup requirements, was being incurred. This extra expense was difficult to justify. All that was really happening, as far as the plant was concerned, was that the same feedback control loops were now being controlled digitally instead of by analog means. It was still basic feedback control, with the same *process* component in each of the loops and the same recovery times from the same disturbances.

DDC was, to some extent, the result of being psyched out by the idea that if something is done digitally, then it has to be better. Numerous computer people and some control people had apparently forgotten the story about the emperor and his new clothes, were carried away with this philosophy, and were promoting DDC. In the end, however, we have to accept the fact that as instrumentation and control people, our job is not to pull off technical stunts. Our mission is to apply the control hardware and systems that will do the most to improve the profitability of the plant in which they are installed.

If we are to make money out of a digital computer application, then we should put the computer to work doing the thing that it does best—computing. The computer that spends most of its time doing trivial jobs such as converting signals from one form to another, generating pictures to be displayed on the screens of monitors, and figuring out whether control valves should be moved or alarms should be turned on or off is really not going to earn its keep. All these functions are minor and should be done in some other way. Yet the DDC computer spent a large part of its time doing these very things.

Process computer control might very well have ended with the demise of phase 2 had it not been for a remarkable technological breakthrough—development of the microprocessor. It was this development which made possible phase 3, which we have today: *distributed computer control systems.*

In a distributed computer control system, the system is separated into discrete modules, each of which performs a specific function. A control module may control a dozen PID loops, give or take a few, and will be able to perform a limited amount of logic, such as linearizing nonlinear measurements or generating an alarm when some process measurement exceeds a safe limit. Another module may generate the displays required for a video monitor.

Other modules regulate the flow of information throughout the whole system. One thing that is common to all the modules, however, is that they can operate independently, because the processing power necessary to carry out their functions is built right into them. This is what the microprocessor made possible. Before microprocessors came on the scene, the intelligence necessary to perform these functions required relatively large circuit boards with their numerous components and relatively large cabinets.

A distributed system consists of the various modules, all linked together by means of a data bus, or highway. A component failure in any one module harms the system only to the extent of losing that one module, at the most. At the least, it may result in the loss of only one function within the module.

An important point not to be overlooked is that since the computer control system is distributed in this way, it also can be distributed geographically. This means that there is now an alternative to the single large operations center concept for a plant that has a number of operating units. It is now practical to have a small control center for each operating unit. All the small centers can be linked together, and the process data from all of them can be sent, via the data highway, to a master operations center for the whole plant.

Decisions as to what should be done for the good of the plant as a whole can be made at the master center, possibly supported by a supervisory computer that is also on the data highway. The results of these decisions, in terms of control action, can be sent back to the appropriate control center and control loop over the same data highway, since two-way information flow is a feature of the distributed computer control system.

The overall benefits of the phase 3 arrangement have only begun to be discovered. This type of control system will likely be around for a long time, if for no other reason than that it will be difficult to improve on its benefits.

18.6 Analog and Digital Data Processing

Before any decision is made on the operating philosophy for a plant, particularly with respect to the choice of instrument panels or consoles, there is a fundamental point that must be taken into consideration. Even though there are now single-loop PID controllers that operate digitally, instrument panels usually have automatic controllers that operate in the analog mode using the standard 4- to 20-mA signals. Thus the merits of operating digital rather than analog may be a factor in the instrument panels versus consoles decision.

Anyone who believes that the ultimate system is one that operates all-digital is a dreamer, because this will never happen. The majority of sensors used for process measurements function in analog fashion, and nothing can be done to change this. Attempts to promote control valves that operate in digital fashion have been unsuccessful so far. Control valves will likely continue to function in the analog mode in the future.

In the physical world, more things happen analog than digital. The planets proceed in their orbits in analog fashion, not stepwise. Everything that grows, grows in an analog way. One of the most basic of measurements, time, advances onward in an analog manner.

The virtues of analog devices show up most clearly in measurements. A measurement made in analog form is a continuous measurement. There is never any loss of measurement information.

Sometimes this is a significant factor in the quality of control. A digital system, on the other hand, samples a measurement at a programmed interval. The measurement information that happens to fall between the measurement points is never acted on.

Proponents of digital systems may argue that the interval between the samples is so short, possibly of the order of 1/10 second, that the digital method is effectively duplicating an analog measurement. The assumption here is that there are no industrial processes that will change to any appreciable extent in 1/10 second. Real authorities on process control, who have seen how fast some operating equipment can react, are unlikely to buy this.

There is also a school of thought that says that digital sampling of measurement data is actually better than analog because the digital method filters out extraneous noise that may be superimposed on the actual measurement. When one thinks about this, the answer to this assertion is that if the measurement information that is rejected because it is between the sample points is bad information, then overall, the measurement is improved. If the information that is rejected is good information, then the measurement is made worse.

The real problem with analog components is that they require calibration periodically to keep their accuracy up to the required level. If this calibration work is not done when it should be, then errors of significant size develop. Digital systems for processing and transmission of data retain their accuracy much better, and in this sense anyway, they are definitely superior. Furthermore, when a measurement is presented in digital form (a display of numbers) rather than in analog form (a scale and pointer), the chance of human error when making a reading is certainly reduced.

18.7 The Physiologic Aspects

The various effects on the human body of sitting throughout an entire shift in front of a video monitor, as opposed to walking up and down in front of an instrument panel, are yet to be evaluated fully. There have been reports publicized about eyestrain, headaches, back and leg problems, and other health disorders associated with too little physical activity and staring too long at video monitors. The main operations center at the Syncrude plant, which contains distributed computer control systems supplied by three different manufacturers, is also equipped with a rowing machine and an exercise bicycle. This is certainly an indication that at that plant anyway the negative effects of video monitors on the physical well-being of the operating staff is a recognized fact.

18.8 Process Computer Project Hazards

Computers that are now running operating plants have received their share of publicity; this is certain. Yet the projects that have been written up and which have been described in technical meetings seem to be those which are brand new and have yet to prove their worth, rather than those which have been through startup and debugging and have been deemed operational. Little, if any, publicity has been given to those projects which never did meet the expectations held for them or which operated for a while and then were abandoned because the time and effort required to get them to attain their objectives made the whole venture unprofitable.

Before embarking on a process computer project, it is only prudent to consider some of the pitfalls that others have encountered and take whatever action may be advisable to protect the project, and oneself. Process computer projects may come to grief for any of the following reasons:

- Failure to set clear, specific objectives for the project from the very beginning and then *stick to them.* Once word is out that the computer project is "go," just about everyone will have his or her own pet task that he or she would like to have added to the computer's functions.
- Failure to realize that the eminent experts from the computer system company, to whom you are introduced when the project is in the bidding stage, will in no way be around to work on your system should their company be awarded the contract. Your computer contractor will have learners on staff who will be real keen to accumulate some experience hours working on your system both in their shop and afterwards in your plant. On the other hand, knowledge of how your plant and its operating equipment function will be critical to the success of the project, and this knowledge can only come from your own people.
- Failure to recognize the prospective contractor who cannot possibly meet your specifications for the price he or she has quoted you.
- Failure to recognize that in many operating plants the quality of key process measurements will have to be considerably upgraded if the computer is to use them effectively.
- Failure to understand, in your enthusiasm, that not everyone will love the computer. On-line computers often add substantially to the workload of the instrument maintenance group. Some operating people may feel threatened. The bottom line is that there may be those who actually hope that the project will bomb.

18.9 Is There Yet Another Unrecognized Hazard?

In the April 1990 issue of *InTech,* the publication of the Instrument Society of America,* there was a letter from a reader who brought up an interesting and possibly disturbing point. This reader thought that he had detected a correlation between the adoption of computer-based process control systems and an apparent increase in the number of large-scale plant catastrophes. The point was that there could be an unexpected weakness in plant monitoring by computer, in that unsafe conditions were not being recognized as quickly as they were under human monitoring.

This observation, if true, would run contradictory to the concept that since a computer can process data faster than a person and it never tires, then it can monitor plant conditions more effectively than can a person. Is it possible, therefore, that when important monitoring functions that were once carried out by people are turned over to a computer, the people become complacent? There has been evidence already, uncovered during the investigations of certain plant accidents, that a computer issued a warning that the operating crew disregarded as being of no consequence. If the news reports were correct, the plants at Three Mile Island and Chernobyl were both monitored by on-line process computers.

If this trend should continue, then more and larger plant disasters will result. We need to know more about the psychological effects of switching to all-out computer process control. The disposition of responsibilities between humans and machines needs to be carefully thought out.

18.10 Distributed Control: Without Video Monitors

The discussion of the relative merits of video consoles versus instrument panels is primarily intended to point out that this is a decision that is too often taken too lightly. There are many aspects to it, and they deserve careful consideration.

Developments in instrument and control hardware and systems have made possible another alternative, namely, to distribute the overall plant control system in a number of small control centers but have instrument panels in the control centers rather than video displays. The various control centers can be linked via a data bus as before, and the whole operation supervised, with an on-line computer

*Now the International Society for Measurement and Control.

if appropriate, from a master control center. The individual automatic controllers on the instrument panel will operate in the digital mode, but since they sample only one measurement, the analog-to-digital conversions will be made with very little loss of measurement information.

This entire question is a graphic example of the basic axiom, when we change to a new system or method to be rid of our old problems, we inevitably find that while the new system does in fact eliminate the old problems, it brings along with it a whole new set of problems that we will have to cope with. So once again the question comes up, "Are we better off with the new way which freed us from the old problems but brought with it a whole batch of new ones, or would we have been better off with the old system and its problems, which we at least understood and knew how to live with?"

Chapter

19

Operations Centers

Someone once said, "A camel is a horse that was designed by a committee." So, it seems, is the situation with a great many operations centers. When it comes to operations centers, everyone is an expert. Consequently, when the time comes to make the important decisions regarding the design of the plant operations center, a meeting is held, and everybody who is anybody attends.

The person whose voice will be heard the loudest is the manager of operations. The operations center is, after all, the domain of the plant operating staff, and the manager is the top person in operations. Therefore, the manager's opinions count for a great deal. Unfortunately, manager of operations turns out to be a rather transient position in some companies, which makes it quite possible that the manager of operations who attends the big decision meeting will not be the manager of operations when the plant is completed and ready to start up.

Thus it is that the regular operating personnel may be left with an operations center designed and constructed to the personal preferences, of the moment, of the operations manager and other VIPs who attend the monumental meeting. Too often the opinions of persons who actually have something to contribute to the design of the operations center—and they are not necessarily process control experts—are not given much weight. Nevertheless, the day-to-day operations staff, with their typical ingenuity, usually manage to find a way around the difficulties and get the plant to run and make money. More power to them.

The people who remain quiet in these meetings, for the most part, are the representatives of the engineering contractor who will eventually build the plant. For one thing, engineering contractors are not usually innovators when it comes to building operations centers, nor do they particularly want to be. Better to let the client come up with his or her unique ideas and then build it that way. Then the contractor is not party to the hassle afterwards if things do not turn out too well. If it is left to the contractor to do the job, the operations center will likely have either flat, vertical instrument panels around the walls of the room or the instrument company's standard console design.

The operations center is the place where information flows from instrumentation systems to people. The people analyze the information and then communicate other information, in the form of requests or commands, back to the instrumentation. The need for a reliable flow of information back and forth between people and instruments has spawned a new area of technology called *ergonomics*. This field, in itself, is one in which people have made careers. It is obviously highly important, especially in those operations which could end in disaster if information is not transmitted or received accurately. This could include everything from operating a nuclear power plant to flying a Boeing 747. Many plant calamities have been attributed to failure to receive, to interpret, or to act on important information.

The operations center is first and foremost a center for communications. Consequently, when decisions are made regarding how it should be designed, accurate communications should be one of the highest-weighted objectives. This chapter identifies some of the important decisions that have to be made, as well as considerations relative to those decisions.

19.1 One Large Center or Several Small Ones

From the viewpoint of overall plant operations, should there be one large operations center from which all the process units are monitored, or should there be a number of small control rooms, one for each process unit in the complex?

Until the advent of distributed computer control systems, the only practical option available was a single central operations center. This was considered necessary because, if a plant had a number of operating units, each with its own crew, but linked together to form one operating complex, it was vital that there be adequate communications between the operators of the various process units. The best form of communication is obviously face to face. Therefore, a single central operations center was thought to be best so that operators could talk to each other without having to rely on telephones or radios.

While the single central operations center is still an alternative today, distributed computer control systems have provided a second possibility. Each of the operating units can have its own smaller operations center, with all the individual centers linked together by means of a data highway.

This arrangement allows process control at the process unit level to be handled out of the individual centers, while supervisory control of the overall plant can be carried out in a separate supervisory control center that is also on the data highway and which may be staffed by operations technical people. Through the data highway, the supervisory center can access process information from any of the individual units. Information also can flow in the other direction. If a change in operating conditions is determined to be advisable based on personal judgment or on computer optimizations, then the new conditions are sent down the data highway from the supervisory center either to the appropriate unit operator or directly to the process control loops involved.

Why would anyone want to distribute the control system for the plant in this way? The answer lies in the geography of the plant. We need to keep in mind that the operators who staff the operations centers are not the only operators on shift. Some members of the crew have the unenviable job of walking the units outdoors. Their responsibilities are opening and closing valves and turning rotating equipment on and off if this cannot be done from the operations center. They also must be on the lookout for things that do not look right, sound right, or smell right. If the plant is a large one and it has only the one central operations center, then some of the field operators may be stationed a hundred meters away.

What happens to these field operators if it rains or if temperatures drop down to −30°C in wintertime? No one can be expected to put in 8 hours outdoors under these conditions. There has to be a shelter provided in the vicinity where a person can escape the weather for a while. This is where the distributed control system has appeal. While the local unit operations center is primarily the place from which the unit is controlled, it also can be the shelter, and perhaps even the change area, for the field operators.

19.2 Building Construction

This book does not deal with building codes, nor with the features of different types of building materials. However, two points should be mentioned. First of all, plants that are processing flammable materials are unfortunately susceptible to explosions. If all the controls for all the plants in the complex are housed in a single operations center, then there must be concern over the damage that an explosion could cause. A single large control center will invariably be located at the geographic center of the complex and would be vulnerable to an explosion shock wave from any direction.

Operations buildings can be blastproofed and have been, at considerable cost. Since blastproofing is really an insurance measure, the decision as to whether the cost is justified is up to the owner of the plant, not to the writers of books. It is, however, a matter on which a decision should be made, one way or another, at the design stage. Limiting the damage that would be suffered, should an explosion happen, is another reason why distributing control in a number of small operations centers could be a good idea.

Plants that are processing hazardous fluids may have also a need for an operations building in which the pressure inside is maintained at a slightly positive value so that any air leakage will be from the inside out rather than in the other direction. It has been found that concrete blocks, one of the popular building materials for operations centers, are actually porous to air seepage if there is any difference in pressure across them. It is unlikely that a concrete block building will be sufficiently airtight to maintain the desired positive pressure in the building.

19.3 Building Layout

Single central operations centers are prone to still another problem: congestion. On unusual occasions, such as the startup of new facilities or possibly in an emergency, the operations center tends to be a gathering place for bystanders who make life difficult for harried shift operators by asking for progress updates and generally getting in the way. Distributing plant control among several centers helps to alleviate this problem also.

When the choice is one large operations center, a prudent move is to install a counter or some other fixed furnishing that can serve as a barrier between the space where an operator works and the open central area where the distinguished observers congregate.

Assuming that the operations center building also will house the change facilities for the operating staff, do not forget that these days many operating crews have women operators.

If the instrumentation and control systems in the operations building have dc electric power, then there will be storage batteries. Storage batteries can generate hydrogen, which is potentially dangerous in an inside area. Hence there may be rules, perhaps enforced by government authority, regarding the housing of storage batteries.

It can be taken for granted that it will not be permissible to stack storage batteries behind an instrument panel. A dedicated battery room will be required. An in-depth investigation should be undertaken to determine just what the local rules are. They may stipulate any or all of the following:

- Special construction for the walls of the battery room
- That one of the walls of the battery room must be an outside wall
- More than one way into or out of the room
- Heavy-duty ventilation

Other requirements are possible, and all of them must be known and allowed for when the building is designed. It can be very expensive to have to modify the building after it is completed to conform with lately discovered regulations for battery rooms.

Electronic instrumentation and computers require an environment that is almost entirely free from hydrogen sulfide; otherwise, equipment can fail, sometimes in incomprehensible ways, from H_2S corrosion. This point is mentioned here because the floor space required to install air-purification equipment can be considerable. Once again, if the need for more or larger air-purification equipment is discovered after the building has been completed, the cost of the modifications will be high.

19.4 Instrument Panels or Consoles

When everyone knows that console desks with video-display terminals (VDTs) are state-of-the-art today, why would anyone even consider an operations center with instrument panels? It should be obvious that the best arrangement is one in which the operator can sit at his or her console and call up and display any needed plant information, simply by pressing the right keys.

All well and good, as long as we do not overlook one point. On an instrument panel there will generally be one instrument for each process measurement. Accordingly, all the plant measurement information is displayed continuously. The operator does not have to ask for anything. He or she just has to go over and look.

Contrast this with the system in which nothing is displayed until it is requested. Everyone who has a PC at home knows that despite the best work of the software designers, there is always some combination of key operations, usually done inadvertently, that causes the computer to hang up. Do we really believe that the same thing cannot occur in a large plant computer, especially if the operator is under duress and does things in a hurry? If it does happen, what are the implications?

A further consideration, already mentioned in Chap. 18, is the possible physiologic problems that may result from sitting for up to 12 hours at a console desk watching a VDT screen.

There will be cases in which the panels versus consoles decision becomes automatic. Electronics, whether digital or analog, cannot tolerate environments where the humidity is very high (such as dairies and water-treatment plants), where there are flammable vapors present (as in gas compression stations), or where there is significant hydrogen sulfide present. These situations require pneumatic instruments. Inasmuch as pneumatic instrumentation comes only in analog form, instrument panels are the only alternative.

19.5 Lighting

The owner of the plant may opt to leave the matter of the operations center lighting up to the contractor or a subcontractor and assume that everything will be done properly. If the person who takes on the lighting job knows nothing about what goes on in an operations center, then management may end up with complaints when the center becomes operational. Generally speaking, control rooms that have instrument panels need a high level of lighting. Operations centers that have consoles and VDTs usually require subdued lighting so that the information displayed on a VDT screen stands out.

The choices for lighting are either fluorescent or incandescent fixtures. Fluorescent fixtures are popular because they provide more light at a lower energy consumption. They also have a relatively low heat output. The problem with fluorescent lighting is that a fluorescent tube is either on or off. It cannot be dimmed.

The only way that variable lighting can be obtained using fluorescent units is to group the individual fixtures into banks and have a switch on each bank. Then the degrees of brightness can be adjusted, in stages, by the number of switches that are on. Staging the lights is easily done provided that the requirement is allowed for when the building is designed. Attempting to rearrange the circuits after the building is in service will be awkward and expensive, and the work going on will disrupt the normal operations center activities.

Incandescent lights can be equipped with dimmers to obtain variable brightness. However, the heat generated by light bulbs will add to the load on an air-conditioning system.

Since most instruments have glass windows or lenses, sometimes curved rather than flat, do not overlook possible complaints about glare if there is a direct path from any light source to the instrument face. Indirect lighting is the only sure preventive measure against this problem. Left to his or her own devices, any lighting contractor will automatically install the lights in the ceiling—not necessarily the best arrangement for an operations center. The owner's own people have to take an interest in the whole matter of lighting to ensure that the decisions that are made will result in lighting that meets everyone's approval.

19.6 Communications

Prior to the advent of portable radios, the standard methods of communicating around process units were telephones and hollering. Portable radios have added a new dimension. While acceptance of radios has been generally good, there are two problem areas that should be recognized. First of all, when there are more than one portable radio on the same channel, only one can be operating at any one time. If two users attempt to transmit at the same time, then only the one that is closest to the base station (usually located at the operations center) will be able to transmit a message. This also means that if a user who is remote from the base station is transmitting and a user who is closer decides to transmit, the second user will cut off the first in the middle of the message. What if the message was important?

The second problem is that when the plant is large and distances are great, or if there are large steel structures in the plant, the tendency is to go to higher-output portable radios to improve reception. Some portable radios can output as much as 1.25 W. This is quite enough to adversely affect the behavior of electronic instruments, electronic controls, and computers, as many plants with portable radios have found. It is not unusual to see signs that say "No Radios Allowed in This Area" in operations centers and computer rooms.

Finally, there is the matter of whose job it should be to take care of recharging the rechargeable batteries. This is one chore that no group or person seems to want. Yet a great many complaints about radios that were apparently not working properly have been corrected simply by replacing the battery with one that was in good condition.

A practical approach to radios is to limit their use to those communications jobs which only a radio can do. This means, among other things, that if a person has a portable radio, it should be because he or she needs a radio, not because a radio has become a status symbol around the plant. Communications jobs that can be done in ways other than by using radios should be done in those ways. The use of sound-powered phones by instrument maintenance personnel is a move in the right direction, as is the use of pagers and cellular phones in nonhazardous areas.

19.7 Air Quality

For operations centers, two aspects of air quality need attention:

- The environment must be correct for the safe use of general-purpose electrical equipment, i.e., the type of electrical and electronic equipment that should be used only in areas which are considered to be nonhazardous.
- Air in the building must not contain contaminants above the levels that will cause corrosion of electronic components, particularly contact points on circuit boards and switches.

When an operations center serves a plant that is handling flammable fluids, precautions must be taken to ensure that any electrically powered devices that are in use cannot become a source of ignition. Explosion-proof enclosures and intrinsically safe circuits are two ways in which this objective is met. Much of the equipment used in operations centers, however, cannot be made either explosion-proof or intrinsically safe. This would apply to some electronic instruments and especially to computers.

Under these circumstances, the only alternative is to make the operations center a nonhazardous area. Flammable vapors that may be outside the building must not be allowed into the building. To achieve this, the pressure inside the building is maintained at a value slightly above atmospheric using forced-draft fans. Then if there is any air leakage, it will be from the inside out, not the other way around.

While the principle of pressure ventilating is simple enough, buildings that were supposed to be pressure ventilated have turned out to be otherwise for one or more of the following reasons:

- Building materials in the walls and ceiling of the building were porous to air seepage so that the required positive pressure could not be maintained. Some concrete blocks have this deficiency.
- Entries into the building for conduits, pipes, and other fixtures were not properly sealed.
- Outside doors or windows did not fit tightly enough. Or doors or windows were left propped open to compensate for problems with the building's air-conditioning system.
- The intake of the forced-air fan was located too low, or it faced in a direction that allowed the fan to draw in air containing flammable vapors.

These potential problems should be guarded against if a pressure-ventilated building is the route taken.

The second part of air quality is the elimination of contaminants that can cause damage to electronic instruments and systems through corrosion. At the top of the list of contaminants to be concerned about is hydrogen sulfide. The points that are most often attacked are switch contacts and the edge contacts on circuit boards, although other parts also can be harmed. Damage created through corrosion can cause problems that are often difficult to locate. A fault may come and go with no discernible pattern to it, except, of course, that it never happens when there is a serviceperson around.

The hydrogen sulfide level can be reduced to a safe value by filtering the air. The onus is on the owner to specify what the H_2S limit is to be so that the filtration system can be designed to meet that specification. However, some difference of opinion regarding just what is a safe limit may be encountered. This could be explained, in part, by the fact that the speed and extent of corrosion attack by H_2S may depend on local factors such as temperature, humidity, and ventilation, as well as on the amount of H_2S in the air. Another factor is the materials used in the circuit boards, gold being considerably more corrosion-resistant than silver or copper.

In any event, there would appear to be a consensus that a hydrogen sulfide limit of 0.005 ppm would constitute a safe environment. H_2S levels in the air that are below this figure will not likely cause any significant harm to silver and copper parts.

Commercial filtering systems are available for cleaning up the air. The process consists of passing the air over an adsorbent that will remove the harmful contaminant. Activated charcoal is an adsorbent that is often used because it is readily available and is therefore inexpensive. The use of activated charcoal should be treated with caution, however, because if the charcoal becomes saturated with hydrogen sulfide, other chemicals that may be present in the air may leach the H_2S out of the charcoal. Then, instead of filtering out H_2S, the activated charcoal will become a source of it.

Many plants have found out that their air-filtration system, designed on the basis of the best information they could obtain at the time, did not turn out to have the capacity to remove the contaminants down to the required levels. The lesson to be learned from this is that it is wise to allow sufficient floor space in the building for a possible expansion of the filtration system at some time in the future.

Finally, while hydrogen sulfide gets most of the publicity when it comes to corrosion damage in electronic components, it is not the only contaminant that can cause trouble. One plant found, after an extensive search, that the corrosion problems they were experiencing were cause by chlorine in the cleaning solution used to clean the floors. Too much or too little moisture in the air also can lead to trouble. High relative humidity accelerates corrosion and creates perverse electrical paths on circuit boards. Very dry air encourages static electrical discharges. Humidity control will likely be unavoidable.

19.8 Electric Power

It goes without saying that the electric power supply for electronic instruments and process computers should be reliable. What needs to be done here is to define what is meant by *reliable* and to describe the means of achieving the desired reliability. A historical review of electric power systems that have been tried and which have failed to meet the reliability objective is an approach that could be taken.

The reliability of an electric power supply is often described in terms of maintaining the required voltage on the power lines. While this is all well and good, it must not be forgotten that in addition to being upset by voltage dips, some electronic equipment is also sensitive to transients in frequency. Designing a power system that ensures the correct voltage while ignoring possible transients in frequency may very well result in an unsatisfactory power system.

Every operations center will be tied in to an emergency lighting system. An obvious move, therefore, would be to connect the power supply for all the electronic equipment into the same emergency system. It will be easily seen, however, that this arrangement will not provide the required reliability, without even starting it up.

Emergency lighting systems are powered by an emergency generator. In an oil refinery, the emergency generator is often driven by a steam turbine, which starts up when a valve in the steam supply line automatically opens on general power failure. Since the emergency lighting generator comes on line from a standing start, there will be a delay of some seconds before power is restored. By this time, every control valve in the plant will have gone to its fail position. Unpredictable problems may develop in a process computer, depending on what it was doing at the moment the power went off. And as if this were not enough, there will be frequency transients as the turbine governor attempts to lock the turbine into synchronous speed.

A variation of this method is to keep the turbine turning at synchronous speed but with no load on the generator. If the general power source fails, the load is shifted automatically to the emergency generator. While this is a better method than the first, it still will not meet the requirement for reliable power in most cases. The difficulties are transients in voltage and frequency as the emergency generator takes over the load, as well as the waste of energy due to operating the turbine-generator set when it is not really needed.

The next step up in reliability, but also in complexity and cost, is the dual-drive generator. The generator that provides the power to the operations center has two drivers, one at either end. The normal driver is an electric motor that is powered from the general electric power system. The motor drives the generator most of the time. The driver at the other end is a gasoline engine.

If the general power fails, the generator drive is taken over, almost immediately, by the gasoline engine. A drop in speed of only a few cycles per second is all that is required to cause the gasoline engine to start. From the viewpoint of minimum loss in voltage and frequency, this is a very good method. It has been used fairly extensively by the telephone companies.

The problem with a dual-drive generator has really nothing to do with its electrical output characteristics. It is that a dual-drive generator set is an expensive piece of machinery. Furthermore, dual-drive generators have been found to have a very high requirement for maintenance. This is a real deficiency, since to do maintenance, it is necessary to shut the generator down and put the operations center on general power, sometimes for days on end. This leaves the center vulnerable to anything that can happen to the general power system.

As far as rotating equipment is concerned, the dual-drive generator is about as far as one can go, even sparing no expense. If the dual-drive generator is not good enough, what conclusion should be drawn? The answer, as disconcerting as it might seem, is that for process instrumentation and computers, *there is no ac power system that can meet the 100 percent reliability requirement.* The reason goes back to a simple fact: ac power cannot be stored.

The power system that is probably considered to be state-of-the-art today is an essentially electronic approach. The key component in the power package is a set of storage batteries. The ac power is generated from dc battery power by an inverter, or inverters. The batteries are kept charged by rectifiers that are ahead of the batteries in the power package. The rectifiers are powered from the general power system.

If general power goes out, the batteries will maintain power to the instruments for a considerable length of time. A reserve of half an hour of battery power is a popular specification. As long as the battery voltage is there, the inverters continue to convert dc to ac, and the instruments and data processors stay on line without any upset.

This system meets fully the no-break power objective, with no moving parts. The problems are in the initial cost and in reliability. In a power package of this sort, it is customary to duplicate both the rectifiers and the inverters. Each of the rectifiers and each of the inverters is made large enough to carry the entire load. This is done so that either rectifier or inverter can be taken out for servicing without having to switch to general power.

At one time there was a general distrust of big inverters. An inverter larger than 5 kVA was considered to put the reliability of the whole power package at risk. Whether or not this situation has changed as a

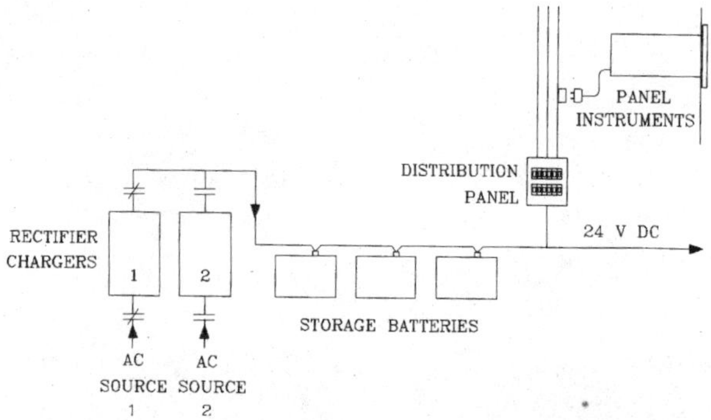

Figure 19.1 A 24-V dc power system for electronic instruments.

result of good work by the manufacturers of electronic power systems is not entirely clear. Accordingly, this point should be investigated before entrusting the security of the whole operations center to an electronic power package that relies on inverters.

Actually, any concern one may have about the reliability of large inverters can be relieved simply by not using them. All the leading instrument companies offer electronic process instruments designed to operate from a 24-V dc power source. Consequently, it would make sense to use them. The power package will then look like that shown in Fig. 19.1. There are a number of features about this power system that require elaboration.

- Commercial 24-V battery chargers, which are designed for charging batteries in heavy vehicles and construction machinery, are not good enough for this power system. There is too much ripple on their dc output. For an instrument power system, an ac-to-dc power supply with no more than 1 percent RMS ripple on the output voltage is required.
- Battery capacity should great enough to maintain the required current flow for 30 minutes with the rectifier chargers shut down and without the battery voltage dropping below 22 V.

Thirty minutes of reserve capacity usually will be adequate for allowing the operations center to ride out any short-term loss of general power. If the general power outage lasts longer than 30 minutes, the battery power system will not crash, but the battery voltage will gradually fall off, and the controlling instruments will lose their ability to stroke the control valves fully. There is also the line of thought that if general power is out for longer than half an hour, then the plant will have bigger problems than just the ones with the automatic controls.

- All storage batteries have an internal resistance, and the current that flows into or out of the battery must flow through this resistance. Suppose that at the rated current output of the battery, the voltage drop across the battery's internal resistance is 0.5 V. This would mean that when a nominal 24-V battery is charging, with current flowing into the battery, the voltage at the battery terminals, and on the dc power lines, would be 24.5 V. Conversely, when the battery is discharging, with current flowing in the opposite direction across the internal resistance, the voltage at the battery terminals would be 23.5 V.

Thus the voltage at the battery terminals could change by 1 V depending on whether the battery was receiving current or delivering it. This situation may not be satisfactory for the electronic instruments that are powered by the dc system. Notice that the dc power buses should be tied right onto the battery terminals, as shown in Fig. 19.1. They should not be strung some feet away from the batteries and connected to the batteries by branch connections. If this is done, the resistance of the branch lines will add to the internal resistance of the batteries, and the voltage change on the dc bus lines from charging to discharging will be even greater.

There are voltage-equalizing devices available that compensate for the charging-to-discharging voltage difference. Figure 19.1 does not show a voltage equalizer, but good practice would require that one be incorporated into the system.

- For a large plant or complex with many measurement and control loops, the dc current flow can be considerable. At the Syncrude tar sands processing plant, it was 500 A at 24 V. While it is practical to expect to draw this kind of current from batteries for 30 minutes, nevertheless, considering the number and size of the batteries involved, some thought has to be given to the space they will take up. The comments on battery rooms in Sec. 19.3 are relevant.
- In Fig. 19.1, the panel instruments are shown with a flexible power cord ending in a plug. This has been done in some plants to make it legal for an instrument technician to disconnect the power to an instrument so that he or she can take the instrument back to the maintenance shop for service. If the electric power supply were hard wired into the instrument, plant rules might require that the disconnect be done by an electrician.
- The best type of plug for the instrument power cord is the type that contains a fuse. If a fault develops in an instrument, the power

break should not happen in the distribution panel, since this may cut off the power to other instruments besides the instrument that is in trouble.

Aside from this, even if the instrument contains its own fuse, it is quite possible that a blown fuse in the instrument will require shop repair, whereas a fuse in the plug can be changed without removing the instrument from the panel. Finally, a fuse in the plug protects against a fault in the power cord as well as in the instrument.

19.9 DC Instrument Power System Specifications

Specifications for a 24-V dc instrument power system can be summarized as follows:

1. *Nominal voltage:* 24 V dc
2. *Tolerance on voltage:* Must be held within 0.50 V of 24 V for
   - DC current load from 0 to 100 percent
   - Batteries charging or discharging
   - AC supply voltage to rectifiers from 103 to 127 V ac
   - AC supply frequency to rectifiers from 58 to 62 Hz
3. *Ripple:* No greater than 1 percent RMS on the 24-V line
4. *Battery capacity:* When batteries alone are providing full power, the line voltage must not drop below 22 V for at least 30 minutes.
5. *Current output:* As required (amperes)

Appendix

Tables, Formulas, and Procedures

A.1 SI Conversion Factors

Measurement	SI units and symbol	Conversion equation
Length	Meter (m)	1 in = 2.540 cm 1 ft = 0.3048 m 1 mi = 1.609 km
Mass	Kilogram (kg) Tonne (1000 kg) (t)	1 lb = 0.4536 kg 1 ton = 0.9072 t
Matter	Mole (mol)	Note 1
Time	Second (s)	1 min = 60 s 1 hr = 3600 s
Area	Square meter (m^2) Hectare (10,000 m^2) (ha)	1 ft^2 = 0.09290 m^2 1 acre = 0.4047 ha
Volume	Cubic meter (m^3) Liter (1 dm^3) (l)	1 ft^3 = 0.02832 m^3 1 std ft^3 = 0.02826 sm^3 1 U.S. gal = 3.785 dm^3 1 Imperial gal = 4.546 dm^3 1 bbl = 159.0 dm^3
Angles	Radian (rad)	1 degree = 0.01745 rad
Force	Newton (N)	1 lb force = 4.448 N
Torque	Newton-meter (N•m)	1 in lb = 0.1130 N•m
Pressure	Pascal (Pa)	1 psi = 6.895 kPa 1 in water = 0.2488 kPa
Temperature	Degrees Celsius (°C)	1°F = 0.5556°C 1°R = 0.5556 K °F = 1.8°C + 32

A.1 SI Conversion Factors (*Continued)*

Measurement	SI units and symbol	Conversion equation
Flow rate	Kilogram/second (kg/s) Cubic meter/second (m^3/s)	1 lb/h = 0.0001260 kg/s 1 ft^3/s = 0.02832 m^3/s 1 U.S. gpm = 0.06309 dm^3/s 1 Igpm = 0.07577 dm^3/s 1 bbl/h = 0.04416 dm^3/s 1 scfh = 0.000007850 sm^3/s
Frequency	Hertz (Hz)	1 cps = 1 Hz 1 rpm = 0.01667 Hz
Energy	Joule (J)	1 BTU = 1.055 kJ 1 cal = 4.186 J 1 kWh = 3.600 MJ 1 ft•lb = 1.356 J
Specific heat	Joules/kilogram•°C (J/kg•°C)	1 BTU/lb•°F = 4187 J/kg•°C 1 cal/g•°C = 4187 J/kg•°C
Power	Watt (W)	1 hp = 0.7460 kW 1 BTU/h = 0.2931 W
Density	Kilogram/cubic meter (kg/m^3)	1 lb/ft^3 = 16.02 kg/m^3 1 lb/ft^3 = 0.01602 kg/dm^3 Specific gravity = (kg/m^3)/999.1
Absolute viscosity	Pascal-second (Pa•s)	1 cp = 1 mPa•s

1. A mole in the SI system is that quantity of a substance which contains the same number of molecules as the number of molecules in 0.012 kg of carbon-12. The quantity of a substance, in moles, is equal to

$$\frac{\text{Grams of the substance}}{\text{Molecular weight of the substance}}$$

1 kmol = 1000 (g) mol.

2. In the petroleum industry, standard temperature is 15°C; standard pressure is 101.325 kPaa.
3. The density of water at 15°C is 999.1 kg/m^3.

A.2 Optimum Controller Setting Formulas

(Recovery with 1/4 Amplitude Ratio)

1. By measuring the *period of oscillation* (P_u minutes) at the ultimate gain setting K_{cu}

Controller response	Proportional only	Proportional + reset	Proportional + reset + derivative
Gain	$0.5K_{cu}$	$0.45K_{cu}$	$0.8K_{cu}$
Reset rate (rep/min)		$1.2/P_u$	$2.4/P_u$
Derivative (min)			$0.1P_u$

2. By running a *process reaction curve.*

   Step input = Δm percent signal to the control valve
   Measured dead time = L minutes
   Measured process reaction rate = R percent per minute

Controller response	Proportional only	Proportional + reset	Proportional + reset + derivative
Gain	$\Delta m/RL$	$0.9\Delta m/RL$	$1.6\Delta m/RL$
Reset rate (rep/min)		$0.3/L$	$0.6/L$
Derivative (min)			$0.4L$

A.3 Control Valve Sizing Equations in fps Units

For *steam flow:*

$$W\ (\text{pph}) = 1.06\sqrt{\rho P}\, C_1 C_v K_c$$

where ρ = density of the steam at valve inlet conditions (pcf)
P = valve inlet pressure (psia)
K_c = valve capacity correction factor for compressible fluids:

$$K_c = \sin\left(\frac{3417}{C_1}\sqrt{\frac{\Delta P}{P}}\right)$$

as explained in Chap. 17.

This equation can be modified for gas flow using the density relation

$$\rho\ (\text{pcf}) = 0.0932\,\frac{MP}{T}$$

For *gas flow:*

$$W\ (\text{pph}) = 0.324P\sqrt{\frac{M}{T}}\,C_1 C_v K_c$$

where P = valve inlet pressure (psia)
M = molecular weight of the gas
T = temperature of the gas at the valve inlet (°R)

These control valve sizing equations are based on the fundamental sizing relationship for compressible fluids that appears in the Fisher Controls Catalog 10.

A.4 Pneumatic Signal Line Tester

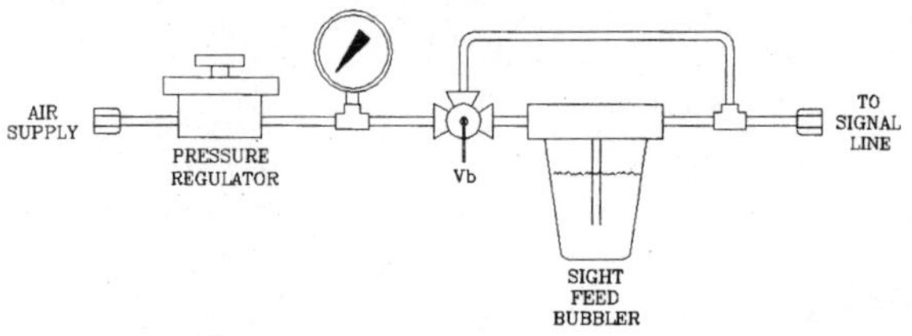

Procedure

1. Disconnect the signal line at the field end (at the transmitter or control valve), and plug it off.
2. Disconnect the signal line at the receiver end, and connect in the signal line tester (right end in the sketch).
3. Set the three-way valve V_b for flow through the bypass around the sight feed bubbler.
4. Connect an air supply with a pressure of at least 50 psig to the tester. Set the air pressure regulator for 50 psig. This will pressure up the signal line to 50 psig.
5. Set the three-way valve V_b for flow through the sight feed bubbler. If any air bubbles up in the bubbler, it is an indication that the signal line is leaking. A leakage rate of more than two or three bubbles per minute can mean a measurement error or a loss of controller performance due to signal-line pressure leaking off.
6. If necessary, trace out the signal line, locate the leak with leak detector fluid (or soap solution or shave foam, etc), and repair it.
7. If the signal line is old, do not assume that there will be only one leak. After a leak is found and repaired, run the test again.

A.5 Dew Point of Air at Elevated Pressures (°C)

Dew point at atmospheric pressure (deg C°)	Air pressure (kPag)										
	0	200	400	600	800	1000	1200	1400	1600	1800	2000
30	30	46	57	65	71	75	78	80	83	86	89
20	20	35	46	53	58	62	65	67	70	73	75
10	10	25	34	41	46	49	52	54	57	59	62
0	0	14	23	29	33	37	39	41	44	46	49
−10	−10	3	11	17	21	24	26	28	31	33	35
−20	−20	−8	0	5	9	11	14	15	18	20	22
−30	−30	−19	−12	−7	−4	−1	1	3	4	7	9
−40	−40	−30	−23	−19	−16	−14	−12	−10	−9	−7	−5
−50	−50	−41	−35	−31	−28	−27	−25	−23	−22	−20	−18
−60	−60	−52	−47	−43	−41	−39	−38	−36	−35	−33	−31
−70	−70	−63	−58	−55	−53	−52	−51	−49	−48	−46	−45

A.6 Thermal Properties of Metals

	Density (kg/m^3)	Coefficient of linear expansion (per °C)	Specific heat (kJ/kg•°C)
Aluminum	2,710	0.000 024 8	0.896
Copper	8,910	0.000 017 6	0.3860
Gold	19,320	0.000 014 0	0.1306
Cast iron	7,200	0.000 010 8	0.498
Lead	11,350	0.000 029 5	0.1281
Nickel	8,910	0.000 013 3	0.440
Tin	7,280	0.000 021 1	0.227
Platinum	21,370	0.000 008 8	0.1356
Silver	10,490	0.000 019 1	0.234
Brass	8,470	0.000 020 2	0.384
Phosphor bronze	8,860	0.000 017 8	0.366
Carbon steel (cast)	7,830	0.000 012 1	0.448
Stainless steel (type 316)	8,030	0.000 015 1	0.500
K Monel	8,470	0.000 014 0	0.544

A.7 Specific Gravity at 15°C and Freezing Point of Ethylene Glycol-Water Solutions

Volume fraction ethylene glycol	Specific gravity	Freezing point (°C)
0.00	1.000	0.0
0.02	1.002	−0.6
0.04	1.005	−1.2
0.06	1.007	−1.9
0.08	1.009	−2.6
0.10	1.012	−3.3
0.12	1.014	−4.1
0.14	1.016	−4.9
0.16	1.018	−5.7
0.18	1.021	−6.6
0.20	1.023	−7.6
0.22	1.025	−8.6
0.24	1.028	−9.7
0.26	1.030	−10.8
0.28	1.032	−12.1
0.30	1.035	−13.4
0.32	1.037	−14.8
0.34	1.039	−16.3
0.36	1.041	−17.9
0.38	1.044	−19.6
0.40	1.046	−21.4
0.42	1.048	−23.4
0.44	1.051	−25.4
0.46	1.053	−27.6
0.48	1.055	−29.9
0.50	1.058	−32.4
0.52	1.060	−35.0
0.54	1.062	−37.8
0.56	1.064	−40.8
0.58	1.067	−43.9
0.60	1.069	−47.2

A.8 Specific Gravity of Petroleum Derivatives at Various Temperatures

Specific gravity	Temperature (°C)								
at 15°C	0	30	60	90	120	150	180	210	240
1.040	1.050	1.031	1.012	0.994	0.977	0.959	0.941	0.922	0.902
1.020	1.030	1.011	0.992	0.974	0.956	0.939	0.921	0.902	0.881
1.000	1.010	0.991	0.972	0.953	0.936	0.919	0.900	0.881	0.861
0.980	0.990	0.971	0.951	0.933	0.915	0.898	0.880	0.861	0.840
0.960	0.970	0.951	0.931	0.912	0.895	0.877	0.859	0.840	0.820
0.940	0.950	0.930	0.911	0.892	0.874	0.857	0.838	0.819	0.798
0.920	0.930	0.910	0.890	0.871	0.853	0.836	0.817	0.798	0.777
0.900	0.910	0.890	0.870	0.851	0.833	0.815	0.796	0.776	0.755
0.880	0.891	0.870	0.849	0.830	0.812	0.793	0.774	0.754	0.733
0.860	0.871	0.850	0.829	0.809	0.790	0.772	0.752	0.732	0.709
0.840	0.851	0.830	0.809	0.788	0.769	0.750	0.730	0.708	0.686
0.820	0.831	0.810	0.788	0.767	0.747	0.728	0.707	0.685	0.661
0.800	0.811	0.789	0.767	0.746	0.725	0.705	0.684	0.660	0.634
0.780	0.792	0.769	0.747	0.725	0.703	0.682	0.659	0.634	0.606
0.760	0.772	0.749	0.726	0.703	0.681	0.658	0.634	0.606	
0.740	0.753	0.728	0.705	0.681	0.657	0.633	0.607		
0.720	0.733	0.708	0.684	0.659	0.633	0.607			
0.700	0.713	0.688	0.663	0.636	0.608				
0.680	0.694	0.667	0.641	0.613					
0.660	0.674	0.647	0.620						

A.9 Atmospheric Pressure Correction for Altitude

[Barometric pressure (kPaa) at sea level should be reduced by the correction in the table.]

	Correction (kPa)									
Altitude (m)	0	10	20	30	40	50	60	70	80	90
0	0.00	0.12	0.24	0.36	0.48	0.60	0.72	0.84	0 .96	1.08
100	1.20	1.31	1.43	1.55	1.67	1.79	1.91	2.03	2.14	2.26
200	2.38	2.50	2.62	2.73	2.85	2.97	3.08	3.20	3.32	3.44
300	3.55	3.67	3.79	3.90	4.02	4.13	4.25	4.37	4.48	4.60
400	4.71	4.83	4.94	5.06	5.18	5.29	5.41	5.52	5.63	5.75
500	5.86	5.98	6.09	6.21	6.32	6.43	6.55	6.66	6.78	6.89
600	7.00	7.12	7.23	7.34	7.46	7.57	7.68	7.79	7.91	8.02
700	8.13	8.24	8.36	8.47	8.58	8.69	8.80	8.91	9.03	9.14
800	9.25	9.36	9.47	9.58	9.69	9.80	9.91	10.02	10.13	10.24
900	10.35	10.46	10.57	10.68	10.79	10.90	11.01	11.12	11.23	11.34

Note: For elevations greater than 1000 m above sea level, the barometric pressure P_e can be estimated from the relation:

$$P_s = P_e(1 + 0.000\,127E)$$

where P_s is the pressure at sea level (same units as P_e) and E is the elevation above sea level (m).

A.10 Critical Temperature and Pressure

The critical temperature and pressure of a substance are relevant to the change of state of the substance between its liquid and gas phases.

The *critical temperature* T_c is the more fundamental point. If the temperature of a gas is greater than T_c, then no amount of applied pressure will cause the gas to change to its liquid state.

The *critical pressure* P_c is the pressure corresponding to T_c on the pressure-temperature equilibrium curve, i.e., the vapor pressure curve for the substance. This point is called the *critical point.* Note that it is really the critical temperature that establishes the critical point. There is also a *critical density* ρ_c for the substance corresponding to the critical point.

Critical temperature, pressure, and density are definable only for pure substances, not for mixtures.

Mixtures of gases

The *molecular weight* of a mixture of gases is equal to the sum of the (mole fraction of each component gas × its molecular weight) for all components.

The *pseudocritical temperature* of a mixture is equal to the sum of the (mole fraction of each component gas × its critical temperature) for all components.

The *pseudocritical pressure* of a mixture is equal to the sum of the (mole fraction of each component gas × its critical pressure) for all components.

The *specific heat* of a mixture is equal to the sum of the (mole fraction of each component gas × its specific heat) for all components.

Note: Mole fractions and volume fractions are equivalent.

A.11 Physical Data for Various Gases

Gas or vapor	Molecular weight	Critical temperature (K)	Critical pressure (kPaa)	Gas law deviation factor at 101.325 kPaa and 15°C
Air	28.97	132.5	3,769	1.000
Ammonia	17.03	405.6	11,300	0.987
Carbon dioxide	44.01	304.3	7,397	0.994
Carbon monoxide	28.01	134.2	3,546	1.000
Chlorine	70.91	417.2	7,711	0.986
Hydrogen	2.02	33.3	1,297	1.002
Hydrogen sulfide	34.08	373.6	9,008	0.989
Nitrogen	28.02	126.0	3,394	1.000
Oxygen	32.00	154.3	5,036	1.000
Sulfur dioxide	64.06	430.4	7,873	0.978
Water vapor	18.02	647.2	22,060	0.991
Methane	16.04	190.7	4,641	0.998
Ethane	30.07	305.3	4,945	0.992
Ethylene	28.05	282.9	5,157	0.994
Propane	44.09	368.8	4,357	0.984
Propylene	42.08	365.5	4,560	0.987
n-Butane	58.12	426.2	3,648	0.969
Isobutane	58.12	407.2	3,749	0.971
Acetylene	26.04	309.2	6,282	0.991

A.12 Certification of Electronic Instruments and Controls by CSA

Certification by the Canadian Standards Association (CSA) for equipment powered by electricity can follow either of two routes:

1. *General acceptance.* This is a contract between CSA and the *manufacturer* of the device, under which a prototype is tested in the CSA laboratories. If it is considered acceptable, that particular model is certified by CSA, after which the manufacturer is permitted to place the CSA certification stamp or sticker on other identical copies of the device tested which are subsequently made and sold.

However, any modifications of any kind to the circuitry of the device will void the certification and require that the modified device be resubmitted for examination.

2. *Special acceptance.* This is a contract between CSA and the *user* of the device, under which CSA examines, and may test, only that item purchased by the user. The examination may be made at the user's plant or at the manufacturer's. CSA bills the user for the service. Special acceptance of any device by CSA applies only to the item or items specifically examined at the time, not to any

identical copies purchased by other users, nor to any copies purchased by the same user at a later date.

A provincial inspector may, at his or her discretion, give a tentative OK to Use to a device which does not yet have official CSA certification, provided that evidence is produced to show that the manufacturer or the user, as the case may be, is in the process of obtaining a general certification or a special acceptance from CSA and that no problems are anticipated.

A.13 Process Instrumentation Facts of Life

1. No plant is constructed to be a place to enjoy instruments and control systems. Rather, it is there to make a useful product which can be sold for profit. The instruments and systems therein, while important, are only one part of the machinery of production.
2. One should diligently strive for an intimate knowledge of the process and its conditions of operation, lest through ignorance one engineers a measurement or control system which never does produce the desired results. Such monuments may well become a millstone around the designer's neck.
3. When we contemplate, starry eyed, some wondrous new instrument or system and the benefits that will be derived from it, the question which must then be asked is, "Who do we have who can make it work?"
4. When anyone expounds on what is good or what is bad about an instrument or system, consider, "Just what axe does this person have to grind?" One's conclusions should then be tempered accordingly.
5. Remember that any instrument or system bears two price tags: the first is its selling price; the second is what it costs to maintain it in good operating order. When judging the merits of a new instrument or system, give equal regard to both.
6. Continuous vigilance must be maintained over engineering specifications. Negligence can lead to failure to benefit from improvements in technology or ways to save money.
7. If any new instrument or system should fail to produce the expected results, ask first, "Are we using it properly?"
8. It is a mistake to presume that because a person once sold you a lemon, instead of an instrument, that he or she will never be able to create a useful instrument from that lemon.

9. When setting out to prove the accuracy of any meter, you are obliged to ensure that the accuracy of the standard to which you are comparing it is at least as good as the accuracy of the meter you intend to prove.
10. It is important to avoid inane comments such as, “That instrument is no good!” Instead, specify the nature of the problem in an intelligent manner. Otherwise, you will probably forfeit the assistance of persons who might provide the remedy you are looking for.

Bibliography

Dresser Industries Inc., *Masoneilan Handbook for Control Valve Sizing*, Dresser Industries Inc., 1987.

Fisher Controls Company, *Control Valve Handbook*, Fisher Controls Company, 1977.

Shinskey, F.G., *Process Control Systems*, McGraw-Hill, New York, 1979.

Spink, L.K., *Principles and Practice of Flow Meter Engineering*, Plimpton Press, 1967.

Ziegler, J.G., and Nichols, N.B., *Optimum Settings for Automatic Controllers*, Taylor Instrument Companies, 1941.

Ziegler, J.G., and Nichols, N.B., *Process Lags in Automatic Control Systems*, Taylor Instrument Companies, 1941.

Index

ABOUT THE AUTHOR

Bob Connell has more than 30 years of experience in the process control industry. He has worked as an instrumentation engineer with the Engineering Division of Imperial Oil Limited, and as a process control engineering specialist with Syncrude Canada Limited. Mr. Connell has also taught instrumentation engineering technology at the Northern Alberta Institute of Technology. He is a fellow member of the International Society for Measurement and Control, he is the author of *Successful Instrumentation Applications*, and has been a frequent contributor to *Chemical Engineering* and other magazines. In 1994, Mr. Connell received the D.P. Eckman Award for contributions to furthering education in the area of instrumentation and application.